Michael Tomczyk

NanoInnovation

For Valerie Thompson,
I hope you enjoy these
nanoinnovation stories
and insights.

Best regards,

Michael
Tomczyk

Related Titles

Fecht, H., Werner, M., Van de Voorde, M. (eds.)

The Nano-Micro Interface

Bridging the Micro and Nano Worlds

Second edition
2014
Print ISBN: 978-3-527-33633-3; also available in digital formats

García-Martínez, Javier (ed.)

Nanotechnology for the Energy Challenge

Second edition
2013
ISBN: 978-3-527-33380-6; also available in digital formats

Vollath, Dieter

Nanomaterials

An Introduction to Synthesis, Properties, and Applications

Second edition
2013
ISBN: 978-3-527-33379-0; also available in digital formats

Allhoff, Fritz / Lin, Patrick / Moore, Daniel

What Is Nanotechnology and Why Does It Matter?

From Science to Ethics

2010
ISBN: 978-1-4051-7544-9; also available in digital formats

Cademartiri, Ludovico / Ozin, Geoffrey A.

Concepts of Nanochemistry

2009
ISBN: 978-3-527-32597-9; also available in digital formats

Balzani, V., Credi, A., Venturi, M.

Molecular Devices and Machines

Concepts and Perspectives for the Nanoworld

Second edition
2008
Print ISBN: 978-3-527-31800-1; also available in digital formats

Michael Tomczyk

NanoInnovation

What Every Manager Needs to Know

WILEY-VCH
Verlag GmbH & Co. KGaA

Author

Michael Tomczyk
Innovator in Residence
Villanova University
Villanova, Pennsylvania
United States

Library of Congress Card No.: applied for

British Library Cataloguing-in-Publication Data
A catalogue record for this book is available from the British Library.

Bibliographic information published by the Deutsche Nationalbibliothek
The Deutsche Nationalbibliothek lists this publication in the Deutsche Nationalbibliografie; detailed bibliographic data are available on the Internet at http://dnb.d-nb.de.

Print ISBN: 978-3-527-32672-3
ePDF ISBN: 978-3-527-65007-1
ePub ISBN: 978-3-527-65006-4
Mobi ISBN: 978-3-527-65005-7
oBook ISBN: 978-3-527-65004-0

Cover Design Grafik-Design Schulz, Fußgönheim, Germany
Typesetting Thomson Digital, Noida, India
Printing and Binding Markono Print Media Pte Ltd., Singapore

Printed on acid-free paper

This book is dedicated to

Fred Kavli (*1927–2013*)
Founder and Chairman, The Kavli Foundation

This book is dedicated to the late Fred Kavli. I was privileged to meet Fred in 2011 at the Wharton School when he was being presented with the Franklin Institute's Bower Award for Business Leadership. He provided the following quotation for this book, although sadly he did not live to see it published:

Figure 1 Fred Kavli and Michael Tomczyk.

"We cannot predict the future and foresee all the revolutionary developments that nanotechnology will bring us, but it gives us confidence that our rate of development and discovery will follow the logarithmic curve and maintain an ever-increasing rate of expansion and discovery to bring us a future that is so revolutionary that we cannot even imagine it today anymore that we could imagine the revolution that the transistor would bring when it was discovered in the 1940's."

Fred was a Norwegian (and naturalized American) entrepreneur, inventor, and philanthropist who used his wealth to fund the Kavli Foundation, Kavli Prize, and Kavli Institutes, including the Kavli Institute for Bionano Science and Technology at Harvard University. As you can see from this photo, the octogenarian retained his enthusiasm throughout his life and is a wonderful example of a world-class innovation champion.

Contents

Preface

In 2000, I was giving a series of presentations to industry and academic groups on radical innovations that have the potential to reshape the future, and one of these innovations was nanotechnology. At the time, there was a lot of hype around "nano," but I had the sense that most people didn't really understand what was really going on in the field. So I began asking my audiences, "Who can name one product that uses nanotechnology?"

To my amazement, most people couldn't name a single product. This happened year after year. In the fourth year, one hand went up and someone said, "carbon nanotubes." At the time, I knew there were already more than a thousand products that used nanotechnology. Obviously, the business community needed to know more about nanoinnovation. At the same time, many of my colleagues in business were expressing frustration over the media hype and constant flow of "breakthrough" announcements that were causing a lot of confusion and misinformation. Finally, I decided to write a book that tells what's "really happening" in nanoinnovation – the book you're reading now.

I started by interviewing nano-insiders in business, government, science, and academia. Thanks to contacts shared by friends like Michael Terlaak and others and by using LinkedIn and other networking resources, I was able to interview more than 150 nano-insiders. Over time, I got to know many of the most prominent nano pioneers. I invited them to provide updates on their research at an annual event I hosted at the Wharton School called the Emerging Technologies Update Day.

Virtually everyone I contacted was eager to participate and to help convey the "real story" of nano. In addition, people in many countries worked behind the scenes to answer questions, provide details, and secure permission for nanoscale images. Some laboratory technicians took nanoscale photos especially for this book. Others provided me with background summaries of nanotech projects and details that are not yet public.

One of my most important challenges was to make this a dynamic publication. When I began this project, I noticed that a lot of nanotechnology books offered snapshots of innovations that quickly grew out-of-date, so I tried to take a slightly different approach. My goals were to make this a starting point, not an endpoint or snapshot. So I tried to design this as a dynamic living document that you the reader can use as a basis to continue your own investigations. Most of the nanoinnovations you'll learn about here will continue to evolve over time, and you can easily track their progress on the Internet and in science and business media.

Another goal of this book is to give credit to some truly impressive pioneers, and tell their stories in their own words where possible – because many of the most significant breakthroughs are the result of extraordinary personal effort. It's fascinating to learn, for example, how Ned Seeman gleaned a breakthrough idea from a woodcut on the wall of a pub while enjoying a beer, or how Tony Atala redesigned inkjet printers to "print" human organs. I also invited business entrepreneurs to discuss how they developed their ventures – including failures as well as successes.

This is not just a book about what's happening in nanotechnology – this is also a book about what you can make happen. You don't have to be a scientist to be a nanoinnovator, or to champion nanotechnology. I know this, because early in my career I was fortunate to play a role in developing and launching the world's first home computer (the Commodore VIC-20). I was not an engineer. I went to college in Oshkosh, Wisconsin and studied literature and journalism. After military service as a US Army officer, I earned an MBA from UCLA, fell in love with home computers, and played a lead role in developing and launching the first home computers at Commodore. My love affair with emerging technologies helped me to become a pioneer in home computing and kept me involved in innovation throughout my career, including 18 years at the Wharton School as Managing Director of the Emerging Technologies Management Research Program, the Mack Center for Technological Innovation, and the Mack Institute for Innovation Management. As my own story confirms, there are many paths that allow you to get involved in innovation.

Most of the nanotech pioneers you'll meet in this book did not start out specializing in nanotechnology. They come from physics, chemistry, biology, engineering, business, environmental technology, and other fields. Most are self-taught. For many researchers, nanotechnology was so new and different that it forced them to think in new and different ways. Their stories are fascinating.

One of the things we learn from their stories is that anyone can be a nanoinnovation *champion*. If you're in a company, you can support and cultivate an innovation culture that includes the manipulation of atoms and molecules in the R&D toolkit. If you're in marketing, think about how nanoinnovations will compete with existing technologies and change your industry. If you're managing a business, think about how nanotechnology will affect your organization, your industry, and your competitive markets. If you're a parent, encourage your children to study science and technology – especially nanotechnology. If you're a teacher, push to integrate nanotechnology in your school curriculum. Learn about nanotechnology and scan the horizon for emerging technologies and applications. If you're in a country that is out of the nanoinnovation mainstream, find ways to educate students and seed the nanoinnovation process. Focus on how nanotechnology can solve problems in your country, especially problems for which there are no other solutions.

As you read this book, keep thinking about how nanoinnovation will change your world. Will a new material made from carbon nanotubes or graphene replace plastic, steel, or aluminum? Will nanosensors create a "sensor revolution" where almost anything can be detected? Will nanomedicine cure diseases that have been stubbornly resistant to cures for decades or centuries? Will we be growing our own hearts and livers to replace failing, damaged, or even aging organs? Will we wrap buildings in nanoskins to regulate environmental conditions? Will we change the structure of materials by mimicking innovations created by Nature, such as creating a new type of dry adhesive by imitating the footpads of a lizard, or engineering a material based on the nanostructure of a butterfly's wing?

The answer is, these wondrous things are already happening. The revolution in nanotechnology is changing our lives, wherever we live on the planet. Many of these nanoinnovations are happening out of sight and are hidden from view in research laboratories. Some innovations are discussed only in specialized science, engineering, or medical journals. This book throws light on hidden corners of science and technology, just as nanoimaging systems reveal nanoscale objects that are smaller than visible light waves. It also gives you a portfolio of ideas and themes that belong on your radar screen, if you want to keep current on what will be happening in the coming decades, in nanotechnology and other areas.

These are exciting times to be involved in science, technology, and business. There have never been so many innovations poised to change our lives, from robots that walk and fly to genetic solutions that will save and prolong our lives. Everything is changing, from how we use mobile social media to communicate, to how we process and package food, to how we use energy. Most of these innovations are visible, but the science and technology that drives them is invisible, and that's why we need to know more about nanoinnovation.

I invite you to join me and millions of others who are helping to drive progress through nanoinnovation. Together, we can make the future happen faster.

Acknowledgments

I am extremely grateful to the many nano-insiders who devoted time to participate in interviews, discuss their research, share their personal stories, and read portions of my book to ensure I "got it right." Many nano-insiders generously provided referrals and introductions to colleagues who allowed me to expand my network of nanotech professionals in science, business, government, and media.

I especially want to thank Michael Terlaak, founder of the Nanotechnology Research Foundation in San Diego, who provided referrals to numerous colleagues at the beginning of my research. This helped jump-start the interview process. Lynn Foster also opened his contact network to me and provided some excellent insights. My friend and colleague Brent Segal, cofounder of Nantero, provided a wonderful entree to critical issues in nanotechnology – by inviting me in 2007 to help organize and serve on the senior leadership of the IEEE/IEC project developing standards for the use of nanomaterials in electronics.

Several university colleagues at the Wharton School and University of Pennsylvania played an important role in my "academic development." I want to thank Jerry Wind, Saikat Chaudhuri, and the Mack Institute's Core Group for giving me an informed vantage point on a wide array of emerging technologies during my more than 18 years at the Wharton School. I joined Wharton in 1995 to help launch the Emerging Technologies Management Program, which in 2001 became the Mack Center for Technological Innovation and in 2013 became the Mack Institute for Innovation Management. Throughout these changes I was privileged to provide managerial leadership as Managing Director, which kept me thinking constantly about radical innovations including nanotechnology.

I started writing this book while studying for my master's degree in environmental studies at the University of Pennsylvania. The research methodologies I learned during my graduate studies were immensely valuable. Yvette Bordeaux, who chaired the Master of Environmental Studies (MES) program at the University of Pennsylvania, helped me tailor my graduate studies to include nanotechnology in my curriculum, including a superb course on nanotechnology taught by Dr. Jody Roberts from the Chemical Heritage Society. My graduate advisor Stan Laskowski was extremely helpful and supportive.

I'm especially grateful to my editors at Wiley-VCH: Heike Noethe, my terrifically patient and absolutely awesome editor, and Martin Preuss who championed the book when I first presented the concept and got me started on the project. Dr. Noethe was patient, encouraging, generous, professional, constantly enthusiastic, and supportive. As an author, I couldn't ask for a better editor/publisher.

Having started my career as a journalist, I greatly appreciated the availability of Google, Google Scholar, LinkedIn, Gmail, and other tools that gave me instant access to emerging innovations and allowed me to contact virtually any nano-insider I wanted to interview, including some of the world's leading scientists and business leaders.

My mega-thanks and gratitude go to the more than 150 nanotechnology insiders who participated in interviews and provided information, insights, and images, graciously sharing their experiences to help make this book accurate, relevant, and "real." Many of these pioneers and champions have already made tremendous contributions to the field of nanotechnology, and continue to help drive nanoinnovation forward. Others are toiling 24/7 in laboratories and offices to turn possibilities into solutions. Their enthusiastic participation in this book project allowed me to include observations and opinions that can only come from insiders who truly know what's really happening in nanoinnovation. I also want to thank those who provided or facilitated the use of the images and diagrams included in this book. Thank you again to these remarkable nanoinnovators and champions, with apologies to anyone I may have inadvertently excluded from this very extensive list or whose affiliations/titles may have changed since this list was updated:

Deshamaya Mahesh Amalean – Chairman MAS Holdings, chairman, SLINTEC.

Paul Alivisatos, Ph.D. – Larry and Diane Bock Professor of Nanotechnology, UC-Berkeley; Director, Berkeley Lab, Lawrence Berkeley National Laboratory.

Peter Antoinette – Cofounder, President and Chief Executive Officer, Nanocomp Technologies.

Anthony Atala, MD – W.H. Boyce Professor and Director of the Wake Forest Institute for Regenerative Medicine and Chair, Department of Urology, Wake Forest School of Medicine; founder, Regenerative Medicine Foundation; Scientific Founder and Chairman of R&D at Tengion, Inc.

David Bachinsky, Ph.D. – Founder, CEO, Chief Scientific Officer at Molecular Creativity.

Joe Bailey – Health Care and Life Sciences Business Strategist, Intel Corporation.

Peter Balbus – Founder and Managing Director, Pragmaxis LLC and a Founding Strategic Advisor, Nanotechnology Research Foundation.

Lajos (Lou) Balogh, Ph.D. – Editor-in-Chief of the journal *Nanomedicine: Nanotechnology, Biology and Medicine*; CEO of AA Nanomed Consulting; former Professor, Roswell Cancer Institute, State University of New York, Buffalo.

Mark Banash, Ph.D. – Vice President of Quality and Regulatory Affairs, Nanocomp Technologies.

Peter Bell, Ph.D. – Director, Cell Morphology Core, Gene Therapy Program, Perelman School of Medicine, University of Pennsylvania.

Michael Berger – Cofounder, Nanowerk LLC; Editor, nanowerk.com.

Parijat Bhatnagar, Ph.D. – Consultant, Biomedical Engineering (Biochips, Microfluidics, BioMEMS Devices); Advisor, Intel Research.

Peter Binks, Ph.D. – Chief Executive Officer, Sir John Monash Foundation; former Chief Executive Officer of NanoVentures Australia.

Dawn Bonnell, Ph.D. – Trustees Chair Professor, Materials Science and Engineering, School of Engineering and Applied Science; and Vice Provost-Research, University of Pennsylvania.

Mike Boyer – Engineering Teacher, North Penn High School, Lansdale, Pennsylvania.

Sam Brauer, Ph.D., J.D. – Principal in the Consulting Group, NanoBiz, LLC; and founder/consultant at Nanotech Plus, LLC.

Stephen Brobst, Ph.D. – Chief Technology Officer, Teradata.

Darren K. Brock – Senior Staff Research Engineer, Lockheed Martin.

Darrell Brookstein – Managing Director, The Nanotech Company LLC; Executive Director of Nanotechnology.com; author of the book, *Nanotech Fortunes* (Nanotech Company, 2005).

Charles Brumlik, Ph.D., J.D. – Principal, NanoBiz, LLC.

Ahmed Busnaina, Ph.D. – William Lincoln Smith Professor; and Director of the National Science Foundation (NSF) Nanoscale Science and Engineering Center for High-Rate Nanomanufacturing, and the NSF Center for Microcontamination Control (CMC); at Northeastern University.

Stuart Cantrill, Ph.D. – Chief Editor, *Nature Chemistry* (former editor of *Nature Nanotechnology*).

Han Cao, Ph.D. – Founder of BioNano Genomics.

Taylor Cavanah – Executive at Zyvex Instruments and DCG Systems (2004–2010); founder and CEO, Locai, Inc.

Thomas Celluci, Ph.D. – Former Chief Commercialization Officer, U.S. Department of Homeland Security, former President/CEO at Zyvex Corporation.

Christopher Chen, MD, Ph.D. – Skirkanich Professor of Innovation, Department of Bioengineering; Director of the Tissue Microfabrication Laboratory; and founding Director of the Center for Engineering Cells and Regeneration at the University of Pennsylvania School of Engineering.

Pearl Chin, Ph.D. – Research Fellow (and *former President*) at the Foresight Institute; Managing Director, Seraphima Ventures.

Daniel T. Colbert, Ph.D. – Executive Director, Institute for Energy Efficiency, University of California, Santa Barbara.

Bill Cook – Principal, Commercialization, Molecular Diagnostics at WECA.

Lance Criscuolo – President, Zyvex Technologies.

Alfred J. Crosby, Ph.D. – Professor, Polymer Science and Engineering and Director of the Crowby Research Group; University of Massachusetts Amherst.

Jeffrey Davis, MD – Director, Human Health and Performance, and Chief Medical Officer, NASA Johnson Space Center.

Joseph DeSimone, Ph.D. – Chancellor's Eminent Professor of Chemistry, University of North Carolina at Chapel Hill.

Pankaj Dhingra – Former CEO, Nanostellar; General Manager-Global Mergers & Acquisitions; Nissan Motor Company.

Robert Dombrowski – President and Principal Scientist, Nanoview Associates, LLC.

Samuel Simon Dychter, MD – Senior Medical Director, Halozyme Therapeutics.

Patrick Ennis, Ph.D. – Global Head of Technology, Intellectual Ventures – former Managing Director at Arch Venture Partners.

Terry Fadem – Former Managing Director, Corporate Alliances, Office of Science and Technology, Perelman School of Medicine, University of Pennsylvania.

Omid Farhokhzad, MD – Associate Professor at Harvard Medical School and Director, Laboratory of Nanomedicine and Biomaterials, Anesthesia, Brigham and Women's Hospital.

Guillermo Foladori, Ph.D. – National Autonomous University of Mexico and Co-Coordinator of the Latin American Nanotechnology and Society Network (ReLANS).

Lynn E. Foster – Emerging Technologies Director of the Law Firm, Greenburg Traurig, LLP; Editor and Coauthor of *Nanotechnology: Science, Innovation and Opportunity* (Prentice Hall, 2005), one of the best "insight books" on nanotechnology.

Robert A. Freitas, J.D. – Senior Research Fellow, Institute for Molecular Manufacturing; Co-Chair, Nanotechnology Track, Singularity University.

Stephanie Getty, Ph.D. – Spectrometer Instrumentation Engineer, NASA Goddard Space Flight Center; researching application of nanotechnology solutions to instrument designs.

Patricia Glaza – Principal, Arsenal Venture Partners, former CEO and Group Publisher of *Small Times*, the pioneering micro and nanotechnology publication and Executive Director/CEO of the Clean Technology and Sustainable Industries Organization.

Anita Goel, MD, Ph.D. – Founder, Chairman and CEO, Nanobiosym, Inc.; MD, Ph.D., Harvard–MIT, Division of Health Sciences and Technology.

Piotr Grodzinski, Ph.D. – Director, Nanotechnology for Cancer Programs, Alliance for Nanotechnology in Cancer, National Cancer Institute.

Mary Gros – Director, Business and Academic Relations, Teradata.

C. William Hanson, III, MD – Director, Surgical Intensive Care, Hospital of the University of Pennsylvania and author of *The Edge of Medicine: The Technology That Will Change Our Lives.*

A. John Hart, Ph.D. – Assistant. Professor of Mechanical Engineering and Art +Design, Director of the Mechanosynthesis Group, University of Michigan, Ann Arbor.

Jason Hartlove – CEO, Nanosys Technologies.

Patti D. Hill – Founder and Managing Principal, Penman Public Relations, Inc.

Damien Hoffman – Editor-in-Chief, Wall Street Cheat Sheet.

Gregory Horowitt – Cofounder and Director of Global CONNECT™, University of California, San Diego.

Peter M. Hughes – Center Chief Technologist, Goddard Space Flight Center; NASA.

Sarah Hurst, Ph.D. – Director's Postdoctoral Fellow, Argonne National Laboratory, Center for Nanoscale Materials and Editor, *Biomedical Nanotechnology* (Springer, 2011).

Duncan Irschick, Ph.D. – Professor of Biology, University of Massachusetts, Amherst.

Matthew Jaffe – Partner in the international practice group of the Washington, DC law firm, Crowell & Moring; his practice includes issues related to regulatory control of nanotechnology and nanoscale materials.

Helia Jalili, Ph.D. – Application Scientist, PANalytical B.V.

A.T. Charlie Johnson, Ph.D. – Professor of Physics, Director, Charlie Johnson Group, Experimental Nanoscale Physics, University of Pennsylvania.

Hai-Yong Kang, Ph.D. – Department of Chemical Engineering and Materials Science, UC-Davis.

Fred Klaessig, Ph.D. – Manager, Pennsylvania Bio Nano Systems.

Jerry Kolbe – Director of Marketing, Murata Electronics N.A.

Joseph Kunze, Ph.D. – President, SI2 Technologies, Inc. (spinoff of Triton Systems).

Stephanie Lacour, Ph.D. – Assistant Professor, Institute of Microengineering at EPFL, Switzerland.

Robert L. Langer, ScD – David H. Koch Institute Professor, Massachusetts Institute of Technology (MIT); cofounder, BIND Biosciences.

David Lashmore, Ph.D. – Vice President and Chief Technology Officer, Nanocomp.

Michael Lefenfeld – Cofounder, President and CEO, SiGNa Chemistry Inc.

Judith Light Feather – President, The NanoTechnology Group Inc.; founder/president, Nano NEWS TV; cofounder, CANEUS.

Wenbin Lin, Ph.D. – Professor of Chemistry and Molecular Pharmaceuticals, Eshelman School of Pharmacy, University of North Carolina at Chapel Hill.

Hod Lipson, Ph.D. – Assistant Professor, Department of Mechanical and Aerospace Engineering, Cornell University.

Michal Lipson, Ph.D. – Associate Professor, Nanophotonics Research Group, Cornell University.

Michael LoCascio – Senior Manager, BASF; Former Senior Analyst, Lux Research.

Dan Luo, Ph.D. – Professor of Biological and Environmental Engineering, Cornell University.

Les Makepeace – Managing Principal, Les Makepeace Consulting.

Kevin Maloney – President, CEO and cofounder, QuantumSphere.

Annette Mason – Director, Strategic Collaboration, BAE Systems.

Michael Mayberry, Ph.D. – Vice President, Technology and Manufacturing Group and Director of Components Research, Intel Corporation.

James McGonigle – Director of Education and Outreach, Nano/Bio Interface Center, School of Engineering and Applied Science, University of Pennsylvania.

John Miller – Vice President, Intellectual Property, Arrowhead Research Corp.; cofounder/former Senior Vice President of Intellectual Property, Unidym; coauthor of *The Handbook of Nanotechnology Business, Policy and Intellectual Property* (Wiley, 2004).

Bevey Miner – Advisor and cofounder, Tenzing Group; formerly President and Chief Operating Officer, Nanotechnology Research Foundation.

Chad Mirkin, Ph.D. – Director of the Nanoscale Science & Engineering Center and the International Institute for Nanotechnology, George B. Rathmann Professor of Chemistry, Northwestern University.

Scott Mize – Consultant and Entrepreneur; former President of the Foresight Institute.

David Model – Acting CFO, ADuro Biotech; Senior VP, Finance at iSonea; Executive VP, Triton Systems.

Tanya Monga – Business Development Manager, Roche Professional Diagnostics; formerly Business Development and Marketing Manager at Intrinsiq Materials and a Research Analyst at Credit Suisse.

Kara Morgan, Ph.D. – Director of Public Health Measurement and Analysis, Office of Planning, U.S. FDA.

Valentina Morigi – Consultant, PricewaterhouseCooper (Milan, Italy); previously Visiting Research Fellow at Methodist Hospital Research Institute, and at the University of Texas Science Center.

Mark Morrison, Ph.D. – Chief Executive Officer, Institute for Nanotechnology, Glasgow, Scotland, UK.

Chris Murray, Ph.D. – Richard Perry Professor of Chemistry and Materials Science and Engineering, University of Pennsylvania School of Engineering.

Mike Nemeth – Director of Commercial and Defense Applications, Zyvex Technologies.

Heike Noethe, Ph.D. – Wiley-VCH Verlag GmbH.

Leo O'Connor, Ph.D. – Vice President of Global Research, Frost & Sullivan's Technical Insights Business Unit.

Geoffrey Ozin, Ph.D. – Tier 1 Canada Research Chair in Materials Chemistry, Distinguished University Professor at the University of Toronto; Founding Fellow of the Nanoscience Team at the Canadian Institute for Advanced Research.

Krsto Pandza, Ph.D. – Professor in Strategy and Innovation; Subject Group Leader Strategy, Innovation and Leadership, Management Division, University of Leeds; Coordinator, Marie Curie Research Training Network, Management of Emerging Technologies for Economic Impact (ManETEI).

Mike Patterson – CEO, Graphene Frontiers, LLC.

Christine Peterson – Cofounder and Vice President, Foresight Institute.

Martin Preuss, Ph.D. – Commissioning Editor, Materials Science, Wiley-VCH Verlag GmbH.

Jaideep Raje – Analyst, Lux Research; previously a researcher in molecular engineering of nanomaterials for biotherapeutic applications at UC-Santa Barbara (UCSB).

Ben Reeve, Theo Sanderson and the other members of the 2009 Cambridge University iGEM team (Anja Hohmann, Bill Collins, Emily Knott, Hannah Copley, Paul Masset, Peter Emmrich, and Will Handley).

Todd Rider, Ph.D. – Senior Staff Scientist, Bioengineering Systems and Technologies Group, Lincoln Laboratory, MIT.

Mihail "Mike" Roco, Ph.D. – Senior Advisor for Nanotechnology, National Science Foundation (*key architect and founding chair of the U.S. National Nanotechnology Initiative*).

Mario I. Romero-Ortega, Ph.D. – Associate Professor, Bioengineering, University of Texas-Arlington; University of Texas Southwestern Medical Center.

Jeffrey H. Rosedale, Ph.D. – U.S. Patent Attorney/Partner and Co-Chair, Nanotechnology and Cleantechnology Practice Groups, Woodcock Washburn LLP.

Paul W.K. Rothemund, Ph.D. – Senior Research Associate, Computer Science, Bioengineering (Systems and Synthetic Biology), Computation and Neural Systems; Caltech.

Albert Sacco, Jr., Ph.D. – George A. Snell Distinguished Professor of Engineering; Director of the Center for Advanced Microgravity Materials Processing; Northeastern University; former U.S. Astronaut (NASA).

John Saiz – Chief Technologist, Johnson Space Flight Center.

Paol Samori, Ph.D. – Visiting Professor and Director of the Nanochemistry Laboratory, Institut de Science et d'Ingenierie Supramoleculaires (ISIS), Universite Louis Pasteur, Strasbourg, France.

Christie Sayes, Ph.D. – Program Manager, Nanotoxicology & Nanopharmacology, RTI International; formerly Assistant Professor at Texas A&M University.

Greg Schmergel, Ph.D. – Chief Executive Officer and cofounder, Nantero, Inc., a leading nanotechnology firm.

Nadrian (Ned) Seeman, Ph.D. – Margaret and Herman Sokol Professor of Chemistry, New York University (NYU).

Brent Segal, Ph.D. – Cofounder, Nantero, Inc.; Director, Lockheed Nanosystems; General Partner, Atomic Venture Partners.

Jaymin Shah, Ph.D. – Research Fellow, Pharmaceutical R&D, Global Research & Development, Pfizer.

Philip Shapira, Ph.D. – Professor of Innovation Management and Policy at the Manchester Institute for Innovation Research, Manchester Business School; and Professor of Public Policy at Georgia Institute of Technology; former Congressional Fellow, Office of Technology Assessment, US Congress.

Kimberly Hamad Shifferli, Ph.D. – Associate Professor of Mechanical Engineering and Biological Engineering, Massachusetts Institute of Technology.

Robert D. "Skip" Rung – President and Executive Director, Oregon Nanoscience and Microtechnologies Institute (ONAMI).

Shay Soker, Ph.D. – Professor of Regenerative Medicine, Cancer Biology, Physiology & Pharmacology and Surgical Sciences; Director of Cellular Therapy Programs, Wake Forest Institute for Regenerative Medicine.

Christopher Stanley, Tatiana Perevozchikova, and Valerie Berthelier – University of Tennessee.

Bill Tai – General Partner, Charles River Ventures; an investor in Nantero, Inc.

Ennio Tasciotti, Ph.D. – Associate Member, Interim Co-Chair, Department of Nanomedicine, The Methodist Hospital Research Institute; and Co-Chair, Regenerative Medicine Director, Spine Advanced Technology Lab; Houston Methodist Hospital.

Michael Terlaak – Founder and CEO/Executive Director, the Nanotechnology Research Foundation and Senior Partner/VP at NEXUSBioMed/NEXUSNanoTec.

Andrew Thompson, Ph.D. – Cofounder, President and Chief Executive Officer, Proteus Digital Health; Managing Member, Spring Ridge Ventures.

Richard Thorpe, Ph.D. – Pro-Dean for Research, Professor of Management Development, Deputy Director of the Keyworth Institute, University of Leeds.

James Tour, Ph.D. – T.T. and W.F. Chao Professor of Chemistry, Professor of Computer Science, and Professor of Mechanical Engineering and Materials Science, Rice University.

Hiran Vedam, Ph.D. – NanoConsulting Pte., Ltd., Singapore.

James R. Von Ehr, II – Founder, Chairman and Chief Executive Officer, Zyvex Corporation, one of the most successful of the early nanotechnology ventures.

Terry Wilkins, Ph.D. – Chair of Expert Advisory Group FP7 Programme, European Commission; CEO, Nanomanufacturing Institute and Professor of Nanomanufacturing Innovation, University of Leeds.

James M. Wilson, MD, Ph.D. – Professor, Department of Pathology & Laboratory Medicine; Director, Gene Therapy Program, University of Pennsylvania.

Josh Wolfe – Cofounder of the NanoBusiness Alliance, a columnist with *Forbes Magazine* and Editor of the monthly *Forbes/Wolfe Emerging Tech Report*.

Shane Woody, Ph.D. and Bethany Woody, Ph.D. – Cofounders of Insitutec.

Tyler Wry, Ph.D. – Associate Professor of Management, the Wharton School.

Peidong Yang, Ph.D. – Professor of Chemistry and Director of the Peidong Yang Group, University of California, Berkeley.

Shu Yang, Ph.D. – Professor of Materials Science and Engineering and Chemical and Biomolecular Engineering; School of Engineering, University of Pennsylvania.

Youseph Yazdi, Ph.D. – Executive Director, Center for Bioengineering Innovation and Design, Johns Hopkins University; formerly Director, Science and Technology, Johnson & Johnson.

Dominick Yee, Ph.D. – Researcher, Department of Chemistry and Biochemistry, University of California-San Diego; Founding Scientific Advisor, Nanotechnology Research Foundation.

Thomas M. Young – Director of Government Affairs, Nanotechnology Research Foundation.

Richard Zhang – Cofounder and CEO of IdentifIED Technologies.

Naiqin Zhao, Ph.D. – Chair of the Department of Metallic Materials, School of Material Science and Engineering, Tianjin University.

Thanks also to various contacts at IBM-Zurich, the European Commission's Joint Research Center, NASA, the National Science Foundation, U.S. Department of Energy, Oak Ridge National Laboratory, University of Leeds, and University of Pennsylvania.

Most important – I want to thank my incredible wife Nancy for making it possible for me to write this book, patiently and encouraging me to keep writing and helping in so many ways too numerous to mention. She drove us to work each morning so I could work on my book during our morning commute, never complained about the tap-tap-tapping and multitasking while watching a movie or TV program, and the countless times she waited patiently while I attended evening courses so I could earn my second master's degree. Nancy continues to be my creative inspiration and is my most passionate and beloved fan.

– Michael S. Tomczyk

Part I
What You Know (or Don't Know) about Nanoinnovation

1
Quick – Name Something "Nano"

Quick – name something "nano."

Name something that occurs at the nanoscale – a structure or process that we measure in nanometers. One nanometer is a billionth of a meter. Nanoscale refers to 100 nm or less. So name something under 100 nm.

As a manager, you should be able to name at least one or two things that are nanoscale. Can you do it? How quickly?

Are you still thinking? Do you have it? Are you *sure* that's a nanoscale example?

Okay. Let's make it easy. Name something nano that occurs naturally in the human body.

What did you say? A human hair? A blood cell?

Wrong.

A human hair is 50 000–80 000 nm thick. A red blood cell is about 2000–5000 nm. A white blood cell is about 10 000 nm in diameter. The period at the end of this sentence may contain 1 000 000 nm of ink.

Think smaller. Much smaller.

Can you name *one commercial product* that uses nanotechnology? This could be something that contains nanosized particles or nanomaterials. Here's a hint: You can find a nanotechnology product in almost any gift shop on the beaches of Cancun, Florida, California, or the French Riviera.

What about man-made nanostructures (Figure 1.1)? Almost any nanotechnology article talks about carbon nanotubes, the popular building blocks of nanotechnology. Can you name one product that uses carbon nanotubes?

Here's one example – most automobiles use carbon nanotubes in engine coatings because their semiconducting properties make nanotubes ideal for controlling static electricity or sparks that might cause a fire or explosion. Nanotubes are also mixed with polymers in seat covers so that you don't get a shock when you slide behind the wheel on a cold winter day. Carbon nanotube fibers are so incredibly strong and durable that they are used to strengthen military body armor.

Nanotechnology is used in an incredible variety of products, processes, and applications. Your tablet computer and smartphone use nanoscale circuits. Your flat screen TV is thin and lightweight, thanks to nanomaterials. Nanoskins protect antique buildings. Edible nanoskins protect apples and other fruits, and

NanoInnovation: What Every Manager Needs to Know, First Edition. Michael Tomczyk.

allow the last drop of ketchup to slide out of the bottle – although most food manufacturers are being highly secretive about their use of nanomaterials, having been spooked by the industry's experience with genetically modified foods.

Nanomaterials protect your pants against stains, liquids, mildew, and wrinkles. A nanofiber pocket in your shirt or jacket can screen the radiation from your cell phone. Athletes can buy no-smell socks infused with nanosilver particles that eliminate odor-causing bacteria. Other applications include clothing that changes color in response to light or body heat.

Many clothing items use technology developed by NanoTex, a US-based company that uses nanotechnology to develop "intelligent fabrics."

(a)
(b)
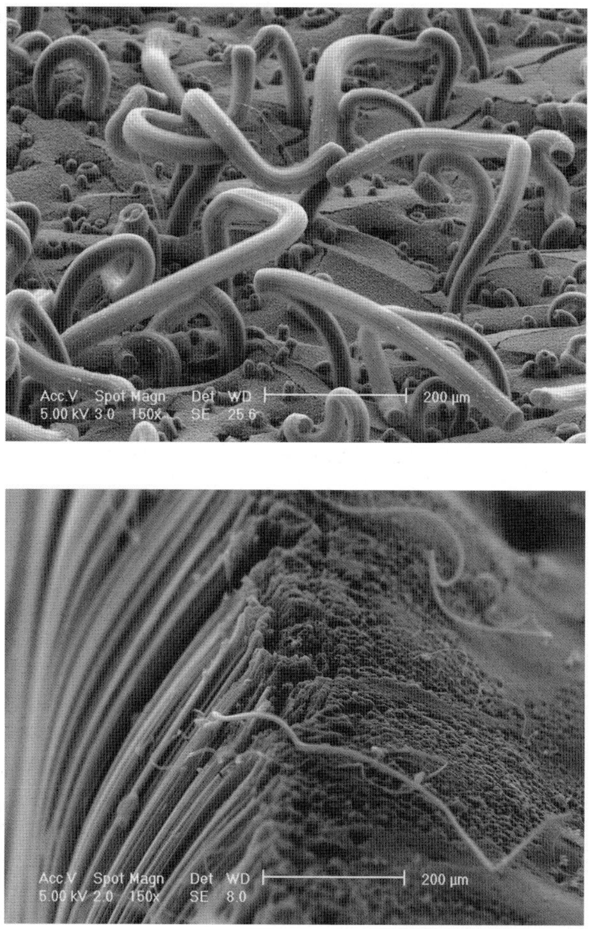

Figure 1.1 Carbon nanotubes that are 1 or 2 nm in diameter can be grown on a substrate as a tangled "forest" (a), made to grow in tight clusters (b) and formed into complex shapes and patterns, using photolithography and catalyst patterning (c). Shown here is the first nanoscale image of a world leader – President Barack Obama. Each Obama face contains 150 million nanotubes (images by Anastasios John Hart, University of Michigan).

(c)

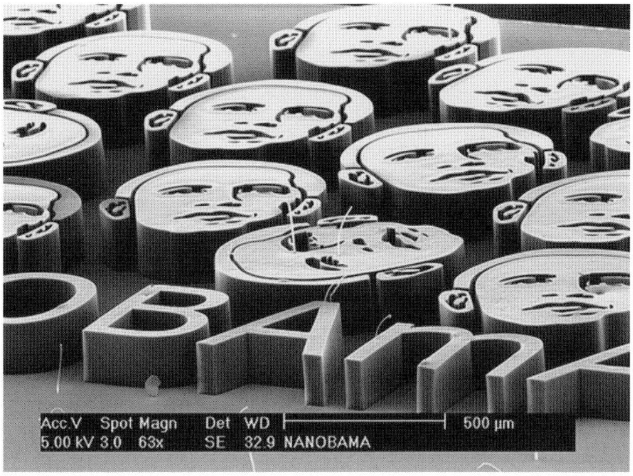

Figure 1.1 *(Continued)*

A few years ago, Philips Lighting created a lot of buzz at the Consumer Electronics Show by demonstrating a jacket with light-emitting diode (LED) optics woven into the fabric. The jacket displayed an animated message – suggesting that wearable computers may be closer than we think.

Another area that we all benefit from involves *catalysts*. Unless you're a chemist, you may not get too excited about catalysts – but think about this: catalysts are a *$100 billion business*.

Historically, catalysts have been used for millennia to strengthen steel, create new materials, and facilitate chemical processes. Today, most metal and plastic products are created using catalysts. Many of these processes have not changed for 100 years, so this isn't a sector where you'd expect to see a dramatic improvement, or radical innovation. However, nanosizing a catalyst can produce dramatic improvements. If you fill a basketball with nanoparticles, those particles will have more surface area than a basketball – that's the secret. Using smaller particles means you need less material to achieve the same catalytic effect – a competitive advantage if the catalytic material is an expensive rare earth element. But what about more common products – maybe a common household product like *ammonia*?

QuantumSphere Inc. (QSI) is a small company based in Santa Ana, California, that makes metal spherical nanocatalysts. Kevin Maloney, president and CEO, explains how a chemical as common as ammonia is benefiting from nanoinnovation. "Most of us think of ammonia as a household cleaner," he explains. "But most ammonia is used for food production and biofuel feedstocks. Fifty percent of all protein consumed comes from ammonia. There are about 400 plants worldwide that produce ammonia – so much that ammonia production and related processes consume as much as 2 percent of the world's annual energy supply and generate 1.6 percent of global CO_2 emissions."

The Haber–Bosch process used to produce ammonia was invented in 1909, and involves combining nitrogen and hydrogen gases over a heated, pressurized bed of iron. There hadn't been a breakthrough in this process for more than a century. QuantumSphere developed a patented process to improve the process using a nanocatalytic form of iron.

"Early in our research we demonstrated a 26 percent increase in yield," Maloney recalls. "We realized that this could translate into a significant improvement in commercial production, as well as financial and energy savings for the ammonia industry." Today, QuantumSphere is working with the leading producer of ammonia reactors to commercialize their process. The "business story" of QuantumSphere is described in more detail in Chapter 6.

This is a great example of a nanoinnovation with enormous potential impact that is happening largely out of sight. An energy-efficient, economical improvement in ammonia production may not be considered "radical" or "disruptive" by most of us, but if you're in the agriculture, biofuels, or ammonia industry, this could be a high-impact, radical innovation.

Radical Nano and Killer Applications

While these applications are important, they are not exactly radical. Most of these innovations have been creeping into products as incremental improvements to enhance functionality, replace or reduce materials, lower production and processing costs, and minimize negative effects. The first waves of nanoinnovations are so embedded in everyday products and processes that we don't give them much thought. There is no "nano inside" label on your car engine or laptop. Most consumer products that use nanoparticles do not list "nano" on their labels (although this is changing).

Most of the "killer apps" from nanoinnovation are still being developed, which makes this an ideal time for you as a manager to start learning about nano.

As Mark Banash, Vice President and Chief Scientist at NanoComp puts it, "There aren't a lot of killer applications yet for nanotechnology, but there are some 'assassin apps' that are replacing existing materials in the form of structural composites, multi-purpose materials, and more efficient catalysts. A lot of substitution is already taking place, and in the future we can expect a lot more."

Mark is in a good position to know. In February 2008, NanoComp set a record for the world's largest sheet of carbon nanotube material (Figure 1.2). Mark observes that carbon nanotubes are less than half the weight of comparable aluminum. Yarns and sheets made from carbon nanotubes possess a unique combination of high strength-to-weight ratio, electrical and thermal conductivity, and flame resistance that exceeds other advanced materials.

The question is, what will product developers actually do with these new materials? What can you wrap in a material that is one or two atoms thin but as strong as metal, as flexible as cloth, and able to provide electrical and thermal conductivity, flame resistance, and more? Will you integrate it into the uniform of a soldier or firefighter, or wrap it around a car, a house, or an office building? Also, will the ultimate "carbon nanosheets" be made from a composite made

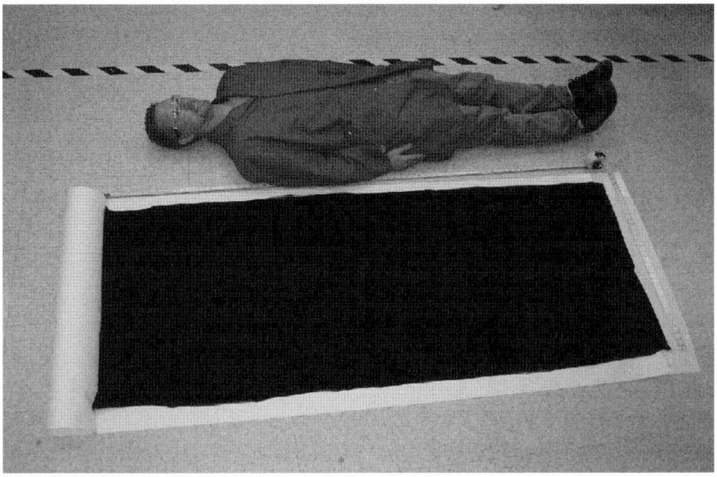

Figure 1.2 Nanocomp's 3′ × 6′ sheet made from carbon nanotubes set a world size record. The company is refining processes to manufacture these sheets, which can be up to 100 times stronger than steel, in commercial quantities. Shown here is Nanocomp technician Jeff Littlefield (who is 6′ tall) lying next to a sheet of the material (photo courtesy of Nanocomp Technologies).

with nanotubes and carbon fibers, entirely from carbon nanotubes, or from single or layered sheets of graphene or graphane?

Professor John Hart's team at the University of Michigan-Ann Arbor predicts a day when nanopores in nanoskins will regulate the flow of air, moisture, and heat in buildings without any mechanical controls. He says, "We know this is possible because we're doing this now in the lab."

There are lots of radical innovations in research laboratories that could be called game changers – digestible nanocomputers that can fit inside a pill, hollow nanospheres that can deliver drugs inside individual cells, and composites that make metals as flexible as plastics, and allow plastics to conduct electricity.

Nanopool GmbH, a family-owned business based in Schwalbach, Germany, has announced the development of liquid glass, an ultrathin form of SiO_2 (the *chemical notation for silicon oxide, which is glass*). Liquid glass can be sprayed on any surface to create a 100 nm coating that is both flexible and breathable. The coating is about 500 times thinner than a human hair, and is invisible to the eye. What is most interesting about this material is that no nanoparticles are added – according to the company, the coatings form and bond due to quantum forces, which suggests that this nanocoating could be safe and inert enough to use to coat hospital surfaces, kitchen sinks, industrial machinery, even textiles, and medical implants. The easy clean nature of a glass coating could greatly reduce the amount of toxic cleaning agents now used to clean and sterilize surfaces in hospitals, homes, and industrial clean rooms. Because it is breathable, it could be sprayed on stone, brick buildings, or statues to help protect against pollution.

What makes nanotechnology so exciting is its potential to solve problems in new and different ways that were impossible and unimaginable just a few decades ago. And that's why you need to know more about nanoinnovation. As a manager, you're in the business of solving problems, creating value, and turning ambiguous situations into something tangible. Nanotechnology holds incredible potential – this can take the form of an improvement, a breakthrough, a competitive advantage or a competitive threat. A breakthrough innovation in one company can be a death blow to another. In a world where open innovation and global networking is already transforming industries and markets, you need to know about nanoinnovations that are looming on the near horizon.

As a manager, you should be able to name one or two examples of nanoinnovations that could change the future. For example, what are scientists working on now that will really shake things up if they succeed? Can you name one area of science where nanoinnovation is playing a critical role?

The Science on "Planet Nano"

My wife Nancy, a medical researcher and nursing student, likes to describe science as a puzzle – all you have to do is solve it, she says.

I like to think of nanoscience as a new frontier. In many ways, innovating at the nanoscale is like exploring an entirely new planet – Planet Nano. The features of this planet are truly "alien." On this planet, objects don't behave like they do in our human world, at our familiar human scale.

At the nanoscale, water becomes sluggish and doesn't flow. A carbon molecule can be made to conduct electricity – or not. Metal has "memory" and can snap back to its original shape. Biological structures warp and flex into odd shapes and sizes. Synthetic DNA can be formed into pyramids and boxes. Stem cells are sprayed on scaffolds using inkjet printers to regenerate human organs. These are only a few examples of what we see when we peer into the nanoscopic world.

The science that drives nanoinnovation is not limited to research laboratories. There are conceptual thinkers like Robert Freitas who conceived – and patented – the artificial blood cell and numerous other innovations, and Eric Drexler, whose seminal concepts of molecular manufacturing generated both enthusiasm and controversy.

Some of the most impressive innovations will come from the *convergence* of unrelated materials, processes, and applications. Convergence has always been a major driver of innovation. In the past, a great deal of convergence came from "eureka" discoveries and serendipity. Nanotechnology is giving researchers the tools to *engineer* convergence (Figure 1.3). By any metric, this is a sea change – one of the most intriguing aspects of nanoinnovation. Instead of accidental convergence, or convergence that comes from decades of experimentation and fortuitous discovery, nanotechnology is allowing us for the first time to manipulate atoms and molecules to create solutions that can only be achieved at the nanoscale.

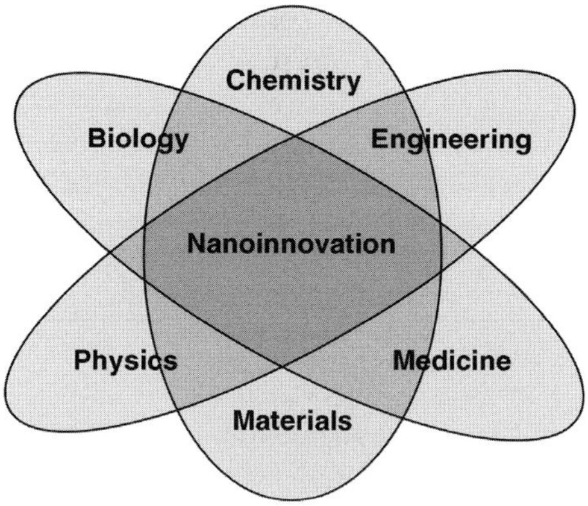

Figure 1.3 Nanoinnovation enables "engineered convergence" at the intersection of the major areas of science and technology.

Many nanoscience innovations sound like science fiction, but they actually do exist. For example, there are nanomaterials that allow electronic devices to twist and stretch and curl – and resume their original shape. Nokia has demonstrated concept phones that change shape when a call is received (Figure 1.4).

The military is developing invisibility cloaks to conceal tanks and submarines from enemy radar – or make soldiers invisible to the naked eye. Researchers have

Figure 1.4 Nokia's Morph concept phone uses nanotechnology to give the cell phone a new form factor that includes "shape shifting" (image courtesy of Nokia).

created mechanical insects that can "fly and spy" and swarms of weapons that resynch and keep functioning even if part of the swarm is destroyed – similar to the nefarious nanoswarms depicted in Michael Crichton's classic novel *Prey*.

In 2013, the US Army's Special Operations Command announced a program called TALOS – Tactical Assault Light Operator Suit – a futuristic combat suit made from lightweight advanced materials and nanoelectronics that the Army hopes to deploy by 2018. TALOS will be worn as an "exoskeleton" that includes such functions as portable power, health monitoring, climate control, biohazard filters, mobile communications, targeting systems, and weaponry.

In the energy field, research teams are working to shrink a complete solar cell to the size of a nanoparticle so that solar cells can be painted on a building or a rooftop. Medical researchers have developed molecular structures made out of gold and boxes made of synthetic DNA that can deliver drugs to cancer cells like nanoscale UPS couriers. This is not science fiction. These are real projects being developed by real people. These nanoscience projects are of special interest because the first wave of nanobusiness ventures came from university science projects and the next wave is being worked on by scientists in every region of the world. The next game changer could come from Silicon Valley or Rice University, from the Biopolis in Singapore, a nano consortium in Europe, or a university in Shanghai or Mumbai. Virtually every large corporation has a nano research laboratory and every major government and industry has a nanotech road map.

Safety Issues: Can Killer Apps Really Kill?

Of course, there is another side of nanoscience that needs to be considered, including the impact of nanomaterials on health, safety, and the environment. There is evidence to show that some "killer applications" could really kill – if nanoparticles find their way into our lungs or other organs where the accumulation of foreign substances can cause cancer. Nanoparticles and fibers have already been shown to sever DNA strands, which can trigger cancer. Longer fibers such as "long carbon nanotubes" have been compared with asbestos fibers that are lethal. Silver ions are natural bacteria fighters used in everything, from medical equipment to washing machines. They can also destroy friendly bacteria and are potentially harmful to aquatic life in streams.

These issues are still being studied and have yet to be resolved. However, as regulatory agencies seek to err on the side of caution, we can definitely expect more "nano inside" listings and warnings on product labels. And of course, safety issues – especially a potential "killer" in some killer applications – raise the stakes for business investors and companies. These risks and uncertainties need to be weighed along with the potential profits of nanoinnovation, which of course is considerable and hopefully worth these risks.

The Business of Nano

Nanotechnology is poised to generate a trillion dollars in revenues worldwide. Economically, the nanoscale business revolution will be as powerful and ubiquitous as the digital revolution. Its impact will be as transformative as electricity and the power grid, the modern highway system, or air travel. But this won't happen overnight. The nanotechnology infrastructure is still forming. Universities and companies are racing to catch up to the need for nanotech scientists and engineers. Companies are still trying to identify their research goals and business strategies.

In the next decade, the most successful companies and institutions will need to become *nanodextrous* – able to work at the nanoscale with the same confidence and proficiency they now work at the macroscale.

Despite strong expectations that nanoinnovation will provide substantial value to customers, companies, and investors, there is also a great deal of uncertainty. Some of these uncertainties are easy to identify, but their outcomes cannot be reliably predicted. To help understand these uncertainties, we can compare nanoinnovation with analogous examples from other technology-driven industries and markets such as the Internet, gene therapy, and genetically modified foods.

Even the most successful technologies and products need to cross the chasm between the laboratory and the marketplace – what some people call the "valley of death." When nanoinnovations first began to appear, there was a concern that some markets might shun nanoinnovations like Europeans initially shunned food products made from genetically modified crops.

Another concern was whether the multibillion dollar investments in nanotechnology research would justify the investment. Gene therapy has soaked up more than a billion dollars in research funding and investment, and generated enormous hype – and hope – but without one major commercial gene therapy available more than a decade after the human genome was mapped. Will nanoinnovation take decades to commercialize like gene therapy?

We know one thing – the commercialization of nanotechnology is already well underway. Many industries have already included nanotechnology in their products, so the question about commercialization is moot. Nanoinnovation will not follow the same course as gene therapy.

Will nanoinnovation emulate the experience of genetically modified foods? It's difficult to predict, but current indications are that nanoparticles and nanomaterials have a few more years of free reign before they are rigorously analyzed, tested, regulated, labeled, and verified as totally safe. Some consumer products have already come under scrutiny; however, there is still no firm consensus on several important safety issues. Will nanoparticles breathed in by workers or customers be treated as harmless elements by the body – or lurk for years or decades as a slow and deadly killer, like cancer-causing asbestos?

Of course, any new technology sector faces business risks. Will business managers, including scientist entrepreneurs who are not schooled in management *per se*, make the same mistakes as the hundreds of Internet ventures that went bust in the 1990s? As a manager, you may be in a position to influence decisions to invest in, develop, deploy, or market nanoinnovations.

Who is Driving Nanoinnovation?

One of the most intriguing aspects of innovation is who's doing the innovation? Most nanopioneers were not trained as nanotechnologists, simply because there were no nanotechnology courses when they got started. Most are self-taught and come from eclectic backgrounds that range from biology and medicine to computer science, from catalytic and crystallogrphy to electronics and mechanical engineering. These pioneers include a generous representation of business entrepreneurs and venture capitalists who were attracted to nanotech because they saw this as the "next big thing" in emerging technologies. Many had to create their own opportunities because there was no infrastructure when they started their research. Here are just a few examples of some notable nanoinnovators you will meet in this book.

Shane Woody and his wife Bethany built their own atomic force microscope while still in college. Today they own a profitable enterprise called InsituTec that manufactures and sells nanoimaging systems, and also does cutting edge R&D.

Nadrian "Ned" Seeman is a bearded, congenial chemistry professor at New York University who founded the field of structural DNA nanotechnology. Ned has been pioneering the use of DNA to build structures for more than 30 years. He recalls that when he started the field more than two decades ago, he was the only researcher doing this. Today, he estimates that more than 50 research teams are working in the field he pioneered.

Another DNA pioneer is Paul Rothemund, a senior researcher at Caltech who is famous for a type of structural DNA called "DNA origami." Dr. Rothemund is famous for creating a smiley face and other familiar images out of DNA. He has written thousands of words explaining and evangelizing DNA origami, so researchers around the world can learn and contribute to this research.

Han Cao, founder and chief scientific officer at BioNano Genomics (formerly BioNanomatrix), is doing something very different – but equally compelling – with DNA. Dr. Cao has learned how to tease single strands of DNA into unraveling and moving through channel on a nanofluidic chip. He watches as a strand fluoresces – able to see the glow with his naked eye. This is the first device that can isolate and image very long strands of DNA molecules.

Dr. Andrew Thompson, the glib English born cofounder and CEO of Proteus Digital Health, is leading a team that pioneered what he calls "intelligent medicine" – including a biochip sensor made of ingestible components that can be embedded in a drug capsule or pill to monitor whether a patient has taken the prescribed medication as required. This "chip on a pill" is depicted on the cover of this book.

Jim Von Ehr is the founder of Zyvex Corporation, the first molecular nanotechnology company. Jim began as a software pioneer. He was the founder of Altsys, a software developer acquired by Macromedia in 1995. Jim leveraged his success into a large personal investment in nanotechnology, where he has been pushing innovation frontiers on several fronts, from instruments to molecular manufacturing.

These are only a few of the more than 150 nano insiders who generously shared their personal stories for this book. Their insights into nanoinnovation and where the future may lead us are truly fascinating and hopefully will help you shape your own strategy.

So how does all this relate to you as a manager?

I started this book by challenging you to "name something nano." If you want to participate in the nanoscale revolution at any level, everything starts with this simple question. But there are lots of other questions that you as a manager need to be able to answer. You may have some of these answers already. Let's start by finding how much you know – or don't know – about nanoinnovation.

Your Nano I.Q

How much do you know about nanoinnovation? Most of us – myself included – don't know as much as we think we do. Here's a short 20-question quiz to test your "Nano I.Q." I.Q. normally means "intelligence quotient," but here it means "innovation quotient." This is not a formal test, just a fun exercise to get you started to see how much you know (or don't know) about nanoinnovation. You'll find the answers to these questions scattered throughout this book – or you can jump to Appendix A that provides the answers along with some additional notes.

1. Which of these consumer products use nanotechnology?
   - ☐ Dockers jeans
   - ☐ Sunscreen
   - ☐ Toyota Prius
   - ☐ Computers and cell phones
   - ☐ All of the above
2. In 2014, approximately how many consumer products used nanotechnologies (according to the Project on Emerging Nanotechnologies of the Woodrow Wilson Center)?
   - ☐ <10
   - ☐ 100
   - ☐ 560
   - ☐ 1300
   - ☐ >1800
3. The term "nanotechnology" was first coined by
   - ☐ Richard Smalley
   - ☐ Norio Taniguchi
   - ☐ K. Eric Drexler
   - ☐ Richard Feynman
4. This Nobel Prize winning scientist gave a lecture at Caltech in 1959 entitled "There's Plenty of Room at the Bottom" – setting the scientific community on a course of discovery that gave rise to the field of nanotechnology.
   - ☐ Richard Smalley

☐ Carl Sagan
☐ K. Eric Drexler
☐ Richard Feynman

5. The scanning tunneling microscope capable of imaging the nanoscale was invented in

☐ Germany
☐ Japan
☐ Switzerland
☐ The United States

6. What was the first recognizable image created by manipulating individual atoms?

☐ The IBM logo
☐ A Mickey Mouse cartoon from 1929
☐ The yin and yang symbol
☐ A yellow "happy face"

7. The first US president to have his image rendered at the nanoscale is

☐ George W. Bush
☐ William Clinton
☐ Barack Obama
☐ Ronald Reagan
☐ None of the above

8. Nanoparticles of gold (below 100 nm) are what color?

☐ Gold
☐ Silver
☐ Red or purple
☐ Black

9. Nanoparticles of iron oxide – called "nanorust" – can be used to

☐ remove bacteria from washing machines
☐ desalinize seawater
☐ remove toxic arsenic from water
☐ kill cockroaches

10. Buckyballs and nanotubes are forms of

☐ gold
☐ silicon
☐ carbon
☐ semiconductor circuits

11. Which country holds the most nanotechnology patents?

☐ Japan
☐ Germany

☐ India
☐ South Korea
☐ The United States

12. Which country is the nanotechnology research leader in Latin America?

☐ Argentina
☐ Brazil
☐ Chile
☐ Mexico
☐ Uruguay

13. Which of these atoms is larger than 1 nm in diameter?

☐ Oxygen
☐ Cesium
☐ Uranium
☐ All atoms are >1 nm
☐ All atoms are <1 nm

14. Which of the following has been created by scientists in nanotech laboratories?

☐ A nanoscale box with a lid, made of synthetic DNA
☐ A paper battery that you can fold or cut
☐ A hollow gold nanosphere capable of delivering cancer fighting drugs
☐ Carbon sheets one atom thick
☐ All of the above

15. Does Europe spend more or less than the United States on nanotechnology research?

☐ More
☐ Less
☐ Approximately the same

16. Which of the following is an example of nanoinnovation?

☐ High-performance baseball bats and golf clubs
☐ Flexible thin-film solar cells
☐ Digestible computers for tracking pills
☐ Cell phones that bend and twist like plastic
☐ All of the above

17. Which of the following are nanoscale (<100 nm in diameter)?

☐ West Nile Virus
☐ Red blood cells
☐ White blood cells
☐ Human hair
☐ *Escherichia coli* bacteria

18. The world's largest carbon nanotube manufacturing facility is located in which country?

- ☐ Germany
- ☐ The United States
- ☐ Japan
- ☐ China
- ☐ India

19. Which of the following companies holds the most nanotech patents?

- ☐ Fujitsu
- ☐ GE
- ☐ IBM
- ☐ Nantero, Inc.
- ☐ Siemens

20. Which products in Europe are required to identify nanosized ingredients with the word "nano" in brackets?

- ☐ Food products sold in supermarkets
- ☐ Cosmetics
- ☐ Candy
- ☐ Clothing
- ☐ All consumer products

2
A Quick-Start Guide: 10 Things You Should Know About Nanoinnovation

> Micro-this, micro-that—how much difference can there be between one kind of tiny thing and another? The answer, of course, is "almost everything."
>
> – *K. Eric Drexler, Radical Abundance: How a Revolution in Nanotechnology Will Change Civilization*

In Chapter 1, I challenged you to "name something nano." In this chapter I challenge you to "learn something nano." This quick-start chapter is designed to help you understand the core essentials of nanoinnovation – including the core principles, properties, scale, industry metrics, jargon, history, and critical issues that you as a manager need to know. This will help prepare you by giving you a grounding in some of the "nano essentials" before drilling down to explore the science and technology that is generating specific products and applications.

Let's start by looking at some of the questions and issues that differentiate nanoinnovation from nanotechnology.

In the field of innovation, everyone talks about getting "out of the box" (Figure 2.1) Nanoinnovation not only gets you out of the box but it also disassembles the box into atoms and molecules. So let's extend this metaphor. Imagine a box. Inside this box is something we call nanotechnology. In the box, the technology is restricted by boundaries. Some of these boundaries are physical, some are mental. Others are contextual. Some involve financial or technical constraints, or various types of resource limitations. Once we take nano out of the box, we enter a realm of endless possibilities, far beyond our real or imagined boundaries – a magical place where materials are one atom thick, computers are digestible, cell phones can be twisted like a pretzel (and bounce back to their original shape), and where DNA molecules can be used to construct a box!

But before we explore these possibilities, let's go over some basics – 10 things you should know about nanoinnovation *before* you open the box.

2.1
10 Things You Should Know about Nanoinnovation

As a manager, you should know how nanotechnology is defined, the properties that make nanotechnologies unique, where we are in the evolution of nanotech,

NanoInnovation: What Every Manager Needs to Know, First Edition. Michael Tomczyk.

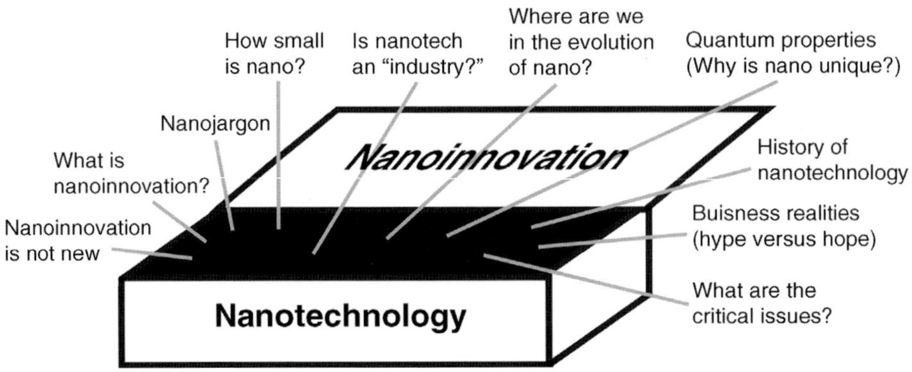

Figure 2.1 Nanoinnovation moves nanotechnology "out of the box" – forcing us to explore questions and issues that managers need to know, to innovate at the nanoscale.

the size and scope of the "industry" – and whether it actually is an industry or something else. You should be familiar with some notable nanoinnovations, including products that use nanotechnology. And most important, you need to have a sense of what's looming on the near horizon. Here are 10 things you should know – starting with a key definition.

2.1.1
What is Nanoinnovation?

In the business world, an **invention** is the *first occurrence of an idea for a new product or process*, while an **innovation** *is the first attempt to carry it into practice* [1]. If we accept this distinction, we can create a working definition of nanoinnovation.

A Definition for Nanoinnovation
Nanoinnovation is the implementation of nanoscale discoveries, ideas, and inventions to create value.

It's important to understand the relationship between an **invention**, the resulting **innovation**, and **commercialization**. For the purpose of this book, an **invention** is a new, previously undiscovered design, method, process, algorithm, composition of matter/material, or device, including new and useful improvements and applications. An **innovation** is the first attempt to carry the invention into practice, and **commercialization** is an attempt to extract value. For business purposes, commercialization may also be considered to be the adoption of the innovation by one or more lead markets. The link between an innovation and the value proposition is critical. Many nanotechnology projects and ventures have crashed and burned because they created something new, but failed to create value.

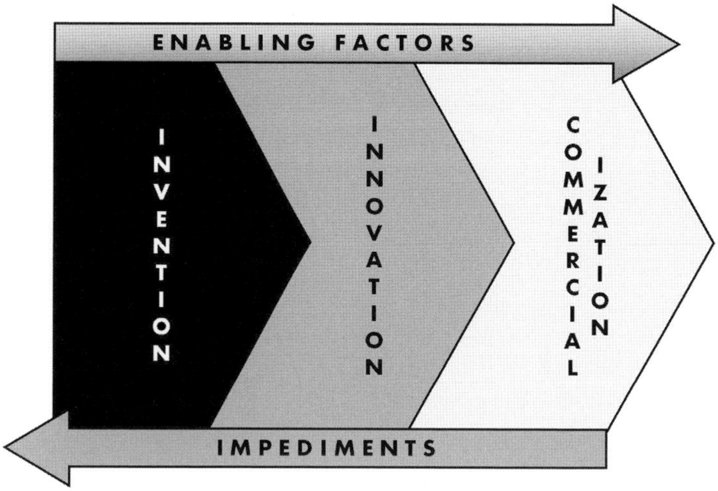

Figure 2.2 A conceptual framework for nanoinnovation (Copyright 2014, Michael Tomczyk).

The innovation process is influenced by *influence factors* that can enable or impede the process. These factors include funding issues, availability of resources, market forces, public sentiment, media attention, government regulation, safety concerns, and so on. These forces create a "push–pull" dynamic that makes the innovation process especially complex, as shown in Figure 2.2, which applies not only to nanotechnology but also to any area of technological innovation.

A Nanoinnovation Example: Untangling DNA

Dr. Han Cao, the founder and chief scientific officer of BioNano Genomics (formerly BioNanomatrix), is both an inventor and an innovator. As a research scientist at Princeton University, he observed that individual DNA molecules tend to form into tangled "balls" when isolated. This phenomenon makes it difficult to analyze individual strands of DNA, which is essential to analyzing individual molecules, or groups of molecules. Dr. Cao and his team *invented* a way to untangle and guide individual DNA molecules into an array of nanochannels on a semiconductor chip. This is an impressive achievement considering that a DNA molecule typically has a width of 2 nm and can be over a meter long. Dr. Cao's invention was granted US Patent No. 7,670,770 in March 2010. The *innovation* – how the invention is implemented – includes a novel method for analyzing strands of DNA on a nanofluidic chip. The *value* of this innovation lies in the possibility to develop genomic tests for individual diseases that could cost as little as $100. BioNano Genomics – founded in 2003 by Dr. Cao – is *commercializing* this innovation in their NanoAnalyzer® system. This level of science is complex, impressive, and game changing. It is a good example of the relationship between the original nanoscale invention, the nanoinnovation, and the commercial application.

2.1.2
Five Categories of Nanoinnovation

It is important for managers to understand what kinds of nanoinnovations we can expect to come from science and technology. There are at least *five categories* where nanoinnovation is likely to impact industries, markets, and our everyday lives. These include (i) new products, (ii) new methods of production, (iii) new sources of supply, (iv) exploitation of new markets, and (v) new ways to organize business [2].

These categories were originally developed by Harvard economist Joseph Schumpeter (1883–1950) to describe innovation in general. Schumpeter is best known for his iconic description of the innovation process as "gales of creative destruction." If we apply these categories to nanoinnovation, we can find strong examples in each:

1) **New Products.** In March, 2011, the Emerging Nanotechnologies Project of the Woodrow Wilson Institute/Pew Charitable Trust listed over 1300 consumer products and product lines that use nanotechnology, estimated to exceed 1800 by 2012,[1)] which extrapolates to over 2000 by 2014 – and these only include products that are reported. The actual number is much higher. Nanomaterials and nanoparticles are used in such familiar items as baseball bats and golf clubs, car tires and engines, clothing, cosmetics, filtration systems, health and fitness products, and LCD screens and virtually any product that uses semiconductors. The most common materials include nanoparticles of silver, carbon, titanium, silicon/silica, zinc and gold. Nano-silver is the most common and widely used commercialized nanomaterial.
2) **New Methods of Production.** Nanotechnology is exerting a major impact on manufacturing in many industries. Nanocatalysts are orders of magnitude more effective than traditional catalysts, which is extremely significant considering that the global market for nanocatalysts is projected to reach $6 billion by 2015. In many industries, carbon nanotubes are being used to manufacture everything from semiconductors and computers to body armor – based on new ways of combining and incorporating composite materials.
3) **New Sources of Supply.** One of the most important benefits coming out of nanotech research is the potential to replace increasingly rare and expensive elements such as gallium and indium with more abundantly available materials such as carbon nanomaterials and metal/plastic composites. Analysts predict we're running out of rare earth elements that are critical to the production of such vital high-tech products as mobile phones, solar panels, lithium ion batteries, semiconductors, optical

1) The Project on Emerging Nanotechnologies, Pew Charitable Trusts and Woodrow Wilson Center, http://www.nanotechproject.org/inventories/consumer/ (March 11, 2011).

technologies, consumer electronic displays, seawater desalination, fuel cells, medical devices, and synthetic fuels. China, which produces more than 90% of the world's rare earth metals, has imposed a tax on rare earth minerals and restrictions on exports. A report by the European Union [3] issued a "supply risk" warning that more than 14 critical minerals and metals will become increasingly scarce. These rare earth metals include antimony, beryllium, cobalt, fluorspar, gallium, germanium, graphite, indium, magnesium, niobium, platinum group metals (PGM), tantalum, tungsten, and a dozen other materials such as yttrium and lanthanum. Many of these could be unavailable as early as 2020. This is a genuine, looming crisis that nanoinnovation is helping to alleviate by finding new ways to replace rare earth metals with nanosized versions of "earth abundant" materials such as silicon and carbon. Jim Tour at Rice University has suggested that conductive nanoribbons could replace indium tin oxide, an expensive material used in flat-panel displays, touch panels, electronic ink, and solar cells. In 2009, Dmitry Kosynkin, a postdoctoral research associate on Dr. Tour's research team, discovered a method for chemically unzipping carbon nanotubes into flat nanoribbons [4]. The IBM solar panel shown in Figure 2.3 is another example of a nanoinnovation solution to the rare earth metals dilemma. The IBM cell uses copper, tin, zinc, sulfur, and selenium – all "earth abundant" materials – to perform functions that are currently provided by costly rare earth elements such as indium, gallium, and cadmium. According to IBM, previous attempts to create solar cells from similar materials did not exceed 6.7% in energy efficiency, which makes the new cell's 9.6% efficiency level a significant leap forward [5]. According to Dr. David Mitzi who leads the research team, the ultimate goal is to develop a solar technology that uses earth abundant materials, compares favorably on a cost per watt basis with conventional electricity generation, and supports the deployment of large systems at the terawatt level.

4) **New Markets.** Visit any consumer electronics store and you can find new markets being created by nanoinnovation. The National Nanotechnology Initiative (NNI) says that almost all high-performance electronic devices manufactured in the past decade use some nanomaterials.[2)] Nano-enabled products are changing the form factors of smartphones, tablet PCs, music players, eBooks, and more. The touchscreen on Apple's iPhone is coated with nanoparticles of indium tin oxide. To give you an idea of the size of this market, Apple has reported that more than half a billion iPhones, iPods, and iPads have been sold.
5) **New Ways to Organize a Business.** Most emerging technologies are accompanied by innovations in business strategy and marketing. Here's one example: nanoinnovations are creating longer-life, fast-charging batteries used in electric cars such as the much-publicized electric cars pioneered by Tesla Motors. At some point, the "business" of fueling

2) National Nanotechnology Initiative; http://www.nano.gov/node/589.

Figure 2.3 This cross section of a thin-film solar cell offers a nanoscale view of an "earth abundant" solution for solar energy. In 2010, an IBM team led by Dr. David Mitzi announced that they had developed a solar cell that uses abundantly available materials in the energy conversion layer to achieve a record energy conversion level of 9.6% – which was 40% higher than energy levels previously achieved with these materials. This was a major advance toward the creation of less costly, more efficient solar cells using commonly available materials (image created by IBM Zurich).

automobiles will shift from gas stations to charging stations and possibly to home garages, parking lots, and electric power outlets on parking meters. In this scenario, electric utilities – not petrochemical companies – will become the new business partners of the world's auto companies – a major business shift.

These are only some of the intriguing issues and prospects for nanoinnovation. Many of these innovations will literally change the future.

An even more exciting development would be the development of a computer that uses carbon nanotubes instead of the array of rare earth materials currently used in semiconductors and other components – which would convey new properties as well as continuing Moore's law using earth abundant materials. There is evidence that this innovation is looming on the horizon. In September 2013, a team of Stanford engineers led by Max Shulaker announced that they have built a computer using carbon nanotubes [6]. This was definitely not an overnight innovation. Stanford faculty and students have been working on this challenge for more than 20 years! This proof-of-concept computer performed 20 different instructions from the commercial MIPS instruction set such as counting, integer sorting, and basic problem solving. This development is a significant step forward for computer technology, and a critical indicator of how nanoinnovation is influencing the design of future computers.

2.1.3
Learning the Jargon of Nanoinnovation

The field of nanoinnovation has given rise to a "swarm" of nano-neologisms – words that begin with the prefix "nano" – and continues to produce some very clever and colorful catchwords and phrases. The jargon of nano can be quirky, fun, and highly descriptive, starting with the core concepts of molecular assembly and molecular manufacturing, and extending to such interesting terms as nanobots, nanoswarms, nanoribbons, molecular torches, DNA origami, and gray goo.

The jargon of nanoinnovation has not only stimulated the public imagination but has also entered the public dialog in a very big way. In December 2007, a Google search for the word "nanotechnology" returned 11.9 million hits, which increased to 14.9 million hits in March 2011 and reached 26.6 million hits in April 2012. In the same 3.5 years, the search results for the abbreviated form "nano" surged from 83.8 million hits in 2007 to a whopping 129 million hits in 2011 and 299 million hits in 2012. By 2013 there were many new terms with "nano" prefixes that expanded the entire search category to more specific nano terms, which actually shrank the number of hits by distributing them among many narrow categories (Figure 2.4).

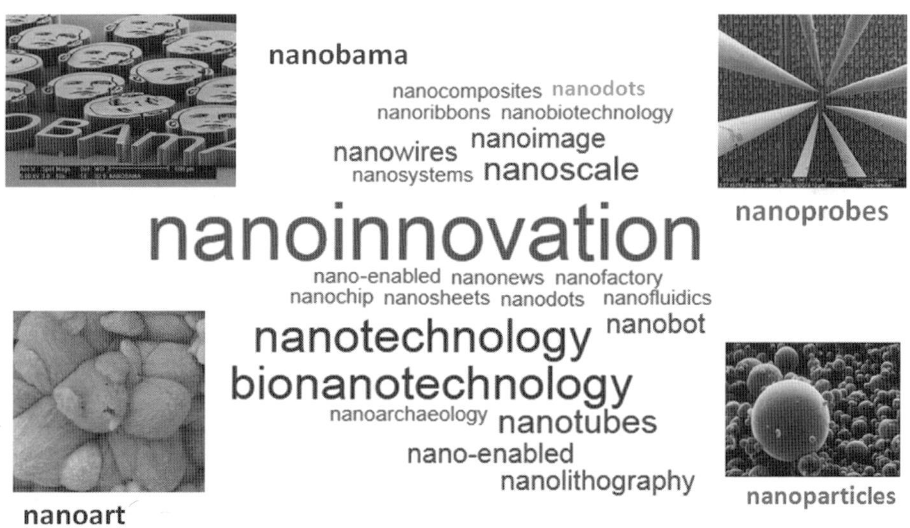

Figure 2.4 A nanoswarm of nano-neologisms. This word cloud includes just a few of the hundreds of nano-neologisms that have entered human language from the realm of nanotechnology. From the nanoimages of President Barack Obama (Nanobama) to a nanoart teddy bear, these terms are becoming a part of modern jargon. The prefix "nano" has been attached to everything from archaeology to zeolites. The term "nano" is used to describe a small car (Tata Motors calls their small car the "Nano") and anything else that is small, compact, and innovative and yet powerful.

If you go to any online dictionary and type a basic term like "nanotechnology," you'll discover more than half a dozen different definitions that tell you that nanotechnology is

- the science of *manipulating materials* on an atomic or molecular scale (*Merriam-Webster*);
- the science of *building devices* such as electronic circuits from single atoms and molecules (www.answers.com);
- a *branch of engineering* that deals with things smaller than 100 nm (*WordNet*);
- the study of *controlling matter* on an atomic and molecular scale (*Wikipedia*);
- the science and technology of *creating nanoparticles and manufacturing machines* in the range of 0.1–100 nm (*Wiktionary*);
- involving *structures, devices, and systems having novel properties and functions* due to the arrangement of their atoms on the 1–100 nm scale (*Foresight Institute*).

So which is it? Science? Engineering? Molecular manufacturing? Most definitions include the same elements: (a) scale (typically 1–100 nm or less), (b) exploitation of size-dependent properties that are unique at this scale, (c) manipulation (including engineering and manufacturing) of nanoscale structures/materials/processes, (d) creation of nanoscale-enabled devices, and (e) at the extreme end of the spectrum, some definitions include the prospects for molecular manufacturing, self-organizing nanomaterials, and nanobots.

Nanotechnology – In Any Language, It's Still "Nano"

Nanotechnology is truly ubiquitous, as revealed by this list of 27 foreign language translations for the word "nanotechnology" – compiled from Wiktionary and other online sources:

- **Arabic:** تقنية النانو (tiqníyya al-naanu)., نانوتيكنولوجي (nanutiknuluji)
- **Armenian:** նանոտեխնոլոգիա (nanotexnologia)
- **Belarusian:** нанатэхналогія (nanatexnalóhija).
- **Bengali:** ন্যানোপ্রযুক্তি bn
- **Bulgarian:** нанотехнология (nanotehnológija)
- **Chinese:** Mandarin: 納米科技 (nàmĭkējì)
- **Dutch:** nanotechnologie
- **English:** nanotechnology
- **Egyptian:** ونانلا ايجولونكت
- **French:** nanotechnologie
- **Georgian:** ნანოტექნოლოგია ka
- **German:** Nanotechnologie
- **Greek:** νανοτεχνολογία (nanotechnología)

- **Hebrew:** ננוטכנולוגיה he
- **Hindi:** नैनोतकनीकी (nainōtaknīkī)
- **Italian:** nanotecnologia
- **Japanese:** ナノテクノロジー (nanotekunorojī)
- **Korean (Hangul):** 나노기술 (nano-gisul)
- **Persian (Farsi):** فناوری نانو
- **Russian:** нанотехнология (nanotexnológija)
- **Spanish:** nanotecnología
- **Swahili:** nanoteknik
- **Thai:** นาโนเทคโนโลยี (naa-noh-ték-noh-loh-yee)
- **Turkish:** Nanoteknoloji
- **Ukrainian:** нанотехнологія (nanotexnolóhija)
- **Urdu:** قزمہ طرزیات
- **Vietnamese:** công nghệ nano

How Organizations Define Nanotechnology

Let's take a quick look at how some credible organizations in the United States and Europe define nanotechnology. This is important because these definitions can determine who qualifies for project funding, inclusion in government initiatives, research parameters, conference topics, and, in general, the allocation of resources.

The NNI is the pace-setting US program established in 2001 to coordinate nanotech R&D across 25 Federal agencies and 8 functional areas.[3] The NNI defines nanotechnology as "the understanding and control of matter at dimensions between approximately 1 and 100 nanometers, where unique phenomena enable novel applications. Encompassing nanoscale science, engineering, and technology, nanotechnology involves imaging, measuring, modeling, and manipulating matter at this length scale."

Mike Roco, who led the development of the NNI, notes that nanotechnology involves the *control* of nanoscale structures and processes. "It is not accidental," he explains. "These are things we can control, which is the real power of nanotechnology."

One of the longest and most comprehensive definitions comes from the National Science Foundation that takes 176 words to define nanotechnology: "Research and technology development at the atomic, molecular or macromolecular levels, in the length scale of approximately 1–100 nanometer range, to provide a fundamental understanding of phenomena and materials at the nanoscale and to create and use structures, devices and systems that have novel properties and functions because of their small and/or intermediate size. The novel and differentiating properties and functions are developed at a critical length scale of matter typically under 100 nm. Nanotechnology

3) National Nanotechnology Initiative Strategic Plan, National Science and Technology Council, Committee on Technology, Subcommittee on Nanoscale Science, Engineering and Technology, February 2011.

research and development includes manipulation under control of the nanoscale structures and their integration into larger material components, systems and architectures. Within these larger scale assemblies, the control and construction of their structures and components remains at the nanometer scale. In some particular cases, the critical length scale for novel properties and phenomena may be under 1 nm (e.g., manipulation of atoms at ~0.1 nm) or be larger than 100 nm (e.g., nanoparticle reinforced polymers have the unique feature at ~ 200–300 nm as a function of the local bridges or bonds between the nano particles and the polymer)."[4]

The definition used by the European Patent Office (EPO) reflects the view that nanotechnology is a bridging technology. The EPO does not set a lower limit of 1 nm: "The term nanotechnology covers entities with a geometrical size of at least one functional component below 100 nanometers in one or more dimensions susceptible of making physical, chemical or biological effects available which are intrinsic to that size. It covers equipment and methods for controlled analysis, manipulation, processing, fabrication or measurement with precision below 100 nanometers."[5]

2.1.4
How Small Is Nanoscale?

Nanoscale and *nanometer* are the core metrics of nanoinnovation. Decision-makers in government, industry, and academia use these nanoscale metrics to define and award nanotechnology research grants, to identify and regulate nanomaterials used in products, to determine quality control standards in semiconductors and filtering systems, study the safety of nanoparticles, determine labeling requirements for consumer products, frame marketing strategies, make investments, and much more. That's why it's important to understand the metrics of nanoinnovation, starting with the distinction between nano*meter* and nano*scale*.

A **nanometer** is one-*billionth* of a meter (1/1 000 000 000) or (10^{-9}), typically abbreviated "nm." This compares to a micrometer (or micron) that is one-millionth of a meter, and a millimeter that is one-thousandth of a meter. There are smaller measurements than a nanometer. A picometer is one-trillionth of a meter (10^{-10}), and an ångström is equivalent to one-tenth of a nanometer (10^{-12}).

The **nanoscale** is typically defined as a range extending from 1 to 100 nm, although some definitions count everything under 100 nm as nanoscale, and some organizations use the phrase "on the order of" 100 nm or less. This is a somewhat arbitrary standard, since the upper limit of 100 nm is not cast in stone. The unique properties associated with nanoscale materials can also occur at larger sizes.

4) National Science Foundation, Nanotechnology Definition, NSET, February 2000.
5) Nanotechnology and patents, European Patent Office, 2009.

It is difficult to wrap our minds around something as small as a billionth of a meter. You can't see individual nanoparticles with the naked eye, or even with a conventional microscope because nanoscale objects are actually smaller than light waves used in most optical imaging systems (visible light corresponds to a wavelength from about 400 to 700 nm). Imagine particles that are smaller than light waves! Nanostructures are so small, they need to be measured by instruments that sense and image beams of electrons instead of beams of light.

We can find nanoscale examples created in Nature, and by human engineers. A DNA molecule is ~2.5 nm wide and up to 1 m long. The smallest virus (parvovirus) is 25 nm across. A typical protein is about 10 nm long. Your fingernails grow about 1 nm in 1 s.[6)]

Manmade examples include a variety of carbon nanostructures that include carbon nanotubes 1 nm in diameter, quantum dots that are clusters of (typically 5–20) atoms, and carbon fullerenes. Fullerenes are formed from carbon atoms arranged in a pattern that resembles the pattern in geodesic domes. Fullerenes can be spheres (buckyballs), ellipsoids, or cylinders. Carbon nanotubes, the best-known nanoscale structures made by humans, are a type of fullerene. Buckminster fullerenes were named in honor of architect Buckminster Fuller, who designed the geodesic dome. Variations on these structures include "carbon onions," which are spherical shells made of concentric layers of graphene.

Some nanoscale issues are subject to interpretation and this can be confusing. For example, should nanoscale as a metric include structures below 1 nm, or stop at 1 nm? Is the upper range 100 nm? What about 101 or 102 nm, or 200 nm? Experts are still debating these issues. While scientists do not favor fuzzy definitions, it is probably most accurate to say that the nanoscale is "approximately" or "on the order of" 1–100 nm.

If we say that the nanoscale includes everything under 1 nm, we're essentially including everything, because all atoms are less than 1 nm. Atoms – the basic units of matter – range in diameter from one-tenth of a nanometer to more than half a nanometer. An atom of gold is about one-third of a nanometer. The smallest atoms (measured by its radius) are helium and hydrogen atoms, at around one-tenth of a nanometer, while cesium, one of the largest, is about seven-tenths of a nanometer. Most atoms are fairly close in diameter because, as the mass of the atom increases, the increased positive charge pulls the electrons closer to the nucleus, so the radius doesn't vary much.[7)]

How nanoscale is specified can present special challenges for nanoinnovators. For example, academic researchers who are working with nanoinnovations in the range

6) Thomas Kenny, Associate Professor of Mechanical Engineering, quoted in a Stanford University news release; July 9, 2001.

7) When most of us were studying chemistry, we were taught that electrons travel in neat orbits around the nucleus of an atom – we now know that electrons actually form "clouds" around atoms. This makes it difficult to measure the physical size of an atom. Also, individual atoms are constantly seeking to bond with another atom to form a molecule, which makes it difficult to segregate and study an individual atom. Consequently, the measurement of the "diameter" of an atom is accomplished by calculating the distance between the nuclei of two atoms and splitting the difference.

of 200–500 nm are often forced to find creative ways to include a 100 nm dimension to their project, to meet the arbitrary definition of nanoscale. A few researchers interviewed for this book reported having projects at the 200–500 nm scale excluded from nanotechnology funding by agencies that adhere strictly to the 1–100 nm definition.

2.1.5 What Are the Unique (Quantum) Properties of Nanotechnology?

In nanoinnovation, it's not the scale that counts, but rather the unique properties at this scale that provide the true value. When we move from the macroscale to the nanoscale, nanoinnovators gain access to an entirely new set of optical, electrical, physical, and chemical properties called "quantum effects" that derive from the interaction of matter and energy at the nanoscale. While these properties are most prevalent from 1 to 100 nm, they can also exist at the subnanoscale (<1 nm) or at larger scales up to 500 nm or more. These properties, which are still being discovered and tested, are an innovator's dream. Suddenly, the alchemist's toolkit contains real magic and the tricks coming out of the nanoscale "magician's hat" are truly amazing.

Using the tools of nanotechnology, scientists are learning that the properties we always assumed were more or less constant for a given substance such as gold or carbon can exhibit radically different properties and characteristics at the nanoscale. Many of these quantum effects involve the availability, movement, or spinning of electrons and the use of light as a force. Some of these effects are tunable, which means their conductivity or reactivity can be tuned along a gradient of temperature, force, light, or other parameters.

Chemistry courses were not teaching modern nanotechnology when most of us were in high school or college, so let's take a look at some of the unique properties that are driving the current wave of nanoinnovation. Here is a partial list of properties that provide opportunities for innovation at the nanoscale, with some brief notes for each:

- **Chemical Effects**
  - Chemical reactivity and reaction rates (*while gold is normally considered an inert metal, gold nanoparticles under 5 nm are catalytically reactive*)
  - Increased surface area (*the surface-to-volume ratio of nanoparticles is significantly greater than one large particle of the same size that is extremely important for catalysts*)
  - Super atoms (*clusters of atoms called "super atoms" can mimic the properties of an entirely different element, for example, a cluster of 13 aluminum atoms can behave like a single iodine atom* [7])
- **Electrical Effects**
  - Electrical conductivity (*materials that are electrical insulators at bulk scale exhibit conductive properties at the nanoscale; for example, electric current can pass nondestructively through liquid glass electrodes that may have applications for labs-on-a-chip* [8])

- **Optical Effects**
  - Transparency (*nanoparticles of zinc oxide or titanium dioxide are used to make sunscreens and cosmetic creams transparent because the small particles let visible light through, while blocking harmful ultraviolet light – in contrast, large particles of zinc oxide appear white and scatter light making the cream opaque*)
  - Color shift (*e.g., nanosized gold appears red or purple in color*)
  - Tunable photoluminescence and electroluminescence (*photoactive quantum dots can be tuned to change color*)
  - Photoactivity (*photoactive nanomaterials can control transmittance of heat and light in solar cells and windows; photoactive nanomaterials include metal oxides such as nanoscale titanium dioxide* [9]*); photocatalytic nanomaterials are used in water purification, and in appliances and medical environments to kill antibiotic-resistant bacteria*)
- **Physical Effects**
  - Hardness (*superhard nanocomposite materials have been synthesized from non-carbon materials that are nearly as hard as a diamond* [10])
  - Boiling point (*nanofluids boil at a higher boiling point; nanomaterials can reduce the energy needed to boil liquids*)
  - Increased strength (*carbon in graphite form is soft and malleable, but at the nanoscale can be stronger than steel*)
  - Metamaterials (*artificially engineered nanomaterials exhibit properties not found in Nature; for example, metamaterials with a negative refractive index have the ability to "bend light," which is being researched to develop invisibility cloaks and new types of optics*)
  - Ductility and malleability[8] (*materials such as silica are brittle in bulk form, but can become ductile at the nanoscale*)
  - Porosity (*aerogels are high-porosity solids where up to 95% of the volume consists of nanoscale pores*)
- **Temperature Variations**
  - *Nanoparticles tend to have a lower melting point than in their bulk size; a 4 nm gold particle melts at 850 °C (1562° F.) compared with bulk gold that melts at 1064 °C (1947.2° F).*

Together, these properties are often characterized as – or have their roots in – the quantum effects that occur when atoms and molecules interact. To illustrate this concept, consider that large objects are influenced by gravitational and magnetic forces that we're all familiar with. However, at the nanoscale, electromagnetic forces and the movement of electrons become a dominant factor. The *mobility of electrons* determines whether a material is conductive or not. Scientists use the movement of electrons to create nanoscale images of

8) Ductility is the ability of a solid material to deform under tensile or stress, such as the ability of a metal to be heated and stretched into a wire; while malleability is the ability of a material to deform under compressive stress exemplified by being able to hammer or roll gold into a thin sheet.

atoms and molecules that are actually smaller than light waves and cannot be imaged using optical microscopes.

Another property that is being exploited by nanoinnovators involves the tendency of electrons to "spin" – which has opened the entire field of spintronics. Real-world applications include using spinning electrons to create quantum computers.

Some of the most exciting and radical innovations come from combining two or more nanoscale properties to produce an entirely new capability or effect. Teams of scientists at the US Department of Energy's Brookhaven National Laboratory and Los Alamos National Laboratory have fabricated thin transparent films capable of absorbing light and generating an electrical charge, which they achieved by blending a semiconducting polymer with carbon fullerenes [11].

2.1.6
The Metrics of Nanoinnovation

In a very short time, nanotechnology has generated a legion of statistics. It's easy to find the numbers on Web sites, in market research reports, press releases, government documents, and academic papers. However, to a nanoinnovator, it's not just the numbers that count, it's what they represent. Many of these numbers are called "industry" metrics that begs the question: Is nanotechnology really an industry?

Is Nanotechnology an Industry?

Government agencies, market analysts, and the media routinely talk about nanotechnology as an industry. However, most of the nano-insiders interviewed for this book agree that nanotechnology is *not* an industry – but acts like one. This can be confusing. For example, there are no SIC or NAIC industrial classifications for nanotechnology. A nanocircuit might be grouped in SIC Code 3674, Semiconductors and Related Devices, but doesn't have an industrial code of its own. Likewise, standards for nanomaterials are still being developed by industry organizations such as IEEE, IEC, and ISO.[9] A separate patent classification for nanotechnology inventions was not established until 2004.

Does this mean nanotech is not an industry? Is this an early-stage industry? Or is it a set of enabling technologies that act like an industry?

Technically, an industry is *a group of companies that operate in the same segment of the economy or share a similar business type.* By this definition, we should be able to find clusters of companies that are entirely focused on nanotechnology – and we can. More than 2000 companies worldwide are involved in nanotechnology. There are companies that manufacture carbon nanotubes

9) I have been privileged to serve on the senior leadership of the IEEE1784 committee led by Dr. Brent Segal, which is developing standards for nanomaterials used in electronics; in 2011 this committee was merged with the IEC (International Electrotechnical Commission) into joint project IEC/IEEE JPT 62 659, Large Scale Manufacturing for Nanoelectronics.

in a variety of form factors as well as quantum dots, fullerenes, and metal oxides. More than a dozen companies make nanoscale instruments such as atomic force microscopes (ATMs), imaging systems, probes, and fabrication equipment. There is a constant stream of nanoventures emerging from universities and government programs, and many corporations have nanotechnology divisions.

Whether you consider nanotechnology to be an industry or not, the same metrics used to evaluate any industry can be used to analyze nanotechnology. Many of these metrics are regularly updated on Web sites such as Nanowerk, AZoNano, the NNI, and Lux Research. Let's take a look at some key metrics that were current during the writing of this book.

Nanotech Companies

In 2010, Nanowerk listed more than 2100 companies from 48 countries involved in nanotechnology research, manufacturing, and applications.[10] By 2013, the number of nanotechnology companies had grown to almost 5000 firms in the United States alone, which is more than half of all nanotechnology companies in the world. There are almost 1000 nanotech companies in Germany and more than 600 firms in Japan.

More than 110 companies that manufacture carbon nanotubes are listed in Appendix B. This list was compiled by Hai-Yong Kang, Ph.D., for a report entitled: "A Review of the Emerging Nanotechnology Industry: Materials, Fabrications, and Applications" [12]. (Author's note: *Kang's report also includes company lists for quantum dots and metal oxides, and the core document includes some very interesting examples of emerging nanotechnologies.*)

Nanotech Revenues and the Value Chain

As early as 2001, the National Science Foundation predicted that the nanotechnology market would reach $1 trillion by 2015 and other sources (Cientifica *et al.*) have projected a multitrillion dollar nanotechnology market. Nanotechnology revenues grew from approximately $380 billion in 2010 to approximately $2.4 trillion in 2015 according to Lux Research (Figure 2.5). The Lux Research forecast was based on a nanotechnology value chain created by Lux that breaks the nanotechnology sector into nanomaterials, nanointermediaries, and nano-enabled products. This is one of several ways that this "industry" can be defined. These projections provide a basis for calculating year-to-year metrics that reveal important growth and sector trends.

These kinds of projections should be read with a critical eye, since they can be misleading. For example, "nano-enabled products" can include such things as automobiles, video displays, paintings and coatings, semiconductors, drugs, and so on – which can inflate the revenue totals if the value of the entire product is counted, since in many of these products, the quantity of

10) Well over 2000 companies worldwide already involved in nanotechnology, *Nanowerk News*, July 1, 2010, www.nanowerk.com.

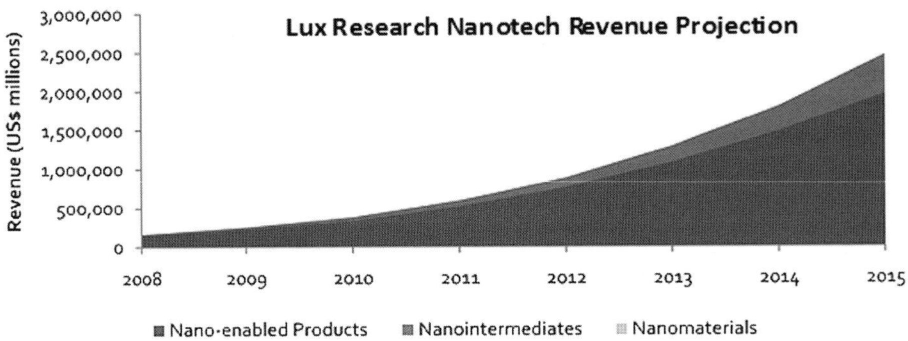

Value chain stage	2008	2009	2010	2011	2012	2013	2014	2015
Nano-enabled products	$145,291	$223,785	$336,062	$519,425	$762,204	$1,081,025	$1,480,928	$1,962,950
Nanointermediates	$18,353	$28,839	$45,592	$75,712	$120,206	$206,823	$322,691	$498,023
Nanomaterials	$812	$1,074	$1,309	$1,540	$1,798	$2,098	$2,462	$2,916
Total	$164,457	$253,699	$382,963	$596,677	$884,208	$1,289,947	$1,806,081	$2,463,890

Source: Lux Research, Inc.
www.luxresearchinc.com

Figure 2.5 This chart by Lux Research shows nanotechnology revenue growth through 2015 – including a downward adjustment that reflects the impact of the economic downturn that began in 2008.

nanomaterials may be a few grams, or a fraction of a gram. Therefore, as awesome as most nanotechnology revenue projections may sound, they need to be considered in context.

Global Investment in Nanotechnology R&D

One way to quantify nanoinnovation is to calculate public and private R&D funding – recognizing that R&D inputs don't always translate directly into commercial outputs. Cientifica estimates that global government investment in nanotechnology is $10 billion a year, with cumulative government funding reaching $100 billion by 2014. When corporate research and venture funding are included, the cumulative total spent on nanotechnology R&D is expected to reach $250 billion by the end of 2015 [13].

More than a dozen countries have invested significant government funds in nanotechnology, and exploited nanoinnovation to achieve institutional and commercial benefits. By most metrics, the leading countries involved in nanoinnovation include the United States; Germany, the United Kingdom, and France in Europe; China, Japan, South Korea, India, and Taiwan in Asia; Brazil and Mexico in Latin America; Israel and Iran in the Middle East; and Russia. Most of these countries have benefited from investing substantial public funds in nanotechnology initiatives.

In the United States, government agencies, corporations, and venture capitalists invested more than $18 billion in nanotechnology research from 2001 to 2014. Until 2010, most of the research in the United States was paid for with government funds (*2010 was the first year that corporate R&D investments exceeded government funding*).[11)]

In Europe, public funding for nanotechnology research was provided under the Sixth and Seventh Framework Programmes (*FP6* and *FP7*). Under FP7, the European Union invested approximately €3.6 billion for nanotechnology research from 2007 to 2013. The next science technology plan of the European Union, called Horizon 2020, does not break out nanotechnology investments as a separate category, but does include a special billion-euro initiative to commercialize graphene.

While the NNI in the United States receives a great deal of publicity in the media, there are significant initiatives underway in many parts of the world. A good example is the Russian Corporation of Nanotechnologies – better known as Rusnano (Роснано) – which was established by the Russian Federation to create a Russian nanotech industry by 2015. In 2008–2009, more than 60 projects were approved for funding by Rusnano. Investments, which include both direct investment and cofunding with partners, are projected to exceed US$8 billion.[12)]

Large ongoing investments by firms such as IBM, Intel, BASF, and Lockheed-Martin in the United States; Fraunhofer, Bayer, and Siemens in Europe; Hitachi, Samsung, and Mitsubishi in Asia – to name a few – drive corporate R&D in nanotechnology. Many of these activities involve industrial secrets and embedded technologies that improve manufacturing processes, make products less expensive, more durable, and lightweight.

In the venture capital sector, nanotechnology took a hit when the Internet and biotech bubbles burst in 1999–2000. The VC sector started to recover in the early to mid-2000s, then saw declines in equity investments after the economic downturn that began in 2008. Lux Research reported that venture capital for nanotechnology dropped as much as 43% in 2009. Venture Capital investments fell from over $1.4 billion to $792 billion, and the average deal size fell by 41% to $8.6 million across 92 deals. These statistics are based on interviews by Lux analysts with 1000 nanotechnology startups and corporate participants, and 15 leading venture capital firms that are investing in nanotechnology [14].

Nano-enabled healthcare investments have accounted for more than half of these investments, followed by energy-related ventures. Jurron Bradley, former senior analyst and lead author of the Lux report, explained that VCs understand that nano-enabled healthcare solutions require long lead times for clinical trials and regulatory approvals; however, these risks are offset by the fact that healthcare is a high-margin, premium-based sector.

11) Party on Nanotechnology, Organization for Economic Cooperation and Development, OECD /NNI International Symposium on Assessing the Economic Impact of Nanotechnology; Background Paper 2: Finance and Investor Models in Nanotechnology; March 16, 2012.

12) Rusnano (www.Rusnano.com); translated using Google translator (accessed March 28, 2011).

To invest in nanotechnology research, VCs need patience and staying power, since the translation of research from laboratory to market has been relatively slow in this sector – although the long-term rewards and potential for commercial breakthroughs are high.

"To invest in nanotech you have to have a passion about it," says Peter Balbus, a business strategist whose experience ranges from CNT yarns and fabrics to medical sensors and nanoparticles. "This isn't just a business decision. It has to be something where you have a belief. It's a matter of faith. Those who get into the industry early on, even if the initial investments don't pan out as originally hoped, have gained experience as the industry evolved and became better defined because they often discovered or created new trajectories. Ironically, in nanotechnology you may have to take greater risks early on and experience some failure in order to figure out how to achieve success later on. This is a real conundrum."

Another challenge facing nanoventures involves persuading their customers – who are often large corporations and manufacturers – to adopt their solutions. "If the nanotechnology solution involves an industrial product or process, getting the industry to migrate from a current technology to another solution takes time and effort, even if the new product has a lot of economic or performance improvement benefits," Peter observed.

How Many Products Use Nanotechnology?

Another way to measure nanoinnovation is simply to count the number of products. As previously noted, the Emerging Nanotechnologies Project of the Woodrow Wilson Institute/Pew Charitable Trusts maintains an online inventory of products and product lines that incorporate nanotechnology (http://www.nanotechproject.org/inventories). Consumer products that incorporate nanotechnology grew 521% from 212 products in 2006 to 1317 products in 2011, exceeding 2000 products in 2013. Note that these are primarily consumer products and do not include many industrial or biomedical products, and only included products in the Woodrow Wilson project that does not include all nanotech products.

Nanotechnology Patents

Patent filings can be a treasure trove of information to nanoinnovators, although some of these treasures may be buried and may require some sleuthing to uncover. During the 1990s and early 2000s, nanotechnology inventions were not classified as "nano." Also, many inventions do not mention nanoscale or refer to nanotechnology, which makes it difficult to classify these inventions, especially in the field of biosciences and medicine. As a result, many nanoinnovations have been underreported or excluded from patent totals.

In October 2004, the US Patent Office (USPTO) established a nanotechnology cross-reference – Class 977 – that has been expanded to 263 subclasses. An early effort by the Patent Office to classify nanoinnovations produced a list of more than 6000 nanotechnology patents, although the classification process is still in progress. Only a few hundreds of these patents included "nano" in the title, illustrating how difficult it is to categorize nanotech patents.

In 2005, the EPO established a tagging system that assigns a "Y01N" tag to nanotechnology patents covering six main groups: Y01N2-Nanobiotechnology, Y01N4-Nanotechnology for information processing, storage, and transmission, Y01N6-Nanotechnology for materials and surface science, Y01N8-Nanotechnology for interacting, sensing, and actuating, Y01N10-Nano optics, and Y01N12-Nanomagnetics.

In 2012, the USPTO issued a record 4000 patents, identified as nanotechnology patents under Class 977 compared with 780 in 2010. In 2012, more than half the nano patents (54%) came from the United States, followed by South Korea (7.8%), Japan (7.1%), Germany (6.2%), and China (4.9%) [15]. In the past few years, the United States, Japan, South Korea, Germany, Taiwan, and China have accounted for most nanotech patents, with the East Asian countries coming on strong. These countries also lead in publications, government funding, and corporate investment. In general, about one-third of nanotech patents originate in East Asia, and recent years have seen a steady increase in patents from Russia and India. In the United States, the states accounting for the most nanotechnology patents were California, New York, Massachusetts, Texas, New Jersey, Pennsylvania, and Illinois.

Some notable examples of US patents include an IBM patent for a self-healing nanomaterial – which includes a self-repair feature provided by encapsulated nanoparticles (U.S. Patent 7,799,849). Another example is Nantero's patent for a high-purity nanotube fabric (U.S. Patent 7,675,766). Nantero, which is profiled in this book, has one of the largest nanotechnology patent portfolios. Other interesting examples include a patent for graphene platelets from Nanotek Instruments Inc. (U.S. Patent 7,790,285); a method for growing uniform diameter nanowires from the Lieber Research Group at Harvard University (U.S. Patent 7,666,708); and a nanostructured silicon lithium battery by Nexeon Ltd. (U.S. Patent 7,683,359). Nexeon had previously announced a world record capacity rechargeable battery based on this technology [16]. In 2010, 230 U.S. patent applications mentioned graphene, more than double the graphene claims submitted in 2009 [17]. Several thousand patents refer to novel methods for purifying DNA.

At this early stage in the development of nanoinnovation, these patent records – which are publicly available – are the best places to scan for research trends. Patent records can also help identify emerging nanotechnologies that may be available for license or purchase, entrepreneurial ventures to invest in, and corporate or university partners who may want to form alliances.

Research Papers

The nanotechnology field has produced a flood of academic papers, which includes research in the field, as well as redundant material as each country's scholars produce papers in their own language. More than 60 countries fund research in nanotechnology, although studies have shown that only 15 countries produce 90% of the research papers.

As part of a project funded by the National Science Foundation, University of Manchester business professor Philip Shapira and USPTO Patent Examiner Jue Wang used data mining techniques to analyze more than 91 000 nanotech-related papers published worldwide in the 1-year period from August 2008 to July 2009 [18]. They found that the top four countries generating nanotech papers (by author affiliation) were the United States (23%), China (22%), Germany (8%), and Japan (8%). About 3% of the papers reported cofunding from the US National Science Foundation and the Chinese National Natural Science Foundation. This sounds like a relatively small number, but actually signals significant and ongoing nanotechnology collaboration between the United States and China. Shapira and Jue suggest that national agencies involved with nanotechnology can improve their outcomes by fostering more international partnerships between domestic researchers and their colleagues overseas. This will be especially important in the coming decade as smaller national budgets and new emerging technologies are expected to reduce the funds available to the global nanotech community.

2.1.7 The Need to Become Nanodextrous

Nanodextrous is a word I coined a few years ago to describe how companies need to develop new capabilities at the nanoscale at the same time they continue to work at the current macroscale. Nanodextrous comes from "ambidextrous," which means being able to do something equally well with both hands. Nanodextrous means being able to work equally well at the nanoscale and at the macroscale (Figure 2.6).

The greatest nanodextrous challenge involves turning nanomaterials into sizes and dimensions that can be used in human scale products. On one hand, we know how to create a carbon material called graphene that is one atom thick, transparent, and superstrong. On the other hand, if we wanted to include graphene in the wings of a space shuttle, we would need to produce sheets that are several meters in length. The need to manufacture large quantities of uniform, defect-free graphene and other materials has posed a serious technical hurdle. While some nanoinnovators see this as a challenge, most see this as an opportunity.

Whether you work in government or business or research/academia, you will need to pursue a nanodextrous approach if you want to cultivate nanoscale

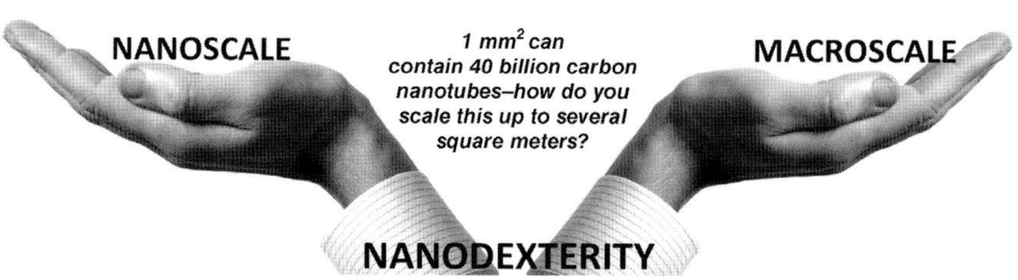

Figure 2.6 Nanodexterity.

capabilities and integrate nanomaterials into your macroscale products and processes. This will involve new strategies and new forms of organizations that are able to work in both worlds. In the decades to come, it will be the nanodextrous organizations that will hit the home runs and score the winning goals in the field of nanoinnovation.

2.1.8
Where Are We Now in the Evolution of Nanoinnovation?

The course of nanotechnology has been moving through five overlapping phases that can be characterized as follows: (1) **Nanoscience**: a period of basic research and discovery (1960–1980), (2) **Nanocore**: development of "core" tools and materials (1981–2000), (3) **Nano-Commerce**: a decade of commercial development (2001–2011), (4) **Nano-Breakthroughs**: a decade of radical/transformative innovation (2012–2022), and (5) **Nano-Future**: horizon innovations (2023–2050 and beyond). These dates are somewhat arbitrary, but are grounded in seminal events and discoveries (Figure 2.7).

Phase 1: Nanoscience (1960–1990)
The field of modern nanotechnology arguably began with Richard Feynman's scientific call to arms on December 29, 1959. Scientists began a serious push to delve into the secrets of molecular interaction and the behavior of atomic clusters. Many of today's nanoscience pioneers were still in school – and beginning to cultivate the ideas and paths of investigation that led them to great discoveries. Companies like IBM and Intel – now famed for their cultures of innovation – were focusing their creativity and resources on the bottom of the scientific pyramid.

Phase 2: Nanocore (1981–2000)
The "tools and materials" phase began in earnest with the development of the scanning tunneling microscope (STM) in 1981, which earned the Nobel Prize in

THE EVOLUTION OF NANOINNOVATION

Phase One	*Phase Two*	*Phase Three*	*Phase Four*	*Phase Five*
1960-1980	**1981-2000**	**2001-2011**	**2012-2022**	**2023-2050**
Nanoscience	**Nanocore**	**Nano-Commerce**	**Nano-Breakthroughs**	**Nano-Future**
Discovery and Basic Research	Tools and Materials	The First Wave of Commercial Innovation	Radical/Disruptive Innovation	Innovations on the Far Horizon

Figure 2.7 The evolution of nanoinnovation. Nanoinnovation has moved from the first discoveries and basic research to a core set of tools and materials, which in turn enabled the first commercial applications. The next wave will produce the first breakthroughs that are truly disruptive, followed by several important "far horizon" innovations that will be implemented in the next few decades – if we can negotiate several technical hurdles.

Physics for its inventors (Gerd Binnig and Heinrich Rohrer at IBM Zürich) in 1986 and allowed scientists to finally visualize, image, and manipulate what was previously based mostly on theory. The discovery of new allotropes of carbon included the discovery of fullerenes in 1985 and carbon nanotubes in 1991. During the Nanocore phase, most major industries and government agencies compiled nanotechnology road maps describing the possibilities and objectives of nanotechnology.

Phase 3: Nano-Commerce (2001–2011)

The launch of the NNI in 2000 and the first NNI research funding in 2001 provide another benchmark – the beginning of nano-commerce. NNI funding jump started a global surge in nanotechnology research and fostered the first wave of commercial nanoinnovation. In its first decade (2001–2011), the NNI invested more than $14 billion in nanotechnology R&D.

During this period, organizations such as the IEEE and IEC began compiling technical definitions and industrial standards. The European Union established the first nano-ingredient labeling requirement. Large corporations such as Lockheed-Martin formed nanosystem divisions. Pharmaceutical firms and biotechs started the first wave of bionano start-ups. Most of the world's leading science and engineering universities established nanoscience centers. Venture capital investments in nanoinnovation exceeded $1 billion a year.

Many nanomaterials such as carbon nanotubes and quantum dots became readily available. In 2011, Mitsui announced that it was producing 30 metric tons of carbon nanotubes each year. Carbon nanotubes are integrated into Mitsui's chemical products that focus on nanoporous membranes and carbon nanotubes. Other companies are producing industrial quantities of nanoscale metal oxides and nanocatalysts – anchoring the first decade of nano-commerce. Other nanotube suppliers include CNano Technology, Nanocyl, Showa Denko, and Arkema.

Until 2013, Bayer in Europe operated a plant and laboratory with a capacity of 260 metric tons per year for the development of carbon nanotube products and processes. However, in May 2013, Bayer announced that it was exiting industrial production of carbon nanotubes primarily due to overcapacity of carbon nanotubes in the market, a steep drop in the price of nanotubes (*from approximately $700/kg in 2006 to $50/kg in 2013*), and fewer uses in Bayer products. While Bayer's exit from carbon nanotube production came as a surprise, the drop in nanotube prices also suggests that carbon nanotubes have become commoditized like many materials, and this may be the first sign of an industry shakeout.

Phase 4: Nano-Breakthroughs (2012–2022)

We are already seeing the first radical nanoinnovations that mark the "breakthrough" decade that began in 2012. Radical innovations can be defined as novel developments that transform markets, consumer behaviors, and patterns of consumption. During this phase, we should expect major improvements in manufacturing processes that will provide the first commercial-scale quantities of graphene and other materials that were previously produced only in laboratories. These breakthrough innovations range from stretchable electronics to nano-enabled cancer therapies. Quite a large number of game-changing innovations have already been reported by researchers and are currently struggling to move from laboratories to markets, a process that typically takes about a decade. Many of these examples are featured in this book.

This Nano-Breakthrough phase also includes a wave of new nanosensors as scientists and engineers refine their capabilities to sense and detect almost any measurable parameter. There are many clues to the future of sensors, such as Google's prototype of a "Project Tango" smartphone that can make over 250 000 3D measurements every second of a surrounding space and update its position in real time to create an indoor 3D map – providing an unprecedented ability to develop novel applications based on space, proximity, and motion. This device could allow users to see and navigate in the dark, help firefighters find their way through smoke-filled buildings, and help vision-impaired users navigate unfamiliar places.

Phase 5: Nano-Future (2023–2050)

The period from 2023 to 2050 is a somewhat arbitrary timeframe designed to encompass the phase that is currently beyond our ability to reliably predict the outcomes. This is the "far horizon" phase of nanoinnovation, when many futurists and technology leaders expect to see nanoinnovations that are true game changers. In this phase, we should expect entirely new computing architectures (quantum computing, biological computers), energy generation and storage, and new form factors for electronic devices. These advances could provide the first implementations of molecular manufacturing and self-assembly, including nanobots and artificial blood cells, as well as commercial applications for plasmonics, spintronics, and light-bending metamaterials. It will be common to manipulate genes to treat and cure diseases. This is an era when entirely new paradigms for energy, healthcare, communication, and transportation will come into use – as science fiction migrates to science reality.

"Nano" Is to Technology What "E" Is to Commerce

To gain a sense of how nanotechnology and nanoinnovation may evolve, we can draw insights from what happened with other emerging technologies in the past. Such analogies offer important clues to what might happen in the future.

For example, in some ways, nanotechnology is analogous to electronic commerce. In the late 1990s, e-Commerce was considered a new kind of commerce. But e-Commerce was not really new – electronic mail order, telecom call centers, and electronic banking were all in place before e-Commerce became a popular term (*actually, o-commerce or online commerce was probably a better term than e-Commerce*). e-Commerce required special skills and capabilities and tools to participate. You needed access to a server, Web site, and web design software. However, in only half a decade, e-Commerce became so accessible and ubiquitous that "electronic" commerce was simply absorbed into commerce. Today, anyone can easily and affordably put a Web site online and process online transactions. Anything can be ordered online and delivered in a few days. Amazon has even boasted that they can predict which products buyers will purchase and put them into the supply chain *before the buyer completes the sale* – with future deliveries made by flying drones!

Nanotechnology is merging into technology the way e-Commerce merged into commerce. Nanotechnology is also following in the footsteps of another analogous example – the microrevolution. In the 1970s and 1980s, the world was talking about "micro" the way we talk about "nano" today. Micro refers to one-millionth of a meter and was a hot buzzword when the first generation of microcomputers appeared in the mid-1970s. The microelectronics revolution gave us microprocessors, microcomputers, microfibers, microwave ovens, and other "micro" technologies that are commonplace today. The term "micro" has blended into the landscape and now micro refers to anything that is "small" or "tightly focused" as in "micromanaging." Today, we don't hear much about microscale technologies or microcomputers, although the term micro has retained its cachet as symbol of everything that is small, friendly, and innovative.

Nano is following a similar course. Nano is already embedded in our jargon, although nano is not as generic as micro, at least not yet. Many companies have called their products "nano" to leverage the image of small and friendly. Notable examples include the two-cylinder, $2500 vehicle called the Nano, introduced in 2009 by Tata Motors, India's largest car company. Apple's Nano, the postage stamp size version of the iPod, used nanotechnology in its chips and circuits. Silver Nano, a trademark of Samsung, is used to describe silver ion technology used in Samsung appliances.

Nano has quickly become ubiquitous and normalized. Many technology companies have nanoscale groups in their R&D and quality control centers. Nanomaterials such as graphene are being commercialized. Nanoimaging and fabrication services are readily available and affordable. The best universities have scanning tunneling microscopes and nanoscience curricula.

Of course, it is still possible that a major safety issue or nano-related tragedy could delay the march of nanotechnology into the future. Assuming we can avoid these pitfalls (which exist with every new technology), the word "nano" should remain as friendly as "micro" and continue to be associated with improvements and innovations at the scale of atoms and molecules. This image will strengthen as more nanoinnovations achieve commercial success. On balance, the evolution of nano will continue to generate exciting and novel solutions to many of society's problems.

2.1.9
A Short History of Modern – and Ancient – Nanoinnovation

Looking ahead to the future is fun. However, we also need to be cognizant of the history of nanoinnovation. The history of nano is actually two histories – modern and ancient. Following are some of the key milestones in the modern history of nanoinnovation, with some concise notes on each.

NANOINNOVATION MILESTONES: 1959–2013

1959 Richard Feynman's Lecture at Caltech: "There's Plenty of Room at the Bottom." This lecture set the scientific community on a course of investigation and discovery that will continue for decades to come (*a more detailed assessment of why this is true is included at the end of this list*).

1965 The Nobel Prize in Physics awarded to Richard Feynman, Julian Schwinger, and Sin-Itiro Tomonaga for their work on quantum electrodynamics.

1974 Norio Taniguchi (University of Tokyo) coined the term "nanotechnology" to describe semiconductor processes – he defined the term as "the processing of separation, consolidation, and deformation of materials by one atom or one molecule" [19].

1978 U.S. Patent 4,107,288 is issued to the Pharmaceutical Society of Victoria (Australia) for technology related to processes for making nanoparticles – later designated a Class 977 patent – making it the oldest nanotechnology patent.

1981 The first scientific paper on nanotechnology is presented –"Molecular Engineering: An Approach to the Development of General Capabilities for Molecular Manipulation" by Eric Drexler, in the *Proceedings of the National Academy of Sciences.*

1981 Invention of the scanning tunneling microscope by Gerd Binnig and Heinrich Rohrer, who developed the historic device at IBM's research laboratory in Zurich, Switzerland.

1985 Discovery of the carbon fullerene by British chemist Harry Kroto, who collaborated with US chemists Richard Smalley and Robert Curl who were studying clusters of atoms. Together they discovered the allotrope of carbon comprised of 60 carbon atoms arranged in 20 hexagons to form a sphere that resembles a soccer ball. The full name of the molecule is buckminsterfullerene, named after architect Buckminster Fuller because the arrangement resembled the hexagons in the geodesic dome designed by Fuller.

1986 The first book on nanotechnology is published – *Engines of Creation: The Coming Era of Nanotechnology* by K. Eric Drexler. Drexler's book includes the first use of the term *nanotechnology* to describe engineering at the nanoscale. Dr. Drexler is credited with introducing and launching the field of nanotechnology as we now know it.

1986 The Foresight Institute is established by Eric Drexler and Christine Peterson, with a mission "to ensure the beneficial implementation of nanotechnology."

1986 Gerd Binnig, Calvin Quate, and Christoph Gerber invent the atomic force microscope (AFM), a very high resolution type of scanning probe microscope.

1986 The Nobel Prize for Physics is awarded (50%) to Ernst Ruska for his design of the first electron microscope and (50%) to Gerd Binnig and Heinrich Rohrer for their design of the scanning tunneling microscope.

1987 Howard G. Tennent of Hyperion Catalysis is issued a U.S. patent for graphitic, hollow core "fibrils" (later known as carbon nanotubes).

1989 First manipulation of atoms – an IBM team led by Don Eigler spells "IBM" using 35 xenon atoms.

1991 Carbon nanotubes are discovered in the soot of an arc discharge by Sumio Iijima at NEC.

1991 First Ph.D. in molecular nanotechnology – awarded to K. Eric Drexler at MIT.

1991 The first "atomic switch" is built by the IBM research team led by Don Eigler.

1992 Publication of the first nanotechnology textbook – *Nanosystems: Molecular Machinery, Manufacturing, and Computation* by Eric Drexler (*based on his doctoral thesis*).

1993 First synthesis of single-walled carbon nanotubes – simultaneously reported by Sumio Iijima and Toshinari Ichihashi at NEC, and by Donald Bethune at IBM's Almaden Research Center.

1996 The Nobel Prize in Chemistry is awarded for the discovery of carbon fullerenes (60 carbon atoms arranged in a ball) – fullerenes are formed when vaporized carbon condenses in an atmosphere of inert gas) – awarded to Robert F. Curl, Harold W. Kroto, and Richard E. Smalley.

(*continued*)

(Continued)

1997	The first nanotechnology company is founded (Zyvex, founded by James Von Ehr, the computer pioneer who previously founded Altsys and helped build Macromedia).
2000	The US National Nanotechnology Initiative is established by the Clinton administration.
2000	The first transistor gates under 100 nm are used in semiconductor chips.
2002	The novel *Prey* by Michael Crichton is published, describing how a runaway research project creates bee-like swarms of self-aware nanodevices that threaten to destroy humanity.
2003	The Drexler–Smalley debate shines a public spotlight on the feasibility of molecular manufacturing, self-assembly at the nanoscale, nanodevices, nanobots, and threats from runaway nanosystems called "gray goo."
2004	First FDA approval and commercial launch of nanoparticulate (nanosized) drugs (developed by Wyeth, Merck, and Abbott using technology developed by Elan Drug Delivery). These drugs include Aprepitant (Emend®) by Merck; Sirolimus (Rapamune®) by Wyeth; and TriCor® by Abbott–resulting in better performance characteristics and bioavailability.
2004	The US Patent Office establishes Class 977 as a new class for nanotechnology patents.
2008	The first Kavli Prize in Nanoscience is awarded by the Norwegian Academy of Science and Letters (which also awards the Nobel Prize) – the $1 million prize is shared by Louis Brus (Columbia University) for his discovery of quantum dots, and Sumio Iijima (Meijo University) for his discovery of carbon nanotubes.
2009	The European Union announces that cosmetics manufacturers will be required to list nanoparticles contained in cosmetics products marketed in the European Union – using the designation (nano) after each ingredient on the label.
2009	Intel becomes the first semiconductor manufacturer to demonstrate 22 nm circuits; the 22 nm chip fits more than 2.9 billion transistors into an area of the size of a fingernail (*by comparison, Intel's 4004 microprocessor introduced in 1971 was based on a 10 000 nm process*).
2010	The Kavli Prize in Nanoscience is awarded to Donald Eigler (IBM) and Nadrian Seeman (New York University) for their development of "unprecedented methods to control matter at the nanoscale" (*Eigler was the first person to pick up an atom and precisely move it to another location; Seeman invented structural DNA nanotechnology*).
2010	The Nobel Prize in Physics is awarded to Andre Geim and Konstantin Novoselov (University of Manchester) for their development of graphene (*which at the time was thought to be impossible*). Both began their careers in physics in Russia.
2011	Intel begins production of 22 nm circuits.
2012	First commercial gene therapy (Glybera) approved by the European Union.
2013	Stanford University team builds first carbon nanotube computer.
2013	The European Community's nano labeling law goes into effect, applicable to all cosmetics that use nanomaterials.
2014	US federal investment in nanoscale research equals $1.7 billion (*$20 billion since 2001*).
2015	Use of 14 nm transistor chips called FinFETs in Apple iPhones and other devices.
2015	Announcement of the winners of the $10 million XPrize medical tricorder contest.

The Roots of Modern Nanoinnovation

The modern history of nanoinnovation is important because the world is still striving to fulfill the vision of the scientific pioneers who opened the floodgates to the nanoworld. It is important for us to pay homage to the pioneers who made nanoinnovation possible. As managers in science, business, government, and academia, we are all fulfilling the legacy they created, which we now call nanoinnovation. Some of the pioneers who have led us to our current advanced level of nanoinnovation include Richard Feynman, K. Eric Drexler, Richard Smalley, and Mihail "Mike" Roco. Their contributions are summarized here.

Plenty of Room at the Bottom

The era of modern nanotechnology began on December 29, 1959 with a lecture at the California Institute of Technology (Caltech) by Richard Feynman who presented his historic vision of a process to manipulate individual atoms and molecules, including some of the unique properties that would prevail at this scale.

Most people do not realize the impact of Richard Feynman (1918–1988), who single-handedly set in motion a chain of investigations and scientific discoveries that laid the foundation for nanoinnovation – more than two decades before the term nanotechnology was coined. Undoubtedly, Richard Feynman is the scientific father of nanotechnology. His 1959 lecture was a call to arms that launched the nanotech revolution.

Dr. Feynman's lecture was entitled "There's Plenty of Room at the Bottom" and subtitled "An Invitation to Enter a New Field of Physics." This scientific call to arms set the nanoscale wheels in motion by challenging scientists to find ways to "arrange the atoms the way we want" by tackling "the problem of manipulating and controlling things on a small scale."

It is remarkable to look back at his comments and see how truly prescient and visionary he was. During his lecture, Dr. Feynman offered several examples of "possibilities" for ultrasmall manipulation. In many ways, this lecture was the first nanotechnology road map.

He discussed placing all the books in the world (which in 1959 totaled about 24 million) on the head of a pin and predicted that by 1970, the Caltech library of 120 000 volumes could be stored on one library card. Dr. Feynman argued that reading this ultraminiaturized information would require at least a 100× improvement in the electron microscopes that were state of the art at the time. He also noted that better microscopes were also needed to solve some of the fundamental problems of biology such as sequencing DNA and RNA. He recognized that the existing room-sized mainframe computers were too large and asked, "Why can't we make them very small?" He then went on to describe how increased pattern matching could provide computerized face recognition.

He told his audience: "In the great future – we can arrange the atoms the way we want" – which is exactly what nanoinnovators are doing today with nanoprobes and optical tweezers.

He cited an idea by Albert Hibbs to create mechanical surgeons that could be swallowed. In this book, I include the story of Proteus Digital Health and their Raisin™ System that includes sensors called "ingestible event markers" that measure the body's response to medications, combined with wearable sensors that communicate medical results. This and other intelligent medical products are direct descendants of the "swallowable surgeons" envisioned by Albert Hibbs and Richard Feynman.

Dr. Feynman described several of the challenges that would need to be overcome to innovate at the molecular and atomic scale. He explained how quantum forces would be more important at the molecular scale than gravity. He described the difficulty of creating billions of structures or devices that would require billions of tiny lathes in billions of tiny factories, to achieve what we now call high-throughput production. He admitted that there are many ways to achieve this, and suggested that one approach might be to make a set of devices that would make smaller devices, which in turn would make smaller devices, until molecular scale devices are achieved.

He chided his audience with this intriguing prediction: "In the year 2000, when they look back at this age, they will wonder why it was not until the year 1960 that anybody began seriously to move in this direction."

He did not stop with this provocative challenge, but was willing to put his money where his mouth was. Dr. Feynman ended his lecture by personally offering a $1000 prize to the first person who could take the information on the page of a book and reduce it 1/25 000 times smaller in a way that could be read by an electron microscope. He offered a second $1000 prize to whoever could make an operating electric motor that is only 1/64 of a cubic inch. This challenge was not surprising, coming from a man who played the bongo drums for fun (you can see him playing on YouTube videos) and whose autobiography is playfully titled "Surely, You're Joking, Mr. Feynman" [20].

Four weeks after issuing this challenge, Caltech graduate William McLellan won the first prize by using (reportedly) a microscope, a watch-maker's lathe, and a toothpick, to create a motor that was 1/64th of a cubic inch. The motor had 13 parts, weighted 250 μg, and rotated at 2000 rpm. The second Feynman prize wasn't claimed until 1985, when a Stanford University graduate student named Tom Newman (working with his colleague Ken Polasko) used electron beam lithography to transcribe the first page of *A Tale of Two Cities* by Charles Dickens in a 6 × 6 μm area, on the head of a pin.

In 1965, Dr. Feynman shared the Nobel Prize for Physics (with Sin-Itiro Tomonaga and Julian Schwinger) "for their fundamental work in quantum electrodynamics, with deep-ploughing consequences for the physics of elementary particles."[13] Dr. Feynman is credited with pioneering the field of quantum computing. Most people don't know that he worked on the atomic bomb. He was also on the panel that investigated the causes of the space shuttle Challenger disaster. He died in 1988 after a long battle with stomach cancer, and despite his health, gave his last lecture 2 weeks before he died. Today, all of us are now engaged in an inexorable march to fulfill the vision espoused by Richard Feynman.

Dr. Feynman's challenge to the scientific community was a true "call to arms" that set in motion an era of scientific investigation that extended from the day of

13) The Nobel Prize in Physics 1965. Nobelprize.org. (April 2, 2011). http://nobelprize.org/nobel_prizes/physics/laureates/1965/.

his lecture to the present. In 2000 – the year Dr. Feynman predicted we will "someday look back on" – the United States launched the NNI, a multiagency to fund nanotechnology research and commercialization.

The National Nanotechnology Initiative

In 2001, the NNI set the standard for government–industry research programs worldwide, by establishing a multiagency research consortium that jump-started research by government and corporate partners. The NNI defined the goals for nanotechnology, set an ambitious research agenda for US scientists and corporate partners, led the establishment of new ventures, enlisted the involvement of leading scientists from many fields, promoted education at all levels, produced road maps and publications, and supported high-risk, high-return activities that would not have been otherwise possible. Today, the NNI spends more than $1.7 billion annually to support nanoscale science and engineering R&D at most US government agencies.

The NNI was conceived and led by Dr. Mihail "Mike" Roco (Figure 2.8) who worked on the initiative for several years before it was implemented in 2000 by President Bill Clinton. Dr. Roco was the founding chair of the National Science and Technology Council's subcommittee on Nanoscale Science, Engineering and Technology.

He recalls how the NNI was developed as a concept in the mid-1990s: "In November 1996, I organized a small group of researchers and experts including Stan Williams (Hewlett Packard), Paul Alivisatos (University of California, Berkeley) and Jim Murday (Naval Research Laboratory). We studied the issues and needs and developed a vision for nanotechnology, despite low expectation of additional funding at that time. More than 150 experts were enlisted to help develop scientific and benchmarking studies and research directions, including milestones for U.S. government investment. Materials were distributed to more than 30,000 organization to rally support." In 1999 at a meeting of the White House's Office of Science and Technology Policy, Dr. Roco formally proposed the NNI with an initial budget of half a billion dollars for fiscal year 2001. "I was given 10 minutes to make the case," he recalls. "Fortunately, nanotechnology captured the imagination of the committee and the discussion stretched to two hours." The first NNI budget was $489 million. "Looking back, it looks like we were off and running but in reality, the work was just beginning."

"When we launched the NNI, six government agencies made nanotechnology an important priority. Over time, this expanded to 26 U.S. agencies, which shows the breadth of nanotechnology. The impact has been profound."

As a nanoinnovation champion, Dr. Roco has had a profound personal effect on the establishment and funding of government programs. As the key architect of the NNI, Dr. Roco has been called the "godfather" of nanotechnology in the United States In 2007, Dr. Roco was awarded the National Materials Advancement Award from the Federation of Materials Society, which described him "as the individual most responsible for support and investment in nanotechnology

Figure 2.8 Mihail "Mike" Roco was a key architect of the NNI in the United States, and continues to play a leadership role in the development of nanoinnovation. His vision and leadership have been responsible for funding and promoting nanoinnovation on an extraordinary scale (photo by Michael Tomczyk).

by government, industry and academia worldwide." Dr. Roco is credited with 13 patents, over 200 articles, and has written or contributed to 20 books. He coauthored (with Hsinchun Chen) the first rigorous assessments of nanotechnology research, publications, and patents [21]. His most recent book, coauthored with Chad Mirkin and Mark Hersam, is entitled *Nanotechnology Research Directions for Societal Needs in 2020: Retrospective and Outlook* (Springer, 2011).

Today he serves as Senior Advisor for Nanotechnology at the National Science Foundation, where he continues to guide the advancement of nanoinnovation.

The Vision of Molecular Manufacturing

The most controversial pioneer in the field of nanoinnovation is K. Eric Drexler, a visionary and futurist who is credited with several notable firsts. He wrote the first nanotechnology book, the first papers and books on molecular manufacturing, and the first nanotechnology textbook. He earned the first Ph.D. in molecular nanotechnology and established the first organization dedicated to advancing nanotechnology.

Eric Drexler began formulating his ideas about the future of nanotechnology as an undergraduate at MIT, where he went on to earn a Ph.D. in molecular nanotechnology, a field he created and has promoted tirelessly throughout his career. Dr. Drexler is credited with popularizing nanotechnology in the United States, which led to the creation of the NNI. His best-selling 1986 book, *Engines of Creation: The Coming Era of Nanotechnology*, presented the first detailed conception of molecular manufacturing and self-assembly of nanodevices. It was also the first book to use the term "nanotechnology" to refer to one-billionth of a meter.

Drexler was also the first public personality to recognize the potential dangers of nanotechnology. He described the possible risks, which range from criminal and military uses to dehumanizing effects. Most famously, he reasoned that if nanodevices could in fact be engineered to self-assemble and replicate themselves, they could also potentially spiral out of control and turn life on Earth to "gray goo" – turning engines of creation into engines of destruction. This meant that nanoscale self-assembly had to have built-in controls.

His use of the term nanotechnology, combined with his many "firsts" and his seminal work in developing the concept of molecular manufacturing, has earned Dr. Drexler an indisputable place in the history of nanoinnovation. If Richard Feynman is credited with *inspiring* the field of nanotechnology, then Eric Drexler earns the credit for *introducing and popularizing* nanotechnology.

Regrettably, Drexler's legacy was somewhat tarnished by a series of public exchanges in the media and scientific journals during 2003–04, called the Drexler–Smalley debates. In a series of exchanges and public letters, Eric Drexler and Richard Smalley argued the feasibility of Drexler's concept of molecular manufacturing and self-assembly of nanodevices.

Richard Smalley was a well-respected nanotechnology pioneer whose discoveries included the discovery of the carbon allotrope that he and his team called the "buckminsterfullerene" nicknamed "buckyballs" – a spherical arrangement of 60 carbon atoms arranged in atomic hexagons that resemble the geodesic domes created by famed architect Buckminster Fuller. Smalley and his colleagues were awarded the 1996 Nobel Prize for Chemistry for the discovery of fullerenes. Dr. Smalley was also the founder and leader of the first nanotechnology research program at Rice University, an early champion of the use of nanotechnology to develop alternative energy solutions, and cofounder of one of the first nanoventures (Carbon Nanotechnologies, Inc.).

An early supporter of Drexler's work, Smalley came to believe that quantum forces and properties would make molecular assembly (as envisioned by Drexler) infeasible. He believed that concepts like "gray goo" could scare children, alarm the public, and influence government funding for the NNI. He maintained that synthesis and replication would be achieved through chemical and biological processes – not molecular assembly as described by Drexler.

Drexler knew that molecular manufacturing and the development of nanofactories would take decades to achieve, and he wanted to get started immediately. He pushed for inclusion of molecular manufacturing and self-assembly in the first NNI research budgets. Smalley maintained that Drexler's approach was technically impossible and argued that the NNI should focus on nanomaterials (including, of course, fullerenes).

As a Nobel Prize recipient and atomic scientist, Smalley was accorded more credence than Drexler. In the long run, it is possible that losing the first battle may have helped win the war. As a result of the debate, combined with some political lobbying, molecular assembly was eliminated from the first NNI budgets and a pall was cast over molecular manufacturing.

As a footnote to this historical episode: Richard Smalley died in 2005 at the age of 62. Eric Drexler continues his work to promote the development of high-throughput atomically precise manufacturing, which he now refers to as *productive nanosystems* [22]

It can be argued that the Drexler–Smalley debates were in many ways healthy for the development of nanoinnovation. As a result of this debate, public fears were calmed and politicians were able to embrace and support nanotechnology – secure in the belief that if molecular manufacturing was not possible, nanobots would not spin out of control and destroy civilization. Scientists did not stop their research. Actually, the argument that Drexler's concept of molecular manufacturing is not feasible was like throwing down a gauntlet to the scientific community. It took a few years for the dust to settle, but eventually the NNI, DARPA, several research agencies in Europe, and large corporations started funding molecular manufacturing and self-assembly research.

Despite the technical difficulties involved, a dedicated group of innovators have picked up the flag and have been working to fulfill Drexler's vision, which in recent years has gained increased acceptance as research groups began to make progress in atomically precise manufacturing of nanoscale structures. In the business community, commercial work on molecular manufacturing is being aggressively pursued by several companies. In May 2010, Zyvex founder Jim Von Ehr said he hopes to have a nanotechnology system that can create block-like objects by 2015, and rudimentary molecular manufacturing by 2020.[14)]

The concept is also supported by a loyal following of technology champions and futurists, including Ray Kurzweil, the Foresight Institute and Battelle Corporation. Foresight and Battelle collaborated on a road map for the development of molecular manufacturing. It is likely that Dr. Drexler's vision of molecular manufacturing will be achieved in some form in the future, but no one knows whether this will occur in 5 or 50 years.

Nano Is Not New – Ancient Nanoinnovations

Nanotechnology seems like a totally modern concept. Actually, nanotechnology has an ancient history that dates back hundreds, even thousands, of years!

A surprising number of ancient mysteries are being solved today by nanoarchaeologists, scientists who study the ancient history of nanotechnology. In recent years, nanoarchaeologists have learned that nanoscale properties played an important role in a wide range of remarkable innovations, from hair dyes used in the time of the Pharaohs to weapons used in the Crusades. Many of these

14) Sander Olson (interviewer), Jim Von Ehr Says Zyvex Will Achieve Digital Matter from Building Blocks by 2015 and Rudimentary Molecular Manufacturing by 2020. nextbigfuture.com (May 26, 2010).

discoveries offer ideas that modern nanoinnovators are using to develop new products and processes.

Ancient Hair Dye

Ancient Egyptians, Greeks, and Romans were darkening their hair using dyes that created nanocrystals, as long as 4000 years ago. French researchers from the Center for Research and Restoration of the Museums of France have discovered that ancient lead–lime dyes used to darken human hair by forming nanocrystals of lead sulfide known as galena, a black pigment known to ancient Egyptians and used by Greeks and Romans in the time of Julius Caesar. Their findings are based on analysis of samples that are 2000 years old. The diameter of the nanocrystals averaged <5 nm, which allowed the crystals to penetrate the keratin microfibers of human hair [23]. This is an ancient example of what modern scientists call *synthetic nanoscale biomineralization.* Ironically, today the cosmetics industry is one of the largest users of nanotechnology.

The Damascus Sword

Another mystery involves the famed Damascus Sword used by ancient Islamic and Hindu warriors – a steel weapon so durable, it provided a military advantage to Islamic warriors during the Crusades, more than 1000 years ago. Damascus swords possessed legendary strength and an ability to retain a razor sharp edge. The blades were characterized by a wavy pattern known as "damask." Saladin himself might have used one of these remarkable sabers. Unfortunately, by the eighteenth century, the secret of the ultrahard steel was lost.

Thanks to modern nanoarchaeologists, we know that the steel in these swords came from high-carbon iron ore steel mined in southern India. The ore was smelted into high-carbon steel cakes, called wootz, and forged into sword blades. During the smelting process, the ore formed carbon nanotubes and cementite nanowires, which made the Damascus Sword one of *the* first documented uses of carbon nanotubes in a commercial product [24].

The Lycurgus Cup

If you visit Room 41 of the British Museum in London, you can view the Lycurgus Cup (Figure 2.9), one of the world's most famous examples of ancient nanoinnovation. The cup, which is depicted on the next page, is a rare example of a Roman cage cup, created in the fourth century. The raised images on the cup depict the mythical King Lycurgus from Homer's *Iliad,* being strangled by vines after attacking Dionysus and Ambrosia. The Lycurgus Cup is a deep olive green, but when light shines through the glass, it turns to a translucent red. The stunning effect comes from the suspension of nanoparticles of colloidal gold (and silver) suspended in the glass, which appear red at the nanoscale [25].

Maya Blue

The turquoise blue pigment known as Maya Blue was used to decorate Mayan buildings, sculptures, murals, and pottery 500–1200 years ago. The

Figure 2.9 The 400 A.D. Lycurgus Cup turns appears to transform from green to red when held up to the light, as a result of nanoparticles of colloidal gold embedded in the glass. The cup is on display in Room 41 at the British Museum in London (used with permission.).

striking blue pigment has retained its vivid color for centuries, despite humid conditions and corrosive elements [26]. The pigment is created by heating organic indigo pigment with palygorskite, a type of moist fibrous clay found in the Yucatan region of Mexico. This is a bit of an oversimplification but essentially, when the materials are heated, the moisture evaporates, allowing the indigo to fill the nanochannels in the clay, which are sealed when the clay cools.

These examples are not merely historical artifacts. Some of these discoveries explain technological mysteries that have eluded scientists for millennia. Nano-archaeologists are revealing ancient uses of nanoscale properties that are still applicable today. They attest to the core principles of nanoinnovation – namely, that it's the *implementation* that provides value, even if the hows and whys of a specific innovation are not completely known.

2.1.10 Critical Issues for Nanoinnovation

As part of the research for this book, more than 150 nano-insiders were asked what they thought managers should know about nanoinnovation. Their responses suggest that first and foremost, managers need an understanding of the critical issues that will impact the strategic decisions of the people and organizations that are driving nanoinnovation. Here are a few insights in the experts' own words:

- In general, managers need to know how to separate the hype from the reality of nanotech – and to recognize that hype still largely dominates most public discourse. – *Peter Balbus, CEO and Managing Director, Pragmaxis*
- Nanoinnovation is not just about size, but how size influences the way materials, forces, and processes can be controlled and exploited at the nanoscale – for example, magnetism, friction, and the responses of the immune system. – *Peter Binks, CEO, The General Sir Monash Foundation; former CEO, Nanotechnology Victoria Ltd.*
- Nanotechnology is going to enable devices that are needed and envisioned, but previously were not realized due to technical limitations, material availability, and so on. Managers should also understand that nanoinnovation is evolutionary and not revolutionary – these changes will not happen overnight. – *Mark Banash, Ph.D., VP Engineering, NanoComp*
- Many nanotechnology developments are at least 15 years away – however, companies that discount these as too far away risk losing control and participation in nanoinnovations that will impact their industries. – *Jaideep Raje, Analyst, Lux Research*
- Many products use nanomaterials today, but are not categorized as nano – in the future, managers will find that some of their existing products contain nanostructures or nanomaterials and are subject to new health and safety standards. – *Fred Klaessig, Manager, Pennsylvania Bio Nano Systems*
- Nanotechnology is not just about going smaller – it's about the ability to build more complex materials and devices. – *Charles Brumlik, Managing Partner, NanoBiz LLC; Consulting Principal, RES Partners*
- Nanotechnology will be a pervasive technology provider that will impact most parts of the economy, including low tech as well as high tech. For example, concrete was used by the Romans 2000 years ago and has not changed much in two millennia, but nanotechnology is changing concrete and improving it. – *Patrick Ennis, Global Head of Technology, Intellectual Ventures; former Managing Director, ARCH Venture Partners*
- Nanotechnology will solve problems at the "bottom of the pyramid" in India, China, and other regions of the world and these solutions will in turn help lower costs in healthcare and other sectors in the industrialized world. – *Dr. Anita Goel, Physicist, Harvard-MIT; Chairman/CEO, Nanobiosym*

- The success of nanoinnovation depends to a large extent on the availability of managers and team members who understand both business and technology. The rapid growth of nanotechnology worldwide is straining the pool of available talent. It will take time to train and cultivate enough resources to meet the demand for talent and leadership. – *Michael Terlaak, CEO/Executive Director and Founder, Nanotechnology Research Foundation, VP-Business Development, NEXUSNanoTech*

These are only a few examples of the types of insights managers need to know in order to understand, help develop, promote, and benefit from the innovations that are expected to emerge from nanotechnology research projects and commercial ventures. From this foundation, let's look next at the "science of nanoinnovation" and find out what nanoscientists are working on now, and how their discoveries will transform industries, markets, and societies.

References

1 Fagerberg, J. (2003) *Innovation: A Guide to the Literature*, Centre for Technology, Innovation and Culture, University of Oslo.
2 Schumpeter, J. (1934) *The Theory of Economic Development*, Harvard University Press, Boston.
3 European Commission. (2010) Critical Raw Materials for the EU: Report of the Ad-Hoc Working Group on Defining Critical Raw Materials.
4 Physorg.com (2009) Rice Researchers Unzip the Future.
5 IBM. (2010) Made in IBM Labs: IBM Sets World Record by Creating High-Efficiency Solar Cell Made from Earth-Abundant Materials, IBM Press Release.
6 Shulaker, M.M., Hills, G., Patil, N., Wei, H., Chen, H.-Y., Wong, P., and Mitra, S. (2013) Carbon nanotube computer. *Nature*, **501**, 526–530.
7 Castleman, A.W., Jr. (2009) Cluster Dynamics: Laying the Foundations for Tailoring the Design of Cluster Assembled Nanoscale Materials. Report on research funded by the U.S. Air Force Office of Scientific Research, http://handle.dtic.mil/100.2/ADA523433.
8 Lee, S., An, R., and Hunt, A.J. (2010) Liquid glass electrodes for nanofluidics. *Nature Nanotechnology*, **5**, 412–416.
9 Brunet, L. *et al.* (2009) Comparative photoactivity and antibacterial properties of C60 fullerenes and titanium dioxide nanoparticles. *Environmental Science and Technology*, **43** (12), 4355–4360.
10 Dumé, B. (2007) Nanomaterial breaks new hardness record. Nanotechweb.org, March 15.
11 Tsai, H., Xu, Z., Pai, R.K., Wang, L., Dattelbaum, A.M., Shreve, A.P., Wang, H.-L., and Cotlet, M. (2010) Structural dynamics and charge transfer via complexation with fullerene in large area conjugated polymer honeycomb thin films. *Chemistry of Materials*, **23**, 759–761.
12 Kang, H.-Y. (2010) A Review of the Emerging Nanotechnology Industry: Materials, Fabrications, and Applications, Appendix II. California Department of Toxic Substances Control, Pollution Prevention, and Green Technology.
13 Cientifica (2011) Global Funding of Nanotechnologies & Its Impact.
14 Bradley, J. *et al.* (2009) Nanotech venture capital: healthcare and life sciences provide life support. Lux Research, 2010.
15 McDermott Will & Emery (2013) Intellectual Property in the Next Technology Revolution: How Does the United States Stack Up?

16 Mouttet, B. (2011) Top Ten Nanotechnology Patents of 2010, TinyTechIP, January 2.
17 Rutt, S. (2011) Graphene patenting soaring in numbers. Cleantech and Nano Blog.
18 Shapira, P. and Wang, J. (2010) Follow the money: what was the impact of the nanotechnology funding boom of the past ten years? *Nature*, **468**, 627–628.
19 Taniguchi, N. (1974) On the basic concept of 'Nano-Technology'. Proceedings of the International Conference on Production Engineering, Tokyo, Part II, Japan Society of Precision Engineering.
20 Feynman, R.P. (1984) *Surely You're Joking, Mr. Feynman*!, Norton & Company.
21 Chen, H. and Roco, M.C. (2007) *Mapping Nanotechnology Innovations and Knowledge: Global and Longitudinal Patent and Literature Analysis*, Springer.
22 Eric Drexler, K. (2005) Productive nanosystems: the physics of molecular fabrication. *Physics Education*, **40**, 339–346.
23 Minkel, J.R. (2006) Ancient hair dye harnessed nanotechnology. *Scientific American*, September 29.
24 Reibold, M. *et al.* (2006) Materials: carbon nanotubes in an ancient Damascus sabre. *Nature*, **444** (7117), 286–286.
25 Freestone, I. *et al.* (2007) The Lycurgus Cup: a Roman nanotechnology. *Gold Bulletin*, **40/4**, 270–277.
26 Chiari, G. *et al.* (2007) Pre-columbian nanotechnology: reconciling the mysteries of the maya blue pigment. *Applied Physics A: Materials Science and Processing*, **90** (1), 3–7.

Part II
The Science of Nanoinnovation

3
What Nanoscientists Are Working On

> The real nanotech pioneers aren't those with just grand ideas, but rather those who make them happen.
>
> – *Chad Mirkin, Northwestern University*

Every discipline and industry has its own set of world-changing nanoscience targets. Some of these are being worked on quietly out of the public spotlight, while others are trumpeted almost daily in the media. Many of these science goals sound simplistic when you hear about them on the Discovery Channel or in the *New York Times*, such as invisibility cloaks, miraculous cures for cancer, or materials that are stronger than steel.

The reality is that all of these so-called miracles get their start in a science laboratory – in government laboratories such as the National Science Foundation, NASA, the US Navy, and DARPA; at corporations like Bayer, Eastman Kodak, Fraunhofer, IBM, Lockheed Martin, Nippon Electric, Nokia, and Samsung; and most notably, on hundreds of university campuses around the world. Many science centers are university–industry collaborations, such as the Binnig and Rohrer Nanotechnology Center in Zurich, Switzerland; part of a $90 million shared research infrastructure developed by IBM and ETH-Zurich (the Swiss Federal Institute of Technology).

Of course, before any science-based nanoinnovation can be implemented and commercialized – long before companies and families can reap the benefits – a scientist in a laboratory, somewhere in the world, has to find a way to bend or slow, or stop and start, the first ray of light. Someone has to place the first drug molecule in a nanoshell, deliver it safely to a human cancer cell, and destroy the first tumor. Someone has to make the first nanotube, and the first few flakes of graphene. Someone has to go first.

The real miracle of nanoscience is the perseverance, dedication, and commitment of the nanotechnology scientists and their teams who are working to make these things happen. These teams include students and laboratory technicians as well as Ph.D.s and research directors. They don't all win awards, but they all push the frontiers of science, even if they fail – because failure helps define the pathways to success. The rewards – from the eureka moments to the commercial successes – don't happen overnight. When they happen, it's

NanoInnovation: What Every Manager Needs to Know, First Edition. Michael Tomczyk.

like someone set off a fireworks display. From one breakthrough, whole new worlds of discovery are unleashed. That's what we mean when we talk about waves of innovation, and the incredible exhilaration that comes from surfing these turbulent waves.

As a manager, you need to keep your eye on nanoscience because this is where most of the nanoinnovations come from. If your job involves emerging technologies, you need to know which of the myriads of science projects have the potential to demonstrate proof of concept, to survive a rigorous testing process, to scale up to production levels, and find safe uses in products and systems. You need to learn to distinguish science hype from science reality so you can evaluate which innovations are looming on the near horizon. You need to get a sense of which innovations are on a fast track to commercial development, and which are going to remain in the laboratory for another decade or more.

Knowing what scientists are working on will help you identify commercial opportunities, potential collaborations, where and when to invest, and how to select nanotechnologies and applications that synch with your interests. In this chapter, we'll look at some nanoinnovations that had their beginnings in research labs. We'll meet some very cool nanoscientists, and look at some science-scanning techniques that will help you stay up-to-date with scientific discoveries and breakthroughs.

3.1
Using Nanoscience to Solve Puzzles and Unlock Innovations

Nanoscience involves chemistry and physics, but you don't have to be a chemist or physicist to understand the subject. My wife Nancy is a medical research technician involved in research to develop improved materials for use in medical implants and devices. I'm deeply impressed with how easily she can draw images of organic molecules, understand protein folding schemes, and analyze nanoscale images for research projects. When I ask her how she wraps her mind around these complexities, she just smiles and says, "It's all about puzzles." She's right. It helps if you like solving puzzles, because the challenges facing nanoscientists are similar to the challenges faced by any puzzle solver.

In the field of nanoscience, much of the early puzzle-solving begins in a university research laboratory. As you will see from the examples in this book, academic research is a starting point and anchor for most of the nanoinnovations that eventually become commercial products. Many of these projects involve partnerships between faculty researchers and corporations. For example, at Caltech, Paul Rothemund has been working with Intel to figure out how to get a tangled forest of nanotubes to line up in an orderly pattern or matrix so they can be used in a new generation of computer chips.

Many nanoscientists are working on innovations that have the potential to provide social benefits that are desperately needed, such as cures for diseases and ways to regenerate failing organs and tissues. At Wake Forest University, Anthony Atala and his team are using modified inkjet printers and nanoscaffolds to create new organs from patient tissue, which could one day replace the need for donated organs. Michal Lipson and her colleagues at Cornell are using metamaterials to bend light, which may lead to the development of an invisibility cloak – or a new type of lens that can treat some forms of blindness.

Some of the problems that scientists are working on are of special importance to people living in Third World countries and remote regions of the world. For example, in Bangladesh, where arsenic contaminates as much as 70% of the drinking water, arsenic poisoning is an endemic health problem. At Rice University in Houston, Texas, Vicki Colvin and her colleagues have demonstrated that nanoparticles of iron oxide (nanorust) can be used as an inexpensive method for removing arsenic from contaminated water. The research team included Dr. Colvin, her colleague Mason Tomson, and their students [1]. When contaminated water is mixed with nanorust, the arsenic clings to the iron particles that clump together and can be removed with an ordinary magnet. This discovery paved the way for a low-cost method to remove arsenic from contaminated drinking water, which may help people living in Bangladesh and other countries where arsenic contamination is a major problem.

It is important to note that scientific research begins from basic or fundamental research that demonstrates what can be done with a particular atom or molecule. There is not always a direct link between a basic research discovery and a commercial innovation, but if there were no basic research there would be no nanoscale microscopes, no carbon nanotubes or graphene, and no flexible electronics or thin-film solar cells.

There are many long-standing goals that scientists have been trying to achieve that can only be unlocked by basic, fundamental scientific investigation and experimentation. For example, you would think that by now we should know everything there is to know about water. But this is not true. Water is difficult to study because it is not easy to isolate and observe individual water molecules. That is because water typically exists as a network of molecules that link and bond in liquid water or ice.

In 2011, a team of Japanese scientists trapped a single molecule of water inside a carbon buckyball, which is difficult considering that water molecules are almost always found attached to other water molecules because of the way they form hydrogen bonded networks, and buckyballs are typically slightly hydrophobic [2]. The scientists used temperatures and water vapor pressures to open a C_{60} carbon molecule (buckyball) and then inserted a water molecule in the hollow cage-shaped structure and closed the carbon framework.

These are only a few of the thousands of innovations being developed by nanoscientists, but they illustrate the enormous technological revolution that is occurring across the incredibly broad spectrum of exploration that we call "nanoscience." Let's look at some more interesting examples.

3.2
Solid Smoke: Catching the Comet's Tail

Scientific discovery is often like trying to catch the tail of a comet, so it is ironic that a nanoscience discovery called aerogel – sometimes called "solid smoke" – allowed NASA to catch and analyze the dust in a comet's tail. In 1999, NASA sent a space probe to collect dust samples from a comet called Wild 2. The probe returned to Earth with the sample in 2006.

When NASA scientists were designing the Stardust space probe, they had to deal with the challenge of collecting the dust particles, which ordinarily would vaporize when they come into contact with a solid material. A set of ultralow-density aerogels was used in the Stardust Interstellar Dust Collector to catch and trap the dust grains from the trail of the comet. A form of silica aerogel is also used to insulate the Mars rovers that have been transmitting stunning photos and data from the surface of Mars. Aerogel is also used to insulate space suits.

Aerogels are comprised of up to 99.8% air, which is trapped in individual pores just a few nanometers in size. Aerogel is translucent and has the least density of any porous solid material, which is why it is sometimes called "solid air" or "solid smoke." If you hold it in your hand it feels light as a feather and has the consistency of spongy gelatin. A block of aerogel the size of a person weighs less than a pound and can support an object weighing half a ton. Aerogels can insulate against extreme temperatures and are about 40 times more effective as an insulator than fiberglass. The unique properties of aerogels come from the nanosized bubbles – the spaces in the material. These air bubbles or gas pockets can range from 2 to 100 nm or more.

The accompanying photos from NASA's Jet Propulsion Lab (Figure 3.1a and b) vividly show the insulating properties of aerogels. These photos show crayons and matches being insulated from the heat of a small blowtorch by a thin layer of a feather-light aerogel made from silicon dioxide. The crayons in the photo are made from a mixture of paraffin wax and color pigment, which melts between ~53 and 64 °C (128 and 147 °F). Safety matches burn at 230 °C (446 °F). The blowtorch flame can exceed 1000 °C.

Aerogel was created in 1931 by Samuel Stephens Kistler (1900–1975), an American scientist from California who (according to popular folklore) developed the material to win a bet to see who could replace the gas in a "jelly jar" without causing any shrinkage. The material languished for several decades until the 1980s when new manufacturing methods and the rise of nanotechnology gave aerogels the impetus needed to become commercially viable. Today, aerogels made of silica, carbon, alumina oxide, and other materials are used in building insulation, skylights, scuba diving suits, tennis racket handles, supercapacitors, and so on. They are used in crime laboratory targets to test ballistics.

The global market for aerogels was estimated at less than $90 million in 2008[1] and has been projected to surge to as much as $1 billion by 2015 [3]. This

1) BBC Research, Aerogels, March 2009.

(a)

(b)

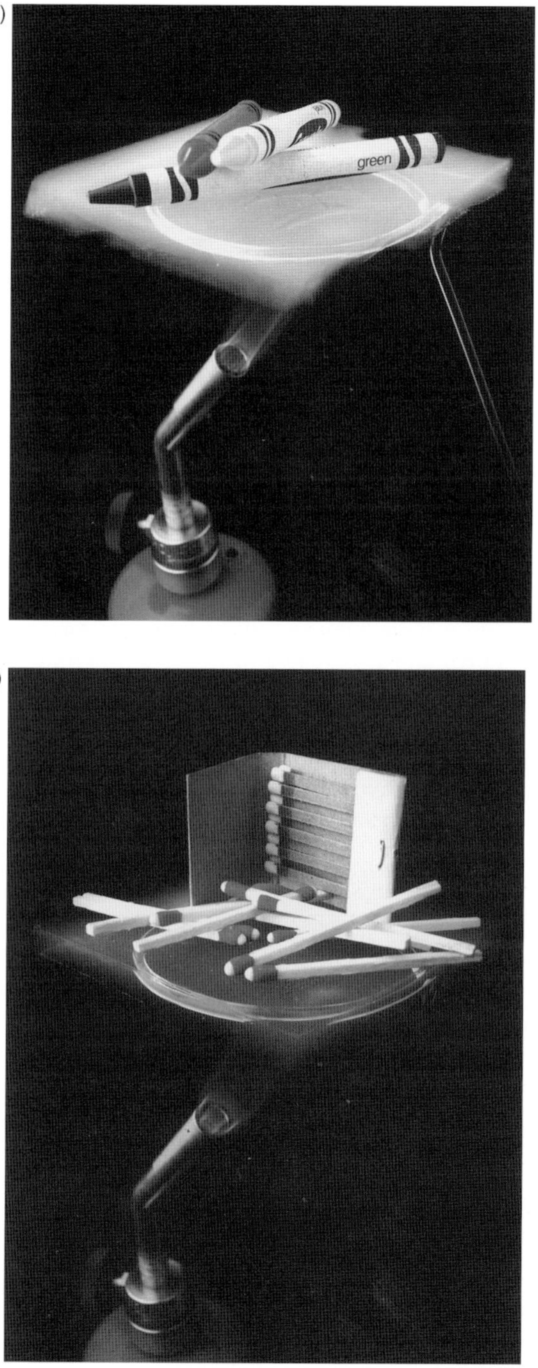

Figure 3.1 (a) Aerogel protecting crayons from melting. (b) Aerogel prevents matches from igniting (*source:* NASA/JPL).

growth is expected to be driven by applications related to thermal and acoustic insulation. This is one example of a science-based nanoinnovation with space-age applications and significant commercial value.

3.3
Turning DNA into Boxes, Lattices, and Pyramids

The following stories are based on interviews with several nanotechnology pioneers. These scientists are representative of the many thousands of researchers who are working 24/7 in laboratories around the world, often without credit and out of the media spotlight, to break down a barrier or unlock a secret that will improve our lives.

Nadrian (Ned) Seeman (Figure 3.2) is an American nanotechnologist and crystallographer who is internationally known as the inventor of DNA nanotechnology – the use of DNA molecules as a structural material to create rigid two- and three-dimensional structures with nanometer precision. Dr. Seeman started his research in the early 1980s and has worked for >30 years in the field that he created and pioneered. He has published >240 research papers and is a recipient of the Feynman Prize in Nanotechnology. In 2010 he was the corecipient of the million dollar Kavli Prize in Nanoscience.

Most of us think of DNA as a carrier of genetic information in living cells. DNA – deoxyribonucleic acid – is the chemical in the nucleus of all cells that carries the code of life. DNA is also a polynucleotide, which is a chemical *polymer* – a polymer is a large molecule composed of repeating structural units. A DNA molecule consists of two strands formed in the shape of a twisted ladder

Figure 3.2 Nadrian (Ned) Seeman, the inventor of DNA nanotechnology (*photo credit:* Mike Summers) (NYU Photo).

(the iconic "double helix" discovered by Nobel Prize winners Watson and Crick). The legs of the ladder are made of two long strands of deoxyribose sugar and phosphate molecules, bound together by four different types of nucleotides or "bases" that are labeled A (adenine), G (guanine), C (cytosine), and T (Thymine). In a DNA molecule, A on one strand pairs with C on the other strand, and G pairs with T. The legs of these ladders can be chemically sliced, separated, and manipulated so that they attach to other "sticky" DNA strands to design objects, lattices, and nanomechanical objects. The DNA strands used in DNA nanotechnology are organic molecules that are synthesized by programming the A, G, C, and T bases; however, it should be emphasized that these synthesized molecules and strands were never part of "life" or living systems and there is nothing "Frankensteinian" about this research.

Dr. Seeman and his colleagues have created DNA crystals that can self-assemble into 3D forms through synthetic sequences. In partnership with Nanjing University in China, Dr. Seeman has explored the potential to build a "DNA assembly line" that uses hundreds of short DNA strands to mold a very long DNA strand into any shape. Programmable DNA "machines" are attached to these shapes, which allow the researchers to direct the assembly using what is essentially a nanoscale robot arm. An impressive achievement from this process is the creation of a DNA "walker" that moves along the assembly line, stopping at the DNA machines to collect and carry the DNA cargo components such as metallic nanoparticles. By switching the devices between "on" and "off" states, the walker can determine which cargo will be transferred, or not transferred. Is this an early demonstration of Eric Drexler's vision of molecular manufacturing? It could be interpreted that way. To date, DNA nanotechnology is one of the few proven techniques for creating and precisely controlling the design of complex nanoscale structures.

Dr. Seeman's research is characterized by a chain of breakthroughs that followed a path as twisting and ladder-like as the DNA molecules he has learned to manipulate. A crystallographer by training, he began applying the principles of crystal formation to the use of DNA in 1981. This led to the creation of one- and two-dimensional structures. In 1991, he and his team announced the synthesis of the first DNA structure – a cube – that was followed by other shapes. However, these structures were flexible and not rigid enough to sustain three-dimensional shapes, which led to the creation of a more rigid tile-based lattice in 1998 (in collaboration with Erik Winfree). These developments led to the development of the field of DNA computing, demonstrated in a 2004 paper by Erik Winfree and Paul Rothemund at Caltech. In 1999, Dr. Seeman demonstrated the first crude DNA nanomachine, and in 2004 and 2005 he and his team demonstrated the first DNA nanomachines that exhibited mechanical motion – in other words, "walkers" that moved. Building on Dr. Seeman's discoveries, Paul Rothemund developed a DNA folding technique now called "DNA origami," which was an easy way to create DNA molecules of virtually any shape. In 2009, Seeman demonstrated the synthesis of a three-dimensional lattice, an achievement he originally conceived as a concept, 30 years earlier. In

2010, Dr. Seeman and his Chinese colleagues described a programmable DNA nanoscale assembly line [4].

In January 2012, a team of researchers from Kyoto University and the University of Oxford reported that they successfully used DNA building blocks to construct a motor capable of navigating a programmable network of tracks with multiple switches. Their research used DNA origami to cause strands of DNA to self-assemble into 2D and 3D structures, thus forming a DNA "rail system" that motor molecules could traverse. The team included Dr. Masayuki Endo and Prof. Hiroshi Sugiyama at Kyoto University's Institute for Integrated Cell-Material Sciences, and Dr. Shelley Wickham at Oxford University. The researchers hope to build on their research to develop programmable molecular assembly lines and more sophisticated sensors [5].

A Pub, Some Fish, and a Nanoscience Eureka Moment

In an interview for this book, Dr. Seeman offered some fascinating personal reflections on the chain of events that led him to his award-winning discoveries.

"By training I'm an X-ray crystallographer. In 1981 I was doing research as a postdoc. My work involved taking biological molecules and getting them to crystallize to make them amenable to using x-ray diffraction, which allowed us to work out the 3-dimensional structure of the molecule that makes up the crystal. This is a complex process, because there are different types of interactions that hold the crystal together – it could be salt, water molecules, charges, etc. Once I read an article that said you have to do 26 things right to hit a home run in baseball. Growing a crystal is very similar. You need to get 9 steps right and if one of them is wrong, the crystal won't grow. I have to admit, I was awful at growing crystals. So I started looking for a way to make the molecules that come out of solution arrange themselves in an exact repeating pattern, which means a 2D would look like rows of headstones in a military cemetery, and a 3D would be like stacking those cemeteries on top of each other. Also, as an assistant professor, I was not publishing on crystallography. So I started modeling branched DNA. I figured out that you can make more than four branches." A smile creeps into his voice as he recalls how he got the inspiration.

"One day in September 1980, I had an epiphany over a glass of beer," he recalls. "I was wrestling with a technical problem, so I went to the campus pub to think about it. I wanted to make six arm junctions – a DNA branch in six directions. While I was thinking about that, I remembered a woodcut by the artist M.C. Escher, called 'Depth.' The woodcut has a bunch of flying fish going from the center. I realized that these flying fish are in principle like a six arm junction. If I took these junctures and put sticky ends on them, I might be able to force a configuration using DNA to create structures."

"During this time, I thought our quest for 3-dimensional crystals was kind of like the quest for the Northwest Passage. As a crystallographer, I didn't know how to work with DNA in solution and we had to learn about that. First, I learned how to do 2-dimension (2D) versions, then did 'floppy' versions of the objects. Then I had to learn how to do 'stiff' motifs. One of my former students worked out how to do a 3-dimensional stiff motif and get them into a 3-dimensional structure. This process took about 30 years. It took us about 10 years just to figure out how to keep the structures from collapsing. Along the way, we learned how to organize nanoparticles in a checkerboard arrangement, and we created some specific devices that were able to capture other species. We have made a variety of mechanical devices from DNA, based on the programmability of DNA, by adding or pulling out some of the strands. About five years ago, we developed a bipedal walker that walks on a track – it moves like an inchworm. This is a nanobot-like bipedal walking device, where the rise and fall of each foot of the biped is controlled by introducing DNA strands with specific sequences into the solution."

"Some of our colleagues have laid the groundwork for nanoscale self-assembly, which is essentially bottoms up chemistry. I may have been a crystallographer, but in essence we're all really chemists. We make stuff."

Ned revealed that his research team is now controlling matter in three dimensions and have scaled up the structures to a quarter of a millimeter, which is large enough to see with the naked eye. "We are now making something that is very big compared to the nanoscale – which involves repeating it somewhere between 50 and 100 trillion times." Everything is still in prototype form, and still very small, but far enough along that it is possible to envision using these repeating units to build memories and for applications in nanophotonics.

In addition to the work by Dr. Seeman and his colleagues, a growing research community has been spawned by his groundbreaking achievements.

"When I started this research 30 years ago," he recalls, "I was the only one working in this field. Today, there are at least 50 research teams working on DNA nanotechnology."

The Next Generation of Structural DNA Scientists

Dr. Seeman's research has spawned an entire generation of DNA nanotechnologists who are pushing the boundaries of the field he created.

In 2009, a team of researchers at the Aarhus University Center for DNA Nanotechnology in Denmark used DNA origami to form a hollow cube with a hinged lid [6]. The technique involved creating a computer program to generate a single-stranded DNA sequence that would self-assemble into the desired shape, using smaller DNA fragments as "staples." The box measured $42 \times 36 \times 36\,nm^3$, large enough to hold enzymes or virus particles. According to lead researcher Jørgen Kjems, the box can be locked shut or opened using DNA latches or "keys." A pair of dye molecules glow red when close and green when separated, which help the researchers detect when the DNA box is open or closed. The team is working on different types of "locks" that are responsive to different conditions, which could be used in a DNA computer, drug delivery, and other applications.

Yamuna Krishnan and colleagues at the National Centre for Biological Sciences in Bangalore have reported an important step toward the vision of a DNA nanomachine that can probe the local pH inside a living cell [7]. Dr. Krishnan

Figure 3.3 These buckyballs were made from self-assembling DNA molecules by Dan Luo and his team at Cornell University. Potential applications range from drug delivery to electronics (image courtesy of Dan Luo. Copyright 2004, Cornell University).

said his team named their DNA nanomachine the "I-switch." It functions as a pH sensor based on fluorescence resonance energy (FRET) inside living cells. The major achievement of this research is that the Indian team has shown that a nanomachine can function inside a living cell, not just in a test tube. This proof-of-principle suggests the feasibility of using nanodevices based on DNA and other nucleic acids for applications in cell biology and biomedical engineering. Krishnan and colleagues suggest that their research could lead to the development of DNA nanomachines that could be used as sensors, diagnostic markers, and targeted therapies in living cells, although they concede that additional technological breakthroughs are needed [8].

In 2004, Dan Luo, Professor of Biological and Environmental Engineering at Cornell University, and Ph.D. grad students Soong Ho Um, Sang Yeon Kwon, and Jong Bum Lee created self-assembling DNA buckyballs that emulate carbon fullerenes (see Figure 3.3). Luo's research group is investigating the properties and unique capabilities of these structures, which measure approximately 400 nm in diameter, compared to carbon buckyballs which average 7 nm in diameter. Potential applications include drug delivery and electronics. Luo's research group has also created DNA barcodes and DNA hydrogels as part of their research, demonstrating the use of synthetic DNA as nanoscale building blocks.

The applications for DNA nanotechnology are complex and exciting. DNA structures can be configured to transport drugs inside cells, especially drugs that are normally rejected and prevented by cell chemistry. A container made from DNA can be accepted into the cell and camouflage the drug until it is inside a tumor cell, where it can then be released.

A collaboration between Oxford University physicists and molecular neuroscientists have shown that molecular cages made from DNA can enter and survive inside living cells until triggered to release their cargo. In a project led by Dr. Andrew Turberfield at Oxford's Department of Physics, synthetic DNA was programmed to form pyramid-shaped cages around protein molecules. The DNA cages were made from four short strands of synthetic DNA formed into a four-sided pyramid about 7 nm tall. The structure resisted attacks by cellular enzymes and remained intact inside human embryonic kidney cells for at least 48 h. This constitutes a convincing proof of the concept that synthetic DNA containers can actually be used for drug delivery, since DNA boxes or cages must be able to enter cells and survive there until needed, and then release its contents in response to some sort of signal or triggering mechanism.

"At 7 nanometers across our DNA tetrahedrons are compact enough to easily enter cells but still large enough to carry a useful cargo," Professor Turberfield has said, noting that more work is needed to understand how DNA cages find their way inside living cells [9].

There are a lot of associated discoveries and processes that still need to be developed by nanoscientists, and the discovery stream is strong and impressive.

3.4 How Nanoinnovation Is Extending Moore's Law

Any discussion of nanotechnology would be incomplete without a discussion of the role of nanotechnology in the evolution of computing, consumer electronics, and semiconductors. As a manager, you don't need to know all the details about the role of nanotechnology in semiconductors. However, you do need to know if Moore's Law is going to keep making computers smaller, faster, with more memory and processing power – and what happens when we reach the physical limits of miniaturization.

Historically, more than 50 years of innovation have taken us from the first transistors, which were over a centimeter in diameter, to the smallest transistors today that are just a few nanometers. Michael Berger, the founder of the Web site Nanowerk.com put this achievement in context: "The first transistor used in a commercial application was in the Regency TR-1 transistor radio, which went on sale in 1954 for $49.95, which is over $375 in today's dollars. Today's world includes transistors that are ubiquitous in miniaturized electronics that drive communications, the world of the Internet, and much more."

Michael observed that transistors have shrunk from over 1 cm in diameter to less than 30 nm, which is *3 million times smaller.* "This feat would be equivalent to shrinking the 509-meter tall Taipei 101 Tower, currently the tallest building in the world, to the size of a 1.6 millimeter long grain of rice." This process of miniaturizing computer circuits follows a pattern known as "Moore's Law."

Moore's Law is based on Intel cofounder Gordon Moore's 1965 observation that the number of components in integrated circuits had doubled every year and would continue to double until at least 1975. His prediction was published in a classic article in *Electronics Magazine* (April 1965). In 1975, he adjusted his prediction to a doubling of circuits and processing speed *every 2 years* and this became the standard metric we call Moore's Law. In practice, Gordon Moore's prediction became a self-fulfilling prophecy when Intel, and the semiconductor industry in general, incorporated his convention into their long-term planning and design targets. This law has been extrapolated to include pixel density and other measures associated with computing and consumer electronics and is now the gold standard for innovation-driven growth in the electronics industry.

In the semiconductor industry, Moore's Law is often expressed in nanometers. Semiconductors are essentially on–off switches that include a source of flowing electrons, a gate that controls the flow through a channel, to a drain where the electrons go. Flowing current is interpreted as a "1" or "on" condition, and when the current is not flowing, that is interpreted as a "0" or "off" condition. The size of the gate has been at the nanoscale since 2000, and in 2014 shrank from 22 to 14 nm. Scientists predict that there will come a point – somewhere ~5 nm – where quantum effects will affect the physical electron flow and new architectures will be required.

In 2010, I was hosting the annual Emerging Technologies Update Day, an event I originated at the Wharton School – the theme was the "Future of Computing" and one of the questions I asked the speakers was whether and how long Moore's Law would continue. When I posed this question to Mike Mayberry, VP-Technology and Manufacturing and Director of Components Research at Intel, he weighed in on the plus side, predicting that the "law" would keep going. Why did he believe this?

"Because we've already hit the wall twice and found ways around it," he said with a broad grin.

In 2011, Intel "broke the wall" again when it introduced a 22 nm microprocessor code-named Ivy Bridge. This device uses a three-dimensional architecture called tri-gate and represented a "reinvention" of transistor architectures. The significance of this innovation is apparent in the side-by-side comparison of the old and new architectures (shown in Figure 3.4), which show a dramatic increase in the circuit density on the new chip.

Of course, it is inevitable that Moore's Law will slow at some point and chip designers will hit a physical barrier. Estimates for when this will happen range from 2021 to 2029.

Why is this important?

For nanoinnovators, finding a way past this wall provides the semiconductor industry with a major research target. There are hundreds of teams working on ways to develop new materials and designs that will maintain the phenomenal growth that has powered the evolution of smart electronics. No one knows if the ultimate solution will involve relatively new physics – such as plasmonics, spintronics, or photonics – or biological computers made from DNA, or something we haven't seen yet.

In 2012, geneticist and genome pioneer George Church and bioengineer postdoc Sriram Kosuri at Harvard University's Wyss Institute announced they were able to store 700 TB of data in 1 g of DNA, which established a new record for

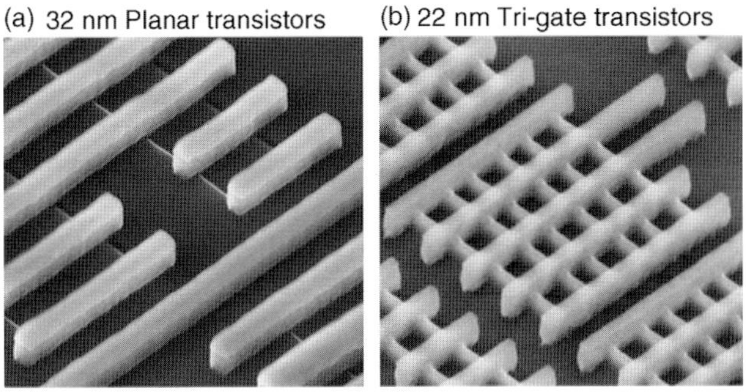

Figure 3.4 It's easy to see the improvement in the architecture of Intel's three-dimensional tri-gate technology (b) over the previous circuit architecture (a). Intel's tri-gate transistor was first announced in 2002, and in 2011 became the first three-dimensional transistor to be put into high-volume manufacturing. The 3D architecture was first used in a 22 nm Intel chip codenamed "Ivy Bridge" and is expected to lead to the development of 10 nm circuits by 2015 (*image source:* Intel Corporation).

biological data storage [10]. The data stored on the molecule were the text of Dr. Church's new book – 53 426 words. This research is supported by the US Office of Naval Research, Agilent Technologies (which makes DNA microchips), and the Wyss Institute. Dr. Church was a leader in the Human Genome Project in the 1980s and 1990s, and is the founder of the Personal Genome Project which hopes to sequence the genomes of 100 000 people.

When it comes to biological computing, the ultimate solution is to develop a circuit that uses just a few atoms. This has been done in the laboratory. In 2012, the world's smallest magnetic storage unit was engineered from just 12 iron atoms by a team of researchers at IBM Research-Almaden. Lead investigator Andreas Heinrich explained that the team posed the question: "How small can we make a magnetic structure and still be able to use it? What is the smallest unit of data storage?"

A bit of data is typically either a 1 or a 0. A hard drive that stores data can use up to 1 million atoms to store each bit, so scaling down the number of atoms used to store data posed an intriguing challenge. They used a form of magnetism called antiferromagnetism, where atoms spin in opposite directions. This allowed them to create an experimental atomic-scale magnet-based memory that is at least 100 times denser than current hard disk drives and solid-state memory chips. Dr. Heinrich's team started with one iron atom on a copper nitrate substrate and used the tip of a scanning tunneling microscope to add atoms until they were able to store 1 byte of data. To store the data, they switched the magnetic information in the bits from a zero to a 1 and back again. They worked at the extremely low temperature of 1 K (272 °C or −458 °F), which allowed them to stabilize the process, build the structure one atom at a time, and eliminate thermal effects [11].

After assembling the single bit of data, they combined 96 atoms to make 1 byte of data (which is used to represent a number or letter in a computer).

Next, they used the process to spell entire words – the first word they spelled was the word "Think," which required 5 bytes of information, or 480 magnetized atoms. In January 2012, Dr. Heinrich told Computerworld, "The ultimate end of Moore's Law is a single atom."

The month after Dr. Heinrich made his prediction, a group of Australian and American researchers reported that they had created a circuit from a single atom. Researchers from the University of New South Wales, Purdue University, and the University of Melbourne reported in *Nature Nanotechnology Letters* that they had created a single-atom transistor using a single atom of phosphorus (*a single atom of phosphorus is approximately one-tenth of a nanometer across*). The atom of phosphorus was positioned in a channel in a silicon crystal. The same team also announced that they created a wire made from phosphorous and silicon, one atom tall × four atoms wide, which exhibited properties of copper wire. To position and functionalize the atom, the researchers had to use temperatures as cold as liquid nitrogen (−391 °F/−196 °C), similar to the IBM team that developed the 12 atom memory storage.

Michelle Simmons, group leader and director of the ARC Centre for Quantum Computation and Communication at the University of New South Wales, noted that the team's single-atom transistor uses the same materials currently used in semiconductors, which makes the process easier to commercialize. She believes their technique is scalable. Professor Simmons credited the many outstanding students and postdoctoral researchers who worked on the pan-Pacific project, and said the process involved numerous systemic studies and "brute determination."

Gerhard Klimeck, Director of the Network for Computational Nanotechnology at Purdue, said much of their collaborative work was facilitated by use of nanoHUB.org – an online meeting place and education hub that provides tools to facilitate international collaboration and community building. NanoHUB content is used in over 450 classrooms at more than 150 universities worldwide.

"To me, this is the physical limit of Moore's Law," said Dr. Klimeck, "We can't make it smaller than this."[2)]

Why Is Moore's Law So Important?

At this point, you might be asking, why are we doing this? Why do we need smaller, faster computers? The reality is, we live in an age of big data. We are doing big things that require massive amounts of information, whether we're analyzing customer behavior on billions of Internet transactions, or trying to predict complex weather patterns that involve a myriad of variables. To engineer supercomputers that are greater in power and smaller in size and energy use, we need to develop smaller, more efficient nanocircuits.

2) One and done: single-atom transistor is end of Moore's law; may be beginning of quantum computing, Purdue University News Service, February 19, 2012.

To put the big data explosion in perspective, your computer measures its memory capacity in megabytes and terabytes. One byte can typically store one letter or number. Conceptually, 1 MB is approximately 1 million bytes; 1 GB is 1 billion bytes; 1 TB is 1 trillion bytes; 1 PB is 1 quadrillion bytes, and so on – although the actual number is different because 1 kB is actually 1024 bytes, not 1000 bytes – so 1 MB is 1024×1024 bytes = 1 048 576 bytes of memory, and so on up the scale.

- The 2009 movie "Avatar" used more than a petabyte of locally stored data to generate its 3D CGI effects. The massively multiplayer online game, World of Warcraft, needs more than 1.3 PB to manage the game that is played by more than 10 million subscribers worldwide. The human brain can store about 2.5 PB of binary data [12]. Google, the search engine giant, processes about 24 PB of data each day. The NSA – whose data mining activities have generated a fair amount of controversy – says the worldwide Internet carries 1826 PB of information per day. The German Climate Computing Center (DKRZ) has a storage capacity that can accommodate 60 PB of climate data. Facebook accounts for about 100 PB of data. The global daily total use of data is estimated to top 500 PB.

How big will "big data" grow? In 2011, IBM built a 120 PB storage array. In January 2012, the supercomputer pioneer Cray started construction of the Blue Waters Supercomputer, which is reported to have a storage capacity of 500 PB. The world's fastest computer at the end of 2013 was the Tianhe-2 supercomputer at Sun Yat-sen University in Guangzhou, China.

Nanotechnology is racing to keep up with the explosion in big data and the need to store and work with these massive amounts of information. Big data is being driven by broadband video, social media, massively multiplayer online games, RFID inventory tracking, Internet searches, medical recordkeeping, genomic libraries, astronomical applications, e-Commerce transactions, language translation, voice technology, massive open online courses (MOOCs), and a surge in cloud computing – a massive list of massive data apps.

Teradata, a pioneer in database warehouses and analytics, created the first data system over 1 TB in 1992, which was needed by its client Wal-Mart to handle the retailer's enormous transaction load. In 2011, Teradata formed the "Petabyte Club" for its largest data users. The first members were eBay, Wal-Mart Stores, Bank of America, and Dell. eBay, which has 100 million active users globally, and 300 million live listings at any point in time, processes 2 billion page views daily, 250 million search queries, and 75 billion database calls, with total data volume ranging up to 20 PB.

It is ironic that we will need increasingly small nanosized computer circuits to provide the memory and speed needed to process increasingly large arrays of big data, and to extend Moore's Law. We are gaining a sense of how small we can make a computer, but at the other end of the spectrum, we are still trying to figure out how intelligent a computer network can become.

A Computer That Wins TV Game Shows? What Next?

At the edge of the computing universe is the intriguing concept of artificial intelligence, which has been moving inexorably forward since the 1980s. In February 2011, an extremely clever supercomputer named Watson designed by IBM was pitted against two of the most successful winners of the TV game show, Jeopardy. One of the contestants had won 20 times and the other had won 74 games, so the competition was stiff. Watson didn't get all the questions right, but after a 3 day tournament the computer beat the two human champions. Watson demonstrated an ability to understand some very subtle and nuanced questions such as "It's a poor workman who blames these" – to which Watson responded, "What are tools?" Watson won \$77 147 (£47 812) and set a new standard for artificial intelligence.

Is Watson the first glimpse of an artificial intelligence android? Migrating Watson's capabilities into an android will require smaller faster circuits to store and process enough knowledge and do the pattern matching required to make "decisions" about what to say and do. Nanosensors will help robots and androids map, understand, and respond to their surroundings. And before these sensors and circuits enable robots, they will enable us humans to extend our reach beyond our grasp.

3.5 Invisibility Cloaks, Optical Tweezers, and Nanophotonics

What do you know about light?

Well, you probably know that light can be visible or invisible, depending on the wavelength (infrared, ultraviolet, etc.). You know that light travels at 186 000 miles/s (300 000 km/s) and moves in a straight line. Light can be absorbed, scattered, reflected, diffused, refracted, and amplified. Light can behave as a wave, but sometimes it acts like a particle. It can also be converted from one form of energy to another, which is how we get solar energy, for example. Light is measured in photons (discreet bundles of electromagnetic/light energy) or lumens (how much illumination a source of light provides).

Visible light has a wavelength that ranges from ~400 to 700 nm (some sources say the range is 380–760 nm). The approximate wavelengths for different colors are violet (400 nm), indigo (445 nm), blue (475 nm), green (510 nm), yellow (570 nm), orange (590 nm), and red (650 nm) (*source:* Atmospheric Science Data Center, NASA). These colors can vary somewhat. For example, the yellow light generated by a sodium vapor streetlight has a wavelength of 589 nm. Ultraviolet light can vary from 10 to 400 nm. The wavelength for a traditional ultraviolet "black light" you might see in an aquarium or a night club is 365–400 nm. Infrared light can range from 700 nm to 300 μm.

A small and extremely talented community of optical physicists are working very hard to develop techniques to slow, trap, manipulate, and even "stop" light, which is creating the foundation for a new generation of optical devices, including optical nanomaterials and optical computers that enable light to be captured, manipulated, modified, and stored.

Some of this groundbreaking research is being done at Dr. Marko Loncar's Laboratory for Nanoscale Optics, at the Harvard School of Engineering and Applied Sciences.[3] To design the next generation of optical chips, Dr. Loncar's group is exploring applications for hybrid active nanophotonic devices, including metallic nanostructures (surface Plasmon-based approach), photonic crystal cavities, and metamaterials. This research has important implications for photovoltaics, photochromatic materials, photocatalysts, photoresponsive gels, and optical computing in general.

It is only a matter of time until we see affordable optical computer systems. One intriguing question is: If we design optical computer systems, including robots and weapons systems, will we also be able to make them optically invisible?

How to Design an Invisibility Cloak and Move Objects with Light

Invisibility has been part of our popular culture since H.G. Wells wrote "The Invisible Man" in 1897, and is the subject of countless movies and television shows. The aliens in the Predator films wear combat suits with an invisibility cloaking feature that makes them look like images from a pixilated video game. More recently, J.K. Rowling's heroic character Harry Potter wore an invisibility cloak made from the hair of a magical creature called a Demiguise. Perhaps the hair of the Demiguise was nanoscale hair, thinner than the wavelength of visible light.

In the military, stealth bombers and naval ships use special materials and construction methods to achieve "invisibility" from enemy radar and missile systems. Of course, it is one thing to conceal something from radar, and quite another to hide something from the human eye. For many years, the technology to make something optically invisible existed only in science fiction. However, science fiction has a habit of morphing into science reality, and so it is with invisibility cloaks.

In recent years, science has begun to accept the notion that invisibility cloaks are something we can actually design. Also, researchers have demonstrated several ways to create "invisibility." One way is to use a network of cameras and projectors – essentially, to shoot a video of what is behind an object, and project that image on the front of the object, to make it appear to be invisible. Some researchers are creating "invisibility" by using optical nanomaterials to induce light to essentially follow a trail of nanoparticles around the cloaked object.

Another approach is to place a crystal in front of the object and refract light in a way that makes an object disappear – similar to how magicians use mirrors to make objects disappear in a Las Vegas stage act. Ironically, mirrors made of

3) Loncar, M. (2011) Laboratory for Nanoscale Optics, Harvard School of Engineering and Applied Sciences, June 3.

Figure 3.5 Dr. Michal Lipson at Cornell heads one of the world's top nanophotonics research teams (Cornell University photo by Jason Koski).

metamaterials have been used to bend light and other types of electromagnetic waves to achieve a form of invisibility – a nanoinnovation that gives new meaning to the phrase "smoke and mirrors."

The Nanophotonics Group led by Michal Lipson (Figure 3.5) at Cornell University in Ithaca, New York, has demonstrated how to hide an object in the "bulge" of a mirrored surface that uses light-bending metamaterials (which are actually nanomaterials) to configure the mirror so that it appears that the surface is flat and the bulge does not exist.

A cloak using nanostructured silicon has been developed by a research team led by Xiang Zhang, a principal investigator with Berkeley Lab's Materials Sciences Division and Director of UC-Berkeley's Nanoscale Science and Engineering Center.

Collaborative teams at the University California, San Diego, Duke University, and ETRI in South Korea have demonstrated that it is possible to control the speed and direction of light using memory metamaterials whose properties can be selectively and repeatedly changed, which enables metamaterials to be tuned. This innovation is an important step toward enabling the manufacture of Gradient Index of Refraction (GRIN) devices [13].

Other researchers have used crystals to cloak objects. Baile Zhang and his colleagues at the Singapore-MIT Alliance for Research and Technology (SMART) center in Singapore have used calcite crystal to shield a 38 mm long steel wedge from visible light underwater, which may be useful for shielding submarines and installations on the floor of the ocean. Collaborators Shuang Zhang at the University of Birmingham and Sir John Pendry from Imperial College in London have demonstrated a calcite cloak that works in air.

Janos Perczel, a Hungarian undergraduate student studying at the University of St. Andrews under the guidance of Professor Ulf Leonhardt, has used an optical device called an "invisible sphere" to slow light as it approaches an invisibility cloak. Perczel calls his technique "partial transmutation." This innovation could enable the wearer of an invisibility cloak to remain invisible while in motion, across a larger range of colors in the visible spectrum, and to move around against different types of backgrounds [14]. It is interesting to note that St. Andrews, which is located in Fife, Scotland, is the third oldest university in the English-speaking world, founded in 1413.

Janos Perczel's innovation is an intriguing example of how researchers are turning science fiction into science reality. In 1967, an award-winning Belfast science fiction writer named Robert Shaw (1931–1996) wrote a short story called "Light of Other Days" that introduced the concept of "slow glass" through which the past could be viewed. Many innovations that are now being enabled by nanotechnology, such as artificial blood cells, swarms of nanosensors – and invisibility cloaks – were first envisioned by science fiction authors and film makers.

Many of the cloaks we read about in science journals use light-bending or light-slowing metamaterials (e.g., man-made nanomaterials) and transformative optics to conceal an object from an observer. Most of the cloaking technologies reported by scientists involve a proof of concept using objects that are only a few microns or millimeters in size. Scaling up to cloak a human, a structure or a vehicle will require a lot more research. The first commercial invisibility cloak will probably be made from configurable nanomaterials that change shape or refractability. This will require a way to manufacture large-scale matrices or "cloaks" made of metamaterials. The current method used to create metamaterials – electron-beam lithography – is expensive and time-consuming.

A group led by John Rogers, a professor of materials science and engineering at the University of Illinois at Urbana-Champaign, has developed a way to "print" sheets of metamaterial that resemble a matrix or fishnet pattern. The group has produced sheets a few square inches in size, and is working to scale up their process to several square feet. Being able to produce metamaterials on a large scale will not only support the development of invisibility cloaks but will also facilitate the development of superlenses, night-vision invisibility cloaks, and more.

In May 2012, a research group from Towson University and the University of Maryland, led by Dr. Vera Smolyaninova, announced the development of an array of 25 000 tiny invisibility cloaks with the capability of "trapping a rainbow" of light. Dr. Smolyaninova says that this innovation could lead to the development of "spectroscopy on-a-chip," which could examine fluorescence at thousands of points simultaneously. This capability has special applications in biochips and biosensors. Arrays of sensors are an important innovation, since they provide the type of parallel processing that will lead to such capabilities as being able to test for multiple genetic conditions in a DNA sample. This array was also demonstrated to have light-slowing properties and may provide a method to test other types of invisibility cloaks. To create their array, the researchers used commercially available microlenses in combination with a thin gold plating. This first-of-a-kind innovation was reported in the Institute of Physics and German Physical Society's *New Journal of Physics* [15].

Some of the most impressive advances in the design of invisibility cloaks are coming from the relatively new fields of *nanophotonics* and *transformation optics.*

The Nanophotonics Group at Cornell University is a great example of how optics researchers are pushing the technology envelope to create innovative technologies that were previously impossible. The group is investigating the physics and applications of nanoscale photonic structures, focusing on light-confining structures that can slow, trap, enhance, and manipulate light.

Dr. Lipson has received considerable media recognition for her team's research on optical cloaking. Her team has demonstrated several designs for an optical cloak that uses nanometer scale dielectric structures to conceal an object under a mirrored surface. The technique involves using transformative optics to reshape the surface of a reflective "carpet" so a bulge or deformation in the surface appears to be smooth to an external observer. This allows objects to be hidden under the deformation without revealing their existence. The device is designed to operate at all angles. According to the team's research paper, these results represent the first experimental demonstration of an invisibility cloaking device at optical frequencies. This research shows how the trajectory of light can be manipulated around a region, and has applications in defense, communications, and other industries [16]. The team also suggests that the process can be reversed to concentrate light in an area, which has applications for efficiently concentrating sunlight in solar energy applications.

Dr. Lipson's research involves innovations that not only bend light but also bend and twist some of the conventional notions of how light has been traditionally thought to behave. She is one of the few physicists in the world who has developed a method to trap and slow light. Her technique involves trapping light in a circular tube, then quickly shutting the gate where the light entered (so it doesn't leak back out) – and reopening the gate. The light loses energy the longer it stays in the trapping device and may need to be intensified after it leaves the device. This type of research has important implications for the development of ultrahigh-speed optical computers and optical switches in broadband telecom systems.

The Cornell group has also used particles of light to move nanoscale objects. That's another thing you probably didn't know about light. At human scale, particles of light from the sun are not strong enough to move objects, but at the nanoscale, particles of light can actually cause nanoscale objects to move, suggesting that light can be used to turn nanoswitches on or off in nanoscale computer chips. Increasingly, scientists are predicting that nanophotonics will soon replace microelectronics in communication and information processing systems.

"Nanophotonics can be used to make things visible as well as invisible," Dr. Lipson explains. She is currently working on a method to use transformative optics and metamaterials to provide a novel solution for some forms of blindness. She noted that this research is very preliminary and the technology is complex, but ultimately achievable.

In 2010, Dr. Lipson was awarded the $500 000 MacArthur Award, one of the most prestigious awards in science, which is sometimes called the "genius" award.

3.6 Nanoscience Wild Cards: Will Tree Lights Replace Streetlights?

In addition to the achievements described here, there are a host of "wild card" innovations that result from unexpected or accidental discoveries. In the field of nanoscience, these serendipitous discoveries are occurring at an unprecedented pace. It is not unusual for a researcher studying a material or process to stumble on a concept or idea that leads to an entirely new innovation.

It is also possible to incentivize the development of wild card innovations through competitions and calls for papers or proposals. In one notable example (see later), MIT has provided a library of biological parts, a competition, and toolkits to promote the development of innovative solutions.

An Accidental Discovery Turns Trees into Lights

A great example of a wild card innovation is the accidental discovery made by Taiwanese researchers who are working on a method for replacing streetlights with trees by implanting their leaves with gold nanoparticles that cause them to glow red – an elegant solution that could provide natural "tree light," which would be especially applicable in rural areas of the world.

Dr. Yen-Hsun Su is a postdoctorate scientist in the Research Center for Applied Science (RCAS) at Academica Sinica in Taiwan. He received his doctorate from the Department of Physics at National Cheng Kung University in Taiwan where he was supervised by professors Wei-Min Zhang and Shih-Hui Chang. Dr. Su and his colleagues were looking for a way to create high-efficiency LED lighting without using toxic chemicals like phosphor powder. In the process of this research, Dr. Su made a remarkable discovery. He found that on injecting gold nanoparticles into the leaves of the *Bacopa caroliniana* plant and exposing them to ultraviolet light, his team was able to induce the chlorophyll in the leaves to produce a reddish glow [17]. Dr. Su is credited with the seminal discovery that gold nanoparticles can induce luminescence in leaves.

This discovery has led to an entirely new course of scientific investigation, to develop what Dr. Su calls "Bio-LED" luminescence, which could be used to induce trees and other plants to emit a luminescent glow strong enough to illuminate streets and roads at night. Dr. Su has observed that this technology could save energy and even absorb CO_2 since the Bio-LED luminescence causes the chloroplasts in trees and plants to conduct photosynthesis, which absorbs CO_2 and produces oxygen.

The notion of using "glowing trees" instead of streetlights raises some interesting possibilities for future research. Imagine the benefits of being able to use "glowing trees" instead of streetlights, especially in rural areas of the world where there is no electricity, or in countries where electricity is too expensive or unavailable.

While some critics have observed that it sounds inefficient to shine ultraviolet light on the trees to induce them to glow, or that using gold nanoparticles would be expensive; the real value of this discovery comes from the concept of using bioluminescent trees to light streets and roads. It is easy to imagine how glowing tree-light will one day supplement electrical streetlights. Over time, more efficient methods are certain to be developed. There are other ways to make trees "glow." For example, one method might involve genetic engineering, using luminescent genes.

iGEM: How Student Nanoinnovators Are Lighting Up the World

Not all nanoinnovators are Ph.D. scientists. Many are students who develop an interest in some aspect of nanoinnovation and wind up creating something really special. Some of the most impressive student achievements have been developed for the International Genetically Engineered Machine Jamboree competition (iGEM) hosted at the Massachusetts Institute of Technology (MIT).

iGEM is the world's premiere synthetic biology competition. iGEM invites student teams to draw from a library of several thousand standard biological parts called "BioBricks." Goals of the BioBricks project include standardizing and automating the development of nanotechnology innovations using standardized biological parts. BioBrick™ parts were originally introduced in 2003 by Tom Knight at MIT. Drew Endy (now at Stanford) and Christopher Voigt at UCSF also participated in developing the project. Randy Rettberg currently maintains the BioBrick registry, which includes several thousand public domain parts that can be viewed online at partsregistry.org. Student teams may also develop their own BioBrick parts as part of their research. This initiative is an ambitious open source effort to help foster safe, ethical development of technology solutions using synthetic biology.

Undergraduate student teams from the Americas, Europe, and Asia work at their own schools over the summer, using biological parts from the Registry of Standard Biological Parts to build biological systems and operate them in living cells. Examples of the student achievements range from an arsenic biodetector to a cost-effective red blood cell substitute. Student teams have participated from Australia, Belgium, Brazil, Canada, China, Colombia, the Congo, Denmark, France, Germany, Hungary, India, Italy, Japan, Kenya, Korea, the Mexico, the Netherlands, Panama, Poland, Scotland, Singapore, Slovenia, South Africa, Spain, Sweden, Switzerland, Taiwan, Turkey, and the United States.

In 2009, a team of nine undergraduate students at Cambridge University won the Grand Prize with an innovation they call "LuxBrick." The students used

Figure 3.6 Ben Reeve reads *The Jungle Book* by Rudyard Kipling by the light of the Cambridge research team's bioluminescent lamp (photo courtesy of Ben Reeve and the Cambridge iGEM team).

genetic material from fireflies and a luminescent marine bacterium (*Vibrio fischeri*) to boost the intensity and activity of light-emitting enzymes. They developed a bacterial culture about the size of a bottle of wine that generated light – in different colors – that was bright enough to read by. The accompanying pictures show one of the students reading The Jungle Book by Rudyard Kipling, using the light from the LuxBrick bacteriological bubble lamp (Figure 3.6) and the other photo shows two of the team members holding flasks of bioluminescent material (Figure 3.7).

Figure 3.7 Cambridge students Ben Reeve and Theo Sanderson illuminate their faces using flasks of bioluminescent material (photo courtesy of Ben Reeve and the Cambridge iGEM team).

The students on the 2009 Cambridge team included Anja Hohmann, Ben Reeve, Bill Collins, Emily Knott, Hannah Copley, Paul Masset, Peter Emmrich, Theo Sanderson, and Will Handley. The diversity of the group is remarkable – each student represents a different discipline. Anja is a biochemist, Ben is a molecular biologist (and the team's "Hacky Sack-master), Bill is a control engineer, Emily is a mechanical engineer, Hannah is a medic, Paul is a systems engineer, Peter is a plant scientist, Theo is a geneticist, and Will is a physicist.

This diversity helped them tackle some of the tricky problems associated with bioluminescent technology. One of the problems with bioluminescence is what happens to the light-emitting compounds after they "glow." Luciferins are compounds that emit light but after the light is generated, the compounds are converted to oxyluciferin, which cannot produce light. The Cambridge students developed a way to recycle oxyluciferin, to maintain the "glow." Their biochemical lights were engineered to produce a variety of colors. Techniques for causing the bacteria to "hibernate" until needed have also been proposed by the team. The students are currently exploring how to apply their discoveries to create bioluminescent streetlights to replace conventional street lamps.

"BioBricks are standardized biological parts," explained Ben Reeve. "One of the BioBricks we produced was the LuxBrick. We used this part to produce both the bubble lamp and fire exit sign" (Figure 3.8) (http://BioBricks.org/faq/). Their "LuxBric" bacterial bubble lamp, illuminated fire exit sign, and simulation of bioluminescent street lamps can be found at their Web site (http://2010.iGEM.org/Team:Cambridge/Tools/Lighting). How the LuxBrick was designed is also described in detail on their site (http://2010.iGEM.org/Team:Cambridge/Bioluminescence/G28).

One interesting aspect of bio-lights is that they have to be "fed" in some way, on a regular basis, which means that bio-lights may need fertilizer or other nutrients in lieu of electricity, to keep operating. Imagine the paradigm shift that would be if we go from powering lights with electricity to powering lights with fertilizer!

This stream of research holds the potential for a range of innovations. For example, in addition to tree lights replacing streetlights or garden lights, it is

Figure 3.8 A standard fire exit sign (a) and the same sign illuminated by bioluminescence (b) (photo courtesy of the Cambridge iGEM team).

easy to envision new types of Christmas trees that glow in the dark, or even change colors biologically. Perhaps trees will be engineered to serve as biological sensors or indicators. Imagine trees or crops glowing when they need water, so they can be irrigated only when needed.

The popular iGEM competition keeps growing – the competition started with a handful of teams in 2005 and increased to 130 teams in 2010. An interesting side note to the iGEM competition is that the top two teams in 2010 were from Slovenia and Peking, which illustrates the global nature of nanoinnovation. Student teams from Slovenia won the iGEM grand prize in 2006, 2008, and 2010.

There are similar student contests and initiatives in virtually every country. A few examples include in2nano (Egypt); the MEMSIC Cup/International Contest of Applications in Nano-micro Technology (iCAN/14 countries); Time for Nano Video Contest (EU); and the Khwarizmi Youth Award Competition (Iran). Many universities hold their own nanotechnology events and contests. A good example is the annual NanoDay sponsored by the Nano/Bio Interface Center at the University of Pennsylvania, which invites Penn students and high school students to "imagine the world at the nanoscale."

Giving Students Access to Nanotechnology Systems

Scanning probe microscopes and other instruments are not yet available to secondary (high) schools, which means that most students need to wait until they enter college to start working with nanotechnology. Fortunately, a growing number of forward-thinking high school teachers are finding ways to teach nanotechnology to their students.

A great example is Mike Boyer, an engineering teacher in the technology and engineering education department at North Penn High School in Lansdale, Pennsylvania. He teaches nanotechnology as part of a program he launched in 2005 called The Future is NEAR (NEAR stands for "Nanotechnology Education and Research") [18].

"The first year I ran the program," he recalls, "we had three teams. At the end of our first year, the students begged to take a presentation on the road to other schools although they were graduating in a few weeks and going off to universities. They wanted other schools to see what could be accomplished."

"Over the years, the student research projects picked up in complexity. I found myself getting farther out of my comfort zone and at the same time the student teams were learning, setting goals, assessing their progress, assessing their Gantt charts and comparing notes with other teams, which challenged me as a facilitator. I've spent many late nights after my family went to bed, researching team projects so I could help guide them, which is important because the field of nanotechnology is constantly growing and our knowledge keeps expanding."

"Because we do not have a scanning electron microscope in class," Mr. Boyer explains, "we collaborate with Drexel University to provide a field trip where our students can run their experiments in the nanotechnology laboratory. This has been an invaluable collaboration. Until the field trip, the students need to work blindly and have to make educated guesses regarding the results of their experiments. They study the results of similar experiments in scientific journals to prepare as much as they can before their field trip. Sometimes the experiments don't work in the first trial, which makes it more difficult to coordinate access and continue the experiments. This is much less hands-on than we would like, but it's difficult to get funding for a $70 000 SEM for one high school, or even a group of schools. Recently, I was able to arrange to get a Hitachi SEM on loan for a week from Angstrom Scientific. One day I hope to have an SEM for my students!"

The list of team projects conducted by Mr. Boyer's nanotechnology class is impressive: portable electrospinning device, electrospinning envelope cone analysis, magnetic polymer fibers, studies of the effects of solvents on nanofiber production, hydrophobic nanofabrics, liquid armor, drug and particle encapsulation in nanobeads, photovoltaic polymers, piezoelectric nanofibers, photochromatic and thermochromic nanofabrics, and the use of bamboo in the creation of antibacterial fabrics. It is remarkable and impressive that such sophisticated experiments are being studied by teams of high school students. Imagine what could be accomplished if all the high schools in the world had nanotechnology programs. Most important, these projects and experiments energize the students to pursue careers in nanotechnology. Many of Mr. Boyer's students have received research positions after graduating high school, and many go on to continue their nanotech research in college (Figures 3.9 and 3.10).

Youth camps and outreach programs that include nanotechnology are being sponsored by government agencies and companies in many countries. Some notable examples include the in2nano program in Egypt and the in2nanotech mentoring program offered by La Trobe University in Australia – similar programs are underway in countries as diverse as Russia, India, and Iran.

James McGonigle, Director of Education and Outreach at the University of Pennsylvania's Nano/Bio Interface Center, discussed the need for more affordable nanoimaging systems for high schools: "For nanoscale science and engineering to really take hold and capture the imagination of pre-college students, we need affordable ways for high school students to experience the authentic process of working at the nanoscale. Students need to be able to be able to synthesize nanoscopic materials in their high school labs and be able to 'see' the materials first hand. Affordable imaging systems that cost thousands of dollars – not tens of thousands – would allow schools and school districts to provide sophisticated and authentic experience on tools that the 21st century technology workforce will use. Solutions might include a simple AFM that is robust enough to travel between classrooms and between schools so many teachers can share the resource."

"More curriculum resources and laboratory activities are needed to prepare students to the enter the nanotechnology workforce of the future. The Nano/Bio Interface Center at the University of Pennsylvania and the A.J. Drexel Nanotechnology Institute have worked together to provide over 100 teachers with summer research experiences in nanoscale science and engineering. This experience translates directly into classroom lessons for their students, helps prepare

Figures 3.9 Students from North Penn High School in Lansdale, Pennsylvania formulate their experiments at school and then complete them using the nanotechnology systems provided by Drexel University in Philadelphia. The technology-oriented high school provides this opportunity under the N.E.A.R. program developed by engineering teacher Michael Boyer. This collaboration is an excellent example of how resourceful teachers and students are finding way to gain access to expensive nanoimaging systems (photos courtesy of Michael Boyer).

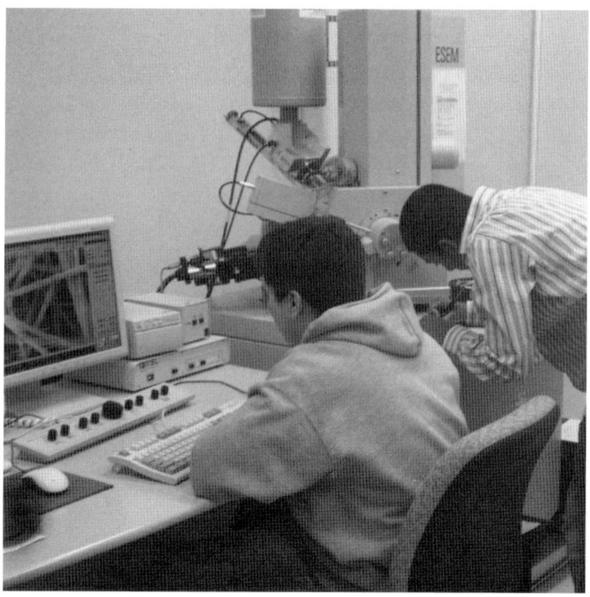

Figures 3.10 High school students in the N.E.A.R. program have a unique opportunity to gain hands-on experience using sophisticated nanotechnology instruments in a university laboratory.

teachers to work with students when they visit university labs, and paves the way for the future when school districts will have access to their own nanotechnology instruments."

3.7
Science Genius versus Commercial Challenge

Translating scientific prowess into commercial success poses a special challenge for nanoinnovators. While it is true that most of the first wave of nanoinnovations came from university research labs, it is also true that many of these commercial ventures have stumbled or failed.

In business, someone who launches and leads several successful business ventures is called a "serial entrepreneur." Someone who develops a string of nanoinnovations can be called a serial nanoinnovator. One of the best examples of a serial nanoinnovator is Chad Mirkin, Director of the Nanoscale Science & Engineering Center and the International Institute for Nanotechnology at Northwestern University and Rathmann Professor of Chemistry (Figure 3.11). Chad leads a

Figure 3.11 Professor Chad Mirkin, Director of the International Institute for Nanotechnology at Northwestern University (photo by Bill Arsenault).

group of 50 researchers that have developed numerous nanoinnovations, including nanoparticle-based biodetection schemes, Dip-Pen Nanolithography, and innovations involving supramolecular chemistry, nanoelectronics, nanooptics, and medical diagnostics.

As a nanoinnovator, Dr. Mirkin is a standout by any standard. He is a world-renowned chemist and nanoscience pioneer, the author of more than 450 manuscripts and over 380 patents and applications. He has been the most cited chemist in the world for the past 2 years and the top-cited nanomedicine researcher, and has been called the world's most-cited nanoscientist. His seminal papers, "A DNA-based method for rationally assembling nanoparticles into macroscopic materials" (*Nature*, 1996) and "Selective colorimetric detection of polynucleotides based on the distance-dependence optical properties of gold nanoparticles" (*Science*, 1997) have both been cited more than 2000 times – contributing to a total of more than 40 000 citations for his published papers. He is the recipient of over 70 national and international awards, including the $500 000 Lemelson-MIT Prize (June 2009), the first NIH Pioneer Award, and the Feynman prize for experimental breakthroughs in nanotechnology (2004), and the American Chemistry Society's Award for Creative Invention (2012). Chad serves on President Barack Obama's Council of Advisors for Science and Technology. He gives 50–70 lectures a year and is constantly in motion, racing to fit as many innovations as possible into one frenetic lifetime.

Dr. Mirkin is the founder of three companies – Nanosphere, NanoInk, and Aurasense – which were some of the earliest and most promising ventures in the field of nanoinnovation. These ventures include both successes and failures, and provide interesting insights for nanoinnovators who want to develop their own ventures and commercial projects.

The Rise and Fall of NanoInk

More than a decade ago, Chad Mirkin recognized the need for a more practical approach to nanolithography. Dr. Mirkin and his research group at *Northwestern University* determined that under the appropriate conditions, molecules rather than energy could be transferred to a wide variety of surfaces to create stable chemically adsorbed monolayers. Their patented lithographic process used atomic force microscope tips as writing tools and diffusible material for ink. They called the process Dip Pen Nanolithography (DPN) [19]. The general process is called polymer pen lithography (PPL). PPL is primarily a fabrication tool, while DPN is used as both a fabrication and an imaging tool.

Their patented process uses an array of millions of polymeric pens that create nanoscale architectures by "drawing" molecules onto a substrate over large areas with extreme precision. The pens are attached to the moving arm of a scanning probe microscope, which traces the designs. As the tips are pressed down on the substrate, the tip is flattened and a larger feature is generated. Different types of substrates, materials, and molecular inks can be used.

In 2001, Dr. Mirkin founded a venture called NanoInk, Inc. to commercialize the technology. Unfortunately, the venture did not fulfill its early promise. In

11 years, NanoInk absorbed more than $150 million in investment capital, which came mostly from Lurie Investments, a Chicago-based investment fund managed by philanthropist Ann Lurie. In 2013, Ms. Lurie withdrew as the company's sole funding source. NanoInk filed Chapter 7 bankruptcy in April 2013 and began liquidating assets – drawing a curtain on one of the very first nanotechnology ventures. This venture had extremely high expectations and was a darling of the press and the nanotech community. With a huge patent portfolio and proven technology, few people expected NanoInk to fail.

At the time of its bankruptcy, NanoInk's assets included 60 US patents, 74 US patent applications, 97 foreign patents, and 230 foreign patent applications. Obviously, the size of a nanoventure's patent portfolio is not always a good indicator of long-term commercial success. In May 2013, a bid process was established and companies interested in purchasing NanoInk's patents were invited to bid on the portfolio.

So how did this venerable nanotech pioneer come to such a sudden end? One observer suggested that NanoInk failed because they were chasing "technology push" rather than "market pull." Tim Harper, CEO of Cientifica, likened NanoInk's dip-pen technology to "replacing the printing press with a bunch of monks." Harper added that "illuminated manuscripts can look good but if you can't mass-produce things there isn't a business [20]." The company's technology used massive arrays of dip pens to "paint" or "draw" nanoscale ink on a substrate, but had trouble finding viable commercial applications for the technology.

Dr. Mirkin has indicated that he had not been active with the company for several years before it failed, having moved on to other scientific projects and ventures. By most accounts, the failure of the company was attributed to the management team, not the scientific team, driven by a lack of commercial markets and applications and overspending.

Nanosphere

Dr. Mirkin is also the inventor of an innovative bioscience technology based on the Verigene® System, an FDA-cleared molecular diagnostics workstation that utilizes patented gold nanoparticle technology to detect nucleic acid and protein targets of interest for a variety of applications. This technology was commercialized through Nanosphere, a public company founded in 1999 by Dr. Mirkin and his Northwestern University colleague Dr. Robert Letsinger. Nanosphere is also an example of how expensive it can be to start and maintain a nanoventure. Nanosphere more than doubled its revenues, to more than $5 million in 2012, but struggled to make a profit. Nanosphere lost an average of $40 million per year since its inception and lost $32.9 million in 2012. By 2013, the company was valued at less than half of its invested capital. In February 2013, Nanosphere's CEO was replaced and in May, Dr. Mirkin left Nanosphere's board. For the fiscal year ended December 31, 2013, Nanosphere reported that revenues nearly doubled from $5.1 million to $10 million, but recorded a net loss of $34.6 million. In January 2014, Nanosphere received FDA approval to market its Verigene *in vitro* diagnostic test for identifying Gram-negative bacteria associated with bloodstream infections. The company's management team expressed confidence and expected revenues to nearly double in 2014.

Aurasense

Dr. Mirkin and his Northwestern University colleague Shad Thaxton formed a venture called "Aurasense" in 2009, to develop nanoparticles for use as therapeutics and intra-cellular biomolecule detection probes. This venture is working to turn gene-regulating nanoparticles into commercial products. The venture is based on scientific developments at Northwestern and continues the process of invention, innovation, and commercialization that Chad Mirkin and his research associates have embarked on, and where they continue to persevere as they work to overcome challenges and setbacks to turn remarkable science into viable commercial ventures.

3.8
Nanoscience Pioneers Are Mapping the Future

Many of the greatest scientific breakthroughs in nanotechnology have come from IBM, which learned to "think small" many decades ago when IBM scientists created the first scanning tunneling microscopes.

IBM scientists have a sense of whimsy – their first nanoscale image was an IBM logo created by manipulating individual atoms. In recent years, IBM has used nanotechnology to create the smallest map of the world, and the world's smallest cartoon animation. IBM's map is the smallest 3D map of the world. It was created using nanoscale tools, structures, and techniques (see Figure 3.12). This map is so small, it can fit in a grain of salt!

In May 2013, a group of IBM scientists announced that they had created the world's smallest movie, a feat that was verified by Guinness World Records. The film cartoon, entitled "A Boy and His Atom" (Figure 3.13) used a scanning

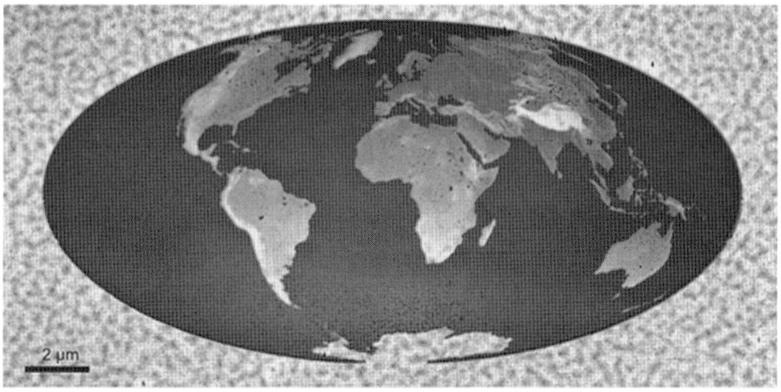

Figure 3.12 To illustrate a breakthrough nanotip-based patterning technique, IBM scientists created the world's smallest 3D map. The device used to create this map is small enough to fit on a tabletop and is accurate to 15 nm or less (photo courtesy of IBM Zurich Research).

Figure 3.13 These scenes are from the world's smallest movie, produced by IBM researchers and entitled "A Boy and His Atom" (image courtesy of IBM Research).

tunneling microscope with a supersharp needle to arrange carbon monoxide molecules on a copper surface to create a $45 \times 25\,\text{nm}^2$ picture of a boy playing with a single molecule. The individual molecules were moved and reimaged for each frame to produce 242 frames. The cartoon is accompanied by a catchy tune. Andreas Heinrich, the principal investigator at IBM Research, said: "This movie is a fun way to open a dialogue with students and others on the new frontiers of math and science."

The group also wanted to focus attention on their initiative to determine the smallest number of atoms it takes to reliably store one bit of magnetic information, which they claim is 12 atoms. This compares to the approximately 1 million atoms it currently takes to store 1 bit of data in a computer or electronic device. According to Dr. Heinrich, the same approach used to create the world's smallest movie is being used to create new computing architectures. One day, he predicts, atomic memory could store all of the movies ever made in a device the size of a fingernail.

The nanoscience innovations described in this chapter are only a few examples of the thousands of nanoinnovators who are pushing the frontiers of nanoscience and redefining what's possible. Many of these scientists work tirelessly with no recognition or publicity, to unlock the complex secrets of the nanoscale world. We owe them a huge debt of gratitude for their dedication, commitment, and perseverance.

References

1 Colvin, V. (2007) Nano-rust removes arsenic from drinking water. *Membrane Technology*, **2007** (3), 6.
2 Kurotobi, K. and Murata, Y. (2011) A single molecule of water encapsulated in fullerene C60. *Science*, **333** (6042), 613–616.
3 Global Industry Analysts, Inc. (2010) Aerogels: A Global Strategic Business Report, February 1.
4 Gu, H., Chao, J., Xiao, S.-J., and Seeman, N.C. (2010) A proximity-based programmable DNA nanoscale assembly line. *Nature Letters*, **465**, 202–205.
5 Kyoto University (2012) DNA motor programmed to navigate a network of tracks, Kyoto University Press Release, January 23.
6 Andersen, E.S., Dong, M., Nielsen, M.M., Jahn, K., Subramani, R., Mamdouh, W., Golas, M.M., Sander, B., Stark, H., Oliveira, C.L.P. *et al.* (2009) Self-assembly of a nanoscale DNA box with a controllable lid. *Nature*, **459**, 73–76.
7 Ishitsuka, Y. and Ha, T. (2009) DNA technology: a nanomachine goes live. *Nature Nanotechnology*, **4**, 281–282.
8 Modi, S., M.G., Swetha, Goswami, D., Gupta, G.D., Mayor, S., and Krishnan, Y. (2009) A DNA nanomachine that maps spatial and temporal pH changes inside living cells. *Nature Nanotechnology*, **4**, 325–330.
9 Oxford University (2011) *DNA Cages 'Can Survive Inside Living Cells'*, Oxford University Press Release, July 1.
10 Church, G.M., Gao, Y., and Kosuri, S. (2012) Next-generation digital information storage in DNA. *Science*, **337** (6102), 1628.
11 Mearian, L. (2012) IBM smashes Moore's law, cutting bit size to 12 atoms. *Computerworld*, January 12.
12 Reber, P. (2010) What is the memory capacity of the human brain? *Scientific American*, April 1.
13 Goldflam, M.D., Driscoll, T., Chapler, B., Khatib, O., Marie Jokerst, N., Palit, S., Smith, D.R., Kim, B.-J., Seo, G., Kim, H.-T., DiVentra, M., and Basov, D.N. (2011) Reconfigurable gradient index using VO_2 memory metamaterials. *Applied Physics Letters*, **99**, 044103.
14 Perczel, J., Tyc, T., and Leonhardt, U. (2011) Invisibility cloaking without superluminal propagation. *New Journal of Physics*, **13**, 083007.
15 Smolyaninova, V.N., Smolyaninov, I.I., and Ermer, H.K. (2012) Experimental demonstration of a broadband array of invisibility cloaks in the visible frequency range. *New Journal of Physics*, **14**, 053029.
16 Gabrielli, L.H., Cardenas, J., Poitras, C.B., and Lipson, M. (2009) Silicon nanostructure cloak operating at optical frequencies. *Nature Photonics*. doi: 10.1038/nphoton.2009.117.
17 Singh, T. (2010) Gold nanoparticles could transform trees into street lights. inhabitat.com (November 30, 2010).
18 Nanotechnology Research at North Penn High School (2014) The Future Is N.E.A.R., www.thefutureisnear.org.
19 Piner, R.D., Zhu, J., Xu, F., Hong, S., and Mirkin, C.A. (1999) Dip pen nanolithography. *Science*, **283**, 661–663.
20 Johnson, D. (2013) Why did nanoink go bust? *IEEE Spectrum*, February 22.

4
Imaging the Unseen: Viewing Structures Smaller than Light Waves

Only he who can see the invisible can do the impossible.

– *Frank Gaines*

When it comes to nanoimaging, seeing is achieving.

Nanoscale imaging – the ability to view structures and processes at the nanoscale – is a core innovation that drives virtually all other nanoinnovations. As a manager, you should have a basic understanding of how nanoimaging works because so much depends on our ability to view this unseen world. Nanoimaging is used to design semiconductors and car engines, to engineer new types of materials, and study the safety of nanoparticles. Nanoimaging helps confirm the quality and consistency of nanopores in fuel injectors and desalination filters. Nanoimages aid the study of bacteria and viruses, and help diagnose and treat disease at the cell level.

Nanoimaging – better known as nanoscale microscopy – is still evolving. Every form of microscopy, from optical and electron microscopes to magnetic resonance imaging (MRI), is advancing rapidly. If you want to understand what is happening in nanoinnovation, it helps to have a sense of the different types of nanoimaging systems, and what is involved in viewing and manipulating the incredibly tiny structures of the nanoworld.

4.1
What Nano Images Reveal

If you surf the Web, you'll see thousands of nanoscale images displayed in online galleries. Many of these images are truly remarkable. They show us what Nature and humans can engineer at the nanoscale, and provide clues to new types of materials that can be developed and mass-produced by nanoinnovators for commercial applications in applications ranging from medical biomarkers to thin-film solar cells.

Nanobama

The nano portraits of President Barack Obama known as "nanobama" were created by Prof. Jon Hart at the University of Michigan (Figure 4.1a and b).

NanoInnovation: What Every Manager Needs to Know, First Edition. Michael Tomczyk.

(a)

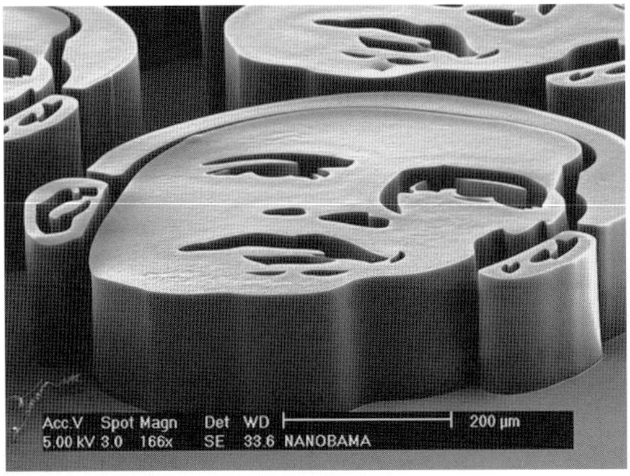

Figure 4.1 (a) John Hart's "Nanobama" image has become an iconic example of nanotechnology and is probably the world's most widely known nanoscale image. Each nanoscale portrait of Barack Obama is comprised of ~150 million carbon nanotubes, which is approximately the number of Americans who voted in the 2008 presidential election. (b) Showing how the images were made, described in detail on the nanobama.com Web site. This is a terrific example of the convergence of nanoart and nanoscience, which is discussed in Chapter 5 (images courtesy of A. John Hart, University of Michigan).

This is one of the most iconic of all the nanoimages produced in the first decade of the current millennium, and was the first nanoscale portrait of a world leader. Nanobama provides an impressive demonstration of how a forest of carbon nanotubes can be transformed into a precise structure – in this case, a presidential portrait.

DNA Triangles

In Figure 4.2a you can see rows of tiny triangles made from strips of synthetic DNA that look like soldiers trying to find their place in a parade formation. These nanosize pyramids were created by a group of IBM scientists led by Greg Wallraff; using a technique invented by Prof. Paul Rothemund at the California Institute of Technology (Caltech). Much denser clusters have been packed into these channels, offering the potential for new semiconductor circuit designs. In Figure 4.2b, DNA origami triangles were self-assembled in a single step by Dr. Rothemund, who developed the DNA origami process. Each triangle consists of over 200 DNA strands, each incorporating 15 000 nucleotides. Each side of the triangle is ~120 nm.

Carbon Onions

One of the best known nanostructures is the buckminsterfullerene, commonly called fullerenes or buckyballs – but what many people do not know is that buckyballs can be nested one inside the other, like Russian matryoshka dolls.

(b)

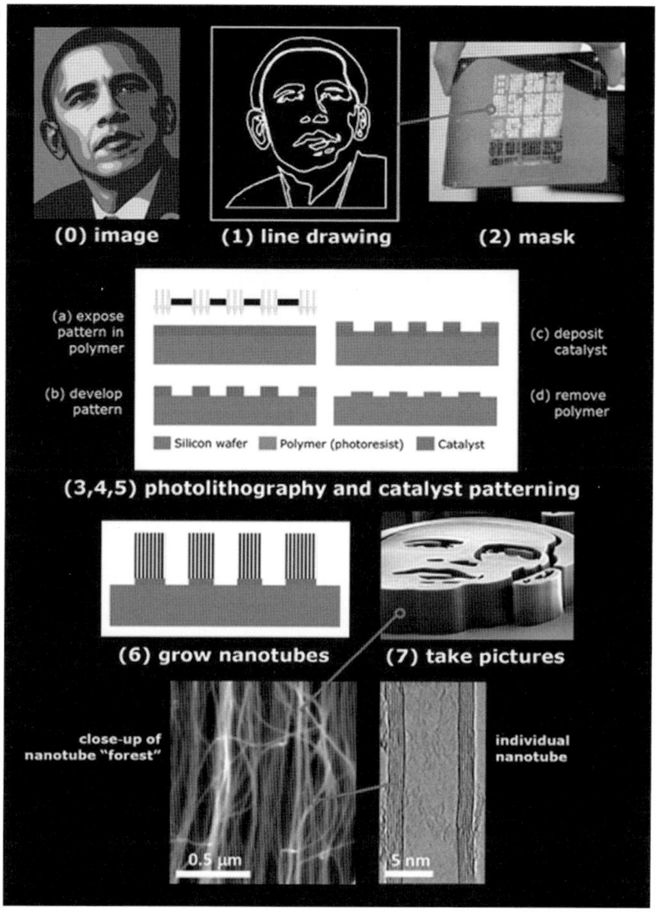

Figure 4.1 (*Continued*)

The nested form of a buckyball is called a "carbon onion." A carbon onion is a multishelled nanoparticle consisting of many concentric graphite layers. Carbon nanotubes and carbon onions were discovered by Sumio Iijima in 1980. The images of carbon onions in Figure 4.3a and b were provided by Naiqin Zhao, an award-winning scientist and Chair of the Department of Metallic Materials at Tianjin University in China. Professor Zhao's research group has developed a method for obtaining pure carbon onion nanoparticles and fabricating them in large quantities [1]. Research on commercial applications of carbon onions was previously limited by the inability to fabricate quantities of this unique form of carbon. In recent years, several fabrication methods have been developed, which is opening the door to more aggressive research on functional applications for carbon onions.

(a)

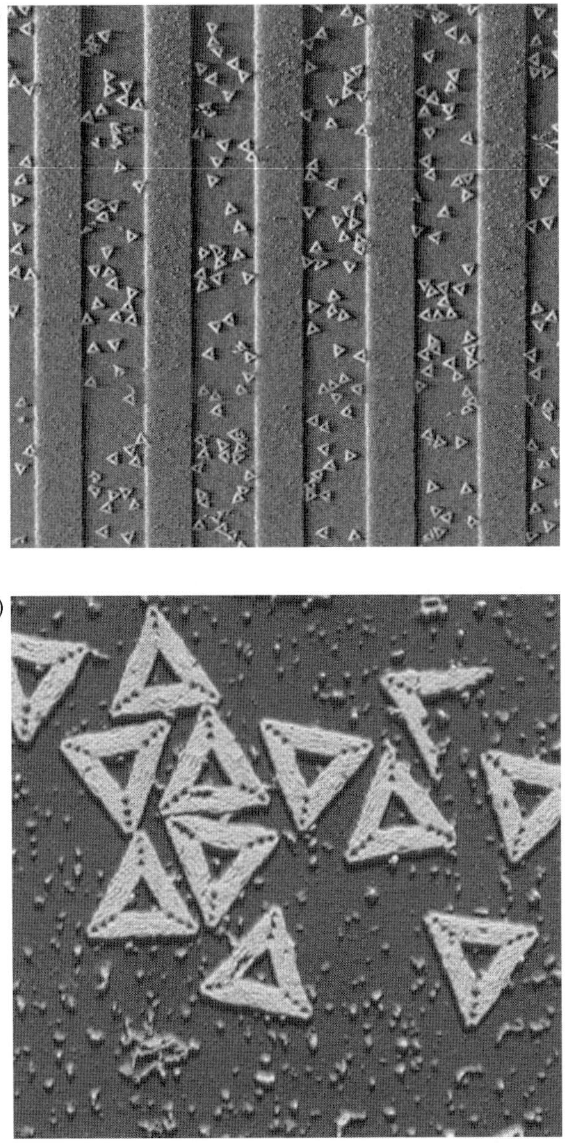

(b)

Figure 4.2 (a) DNA origami techniques pioneered by Paul Rothemund at Caltech have been used to create an array of triangles made from DNA strands. The triangles are in the process of binding to the wide channels that were lithographically etched in the substrate. This technology holds promise for developing new types of semiconductors (image courtesy of IBM). (b) A close-up image of more complex self-assembled DNA triangles created by Rothemund – each triangle is only about 120 nm on each side (image courtesy of Paul Rothemund, Caltech).

(a)

(b)

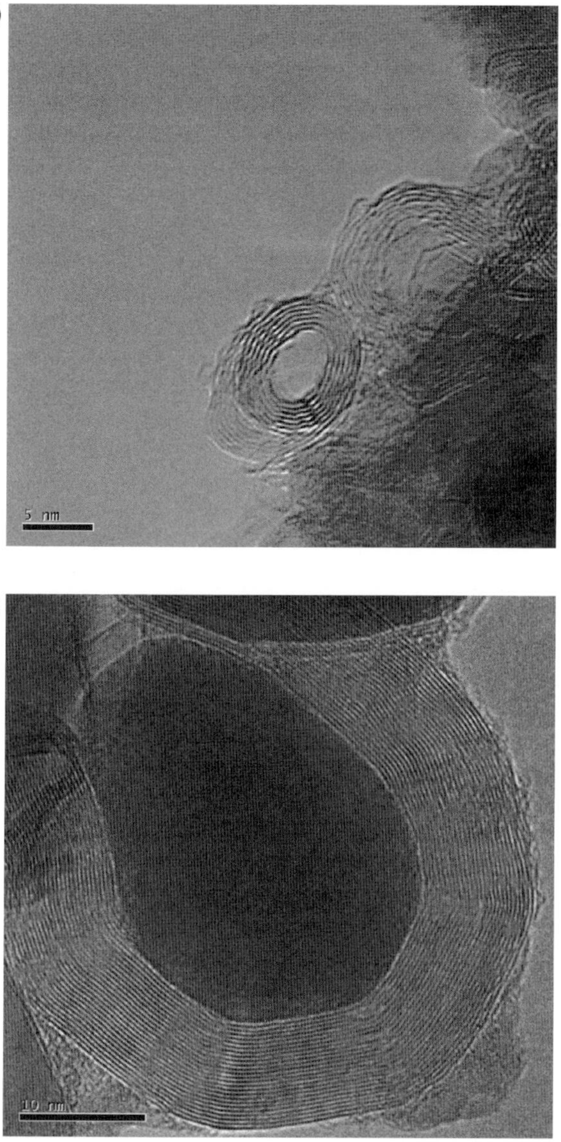

Figure 4.3 (a) The concentric shells formed by the multiple "skins" on this hollow carbon onion, which according to the scale measures only about 10 nm in diameter. (b) Showing even more nested layers on this larger carbon onion, which also shows how this nanoparticle is not perfectly spherical, but is typically distorted. This level of detail is revealed or verified only in nano-images that allow us to actually see the particles (images courtesy of Professor Naiqin Zhao, School of Material Science and Engineering, Tianjin University).

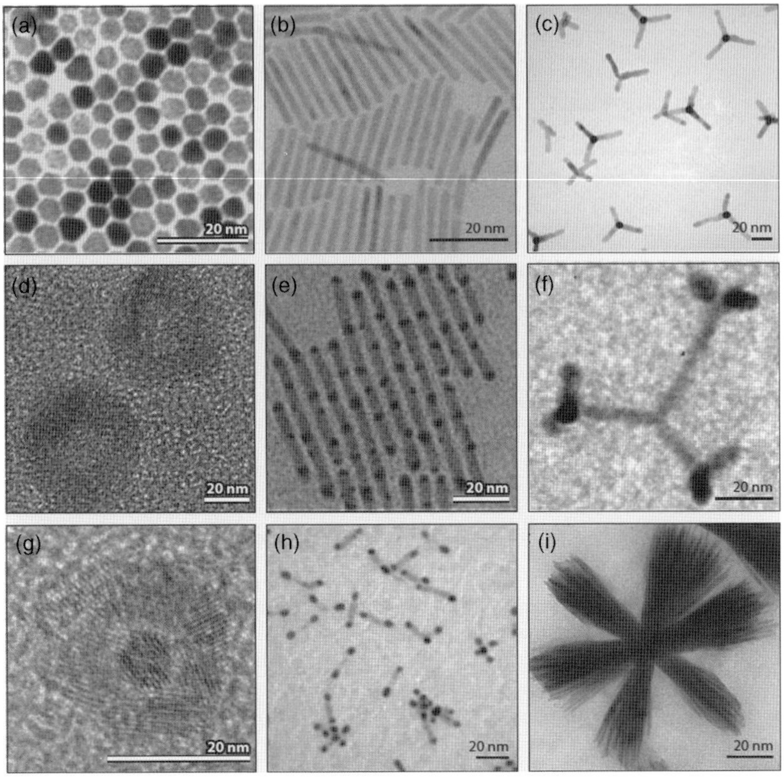

Figure 4.4 These TEM images of nanocrystals synthesized by Paul Alivisatos' research group illustrate just a few of the many types of structures and form factors that nanoinnovators are experimenting with, to learn their properties and apply them to commercial applications. Shown here are (a) quantum dot nanocrystals of cadmium selenide (CdSe), (b) CdSe nanorods, (c) CdSe tetrapods, (d) hollow nanospheres, (e) striped binary nanorods made of silver sulfide (AgS) –cadmium sulfide (CdS), (f) CdSe tetrapods with cadmium telluride (CdTe) dendrimer branches, (g) nested Pt@CoO yolk-shell particles, (h) gold-tipped CdS nanorods, and (i) bismuth selenide (Bi_2Se_3) nanoflowers (images courtesy of Paul Alivisatos, Berkeley National Labs/UC Berkeley).

Myriads of Architectures

Nanostructures come in all shapes and geometries. The gallery of images in Figure 4.4 was provided by Dr. Paul Alivisatos' research group at Berkeley, and shows several semiconductor nanocrystals created in their laboratory. These are only a very few of the many thousands of intriguing images produced by nanoimaging systems. Let's take a quick look at how these images were produced, and the technologies that make it possible to view these structures, which are smaller than visible light waves.

4.2
Using Electrons Instead of Light to View Nanoscale Structures

Most scientists today use microscopes that capture electrons instead of light waves, because nanoparticles are actually smaller than light waves and can't be viewed with conventional optical microscopes. Electron microscopes produce images that are as much as 100 000 times more detailed than an optical microscope. Nanoscale imaging systems come in dozens of configurations and variations.

The first generation of electron imaging systems required the samples to be sliced ultrathin and held rigid in a vacuum, or coated with a material such as gold. This limited the types of samples that could be imaged. Today, high-performance nanoscopes can image a virus that is suspended in the pore of a carbon matrix. A probe sharpened to one atom at the tip can be used to reposition a single atom. With a small charge applied, the same probe can pluck a single nanowire from a forest of tangled nanowires. Some types of nanoscopes can gently move the tip of the probe across the surface of a soft and delicate sample such as a cell. Nanoscope operators can even use light probes – called optical tweezers – to move nanoparticles.

Compared with what we could do with microscopes a century ago, we live in an imaging renaissance – and yet, if we peer ahead and consider what we will be able to image in 10 or 20 years, we may find that we are actually living in the nanoscopic Dark Ages.

Today, most nanoscale microscopes are generally known as scanning probe microscopes (SPM) because the same tips used to image nanostructures can also be used to manipulate them, although scientists use more than a dozen major nanoscale imaging technologies. There is a veritable alphabet soup of nanoscale microscopes that can be confusing to someone who is not working in the industry – atomic force microscope (AFM), SEM, scanning tunneling microscope (STM), scanning transmission electron microscope (STEM), and transmission electron microscope (TEM) – so let's take a moment to sort out what you need to know, starting with a brief history.

4.3
A Short History of Nanoscale Imaging

The concept of using electrons instead of light waves to image nanoscale samples is not new. The first electron images were microscale, not nanoscale. When the first electron micrographs began to appear in magazines like *Life* or *Scientific American*, readers were captivated by the bizarre images of rhinoceros-horned dust mites that live in our homes and close-ups of insect eyes or convoluted grains of pollen. These intriguing and often astonishing images caused us to wonder, "What else is down there?"

The first TEM was developed in Germany by Ernst Ruska and Max Knoll in 1931 and the first commercial TEM was produced by Siemens in 1939. The first electron micrograph of an intact cell was published in 1945. After World War II, the electron microscope was refined into the current form used today.

After the electron microscope was developed, the next major advance was the STM, which was developed to address some of the limitations of the electron microscope, such as having to work with rigid samples in a vacuum.

During 1965–1971, Russell Young and his colleagues at the National Bureau of Standards created an instrument he called the "Topografiner," which was a precursor to the STM. This led to the invention of the STM by Gerd Binnig and Heinrich Rohrer at IBM-Zurich in 1981. In 1986, Binnig and Rohrer shared the Nobel Prize in Physics with Ernst Ruska, who received half the Prize in recognition of his pioneering work on the electron microscope. Russell Young's contribution was acknowledged in the Nobel Prize citation.

By the late 1980s, large corporations were experimenting with electron microscopes. Robert Dombrowski, a materials scientist and expert in nanocharacterization, worked with the first electron microscopes at Colgate Palmolive. "Back then, atomic force microscopes were still considered Rube Goldberg contraptions. I remember when we first got started, there was so much vibration in the lab, we had to put the microscope on bungi cords. Our first scanning electron microscope cost $400,000 and had vacuum tubes that you had to change like a TV repairman." Robert was involved in early nanotechnology research involving skin products, electrospinning of nanofibers, development of biodegradable starch-based polymers for packaging applications, and medical applications that included using liposomes to deliver drugs.

How a Scanning Tunneling Microscope Works

The STM captures images by moving a probe across the surface of the sample, close enough to allow electrons to jump from the surface, which allows them to be measured. The tip of the probe does not actually come into contact with the sample. As the probe detects electrons, it sends signals to a computer, which creates an image.

The STM mechanism is both elegant and complex. A sharpened tip only one atom thick at the point is attached to a movable arm, or cantilever. The probes are typically sharpened to one or two atoms at the tip and may be made from silicon, tungsten, diamond, or some other material. The tip of the probe is created by physically or chemically etching away the atoms until one or two atoms remain at the end of the probe. Some probes are tipped with a carbon nanotube.

The cantilever holding the probe is attached to a set of crystals that are each connected to a nanowire. When a minute current is passed through the nanowire, the crystals move. This movement is possible because of a phenomenon called the *piezoelectric effect*, which causes materials such as crystals or ceramics to move when electric current is applied. By sending electricity (electrons) to the crystals, the crystals can be made to move the cantilever back and forth along the X–Y-axis. By sending current to all the crystals, the cantilever can move up or down along the Z-axis. More detailed descriptions and diagrams are available online, but this is basically how it works.

4.4
Different Types of Nanoscale Microscopes

When you read about nanoinnovations, or if you work with colleagues involved in nanotechnology, you will come across references to a variety of nanoscale imaging systems. Here are brief descriptions of some of the most common imaging technologies.

The Atomic Force Microscope

In 1986, Gerd Binnig (IBM), Calvin Quate (Stanford University), and Christoph Gerber (IBM/University of Basel) developed the AFM. "Atomic force" refers to the microscope's ability to record forces between individual atoms on the surface and at the tip of the probe, such as the van der Waals force. A primary distinction between STM and AFM is that the STM probe is kept a short distance from the surface of the sample and measures the tunneling current but does not touch the surface, while the AFM probe can gently touch the surface of the sample, or can hover above the substrate sensing vibrations in the probe/cantilever. STMs maintain a constant tunneling electrical current and are limited to conducting materials. STMs offer very high resolution (0.1 nm) compared with AFMs, which offer slightly less resolution (2–10 nm), but can be used with more types of materials and surfaces. Systems with probes that come into contact with the sample have greater wear of the probe tips.

Scientists use scanning probe microscopes to determine the stiffness of nanomaterials, to detect defects in carbon nanotubes or nanopores, to determine the geometries of nanoparticles, and to process variations in electrical, magnetic, or mechanical signals that are continuing to reveal a wide variety of quantum properties and effects (Figure 4.5).

While scanning tunneling microscopes still provide the most detailed images at the nanoscale, there are several other methods for imaging nanoscale objects. Other forms of scanning probe microscopy include magnetic force microscopy (MFM), chemical force microscopy (CFM), Raman microscopy, scanning force microscopy (SFM), electrostatic or electric force microscopy (EFM), and total internal reflection fluorescence (TIRF).

TIRF is a type of microscope that uses an evanescent wave to image a thin region of a sample, usually $<$200 nm. TIRF was developed by Daniel Axelrod at the University of Michigan in the 1980s. With TIRF, the fluorescence of a single molecule can be observed, which is especially useful for imaging biological structures.

One of the most intriguing and advanced areas of research in nanoscale metrology involves the use of optical imaging systems to image objects that are smaller than light waves, that is, using optical microscopes to image suboptical structures. As paradoxical as this may sound, this approach is helping to overcome some lingering limitations in our ability to view the nanoworld, especially in biomedicine and organic chemistry where structures change shape and interact differently at different stages, and over time.

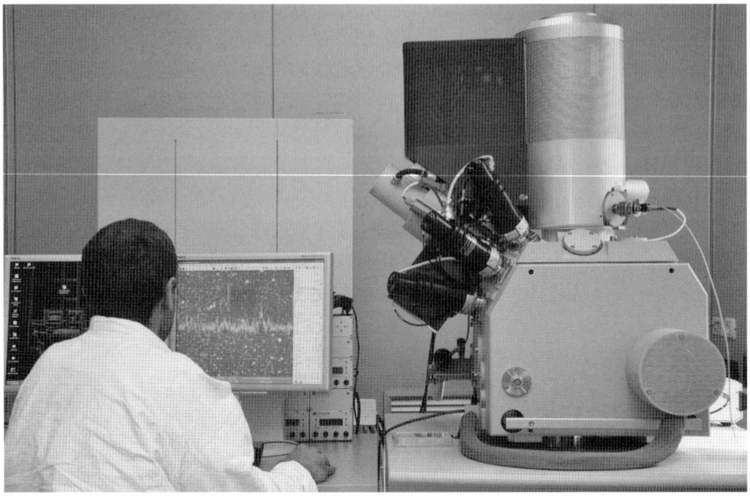

Figure 4.5 A researcher at the European Commission's Joint Research Centre (JRC) uses a scanning probe microscope to analyze nanoparticle samples. The JRC supports nanotechnology research and maintains an inventory of nanoparticles used in commercial products (photo courtesy of the European Commission, Joint Research Centre, Copyright 2010, European Union).

Measuring Neutrons

A neutron is a subatomic particle found with protons in the nucleus of the atom. Neutrons are part of the nuclear glue that helps stabilize the atom. Neutrons also play a role in nanoscale imaging. Neutron scattering is one of the most complex and interesting methods for analyzing nanoscale structures – especially effective for imaging and time-series analysis of biological structures such as proteins. A sample is bombarded with a beam of neutrons that moves through a series of mirrors to focus on the sample. The neutrons interact with the sample, then scatter and are read by a detector that captures the data. This system allows researchers to deduce the size and shape of the sample, including elements that are less than a nanometer in size. The most sophisticated systems can also produce time series showing the stages of development of biological structures associated with causing or curing a disease.

Most textbooks, articles, and online Web sites depict a protein as an artist's rendering of a spaghetti-like tangle of fibrils. To understand protein behavior – such as how a mutant protein can cause or cure a disease – it is important to see the actual development of the protein from inception to maturation. The only way to capture this is to view actual images of the protein during formation. In May 2011, researchers at the Department of Energy's Oak Ridge Laboratory and the University of Tennessee Medical Center used a neutron-scattering instrument called Bio-SANS (biological small-angle neutron scattering instrument) to image and study the formation of the Huntingtin protein, a spaghetti-shaped mutant protein that causes Huntington's disease. The team's groundbreaking

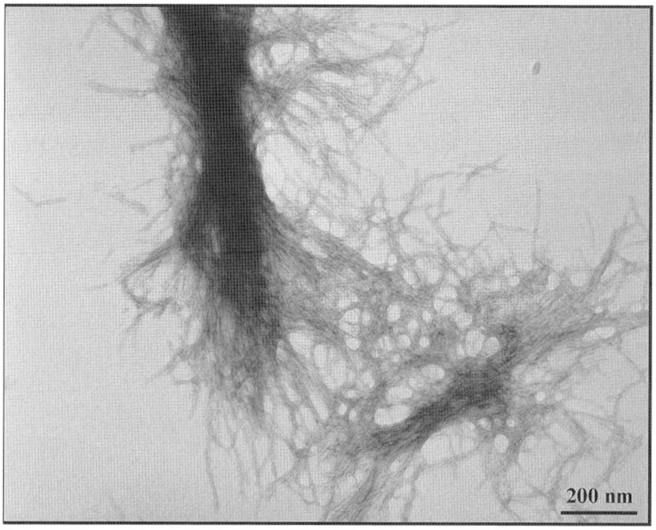

Figure 4.6 A mutant form of the Huntingtin protein thought to cause Huntington's disease. The individual fibrils of the spaghetti-shaped protein are clearly visible in this image captured by the Bio-SANS neutron-scattering system at Oak Ridge National Laboratory in Oak Ridge, Tennessee. This is a rare detailed image showing a protein and its subnanoscale fibrils (image courtesy of Christopher Stanley, Tatiana Perevozchikova, and Valerie Berthelier, University of Tennessee).

research produced the first images of the protein at the various stages of its formation (see Figure 4.6) [2]. This image clearly shows the spaghetti-like fibrils that are as tiny as one or two-tenths of a nanometer in diameter. More details are included in Chapter 9.

Raman Scanning and Microscopy

The Raman microscope is a popular type of nanoscale microscope that measures the Raman effect. It is named after the distinguished Indian scientist Sir C.V. Raman who won the Nobel Prize in Physics in 1930 for his discovery that a change in the wavelength of light occurs when a light beam is deflected by molecules. When you aim a beam of light at an object, light scatters in a pattern that is unique to each molecule, which is useful in identifying the chemical composition of a substance. Each molecule has a unique Raman pattern. Raman testing can be used to measure the vibration, temperature, orientation, and composition of a sample.

The first Raman microscope was invented in 1955 by Harvard graduate Marvin Minsky. Minsky was frustrated by the blurred images that resulted in optical microscopes, caused by a large amount of scattered light. The solution was to use pinhole apertures (nanopores) to focus the light and eliminate the scattering without reducing the brightness. The development of economical lasers in the 1980s facilitated commercial use of Raman technology, which uses lasers to focus the light and amplify the signal.

Raman technology has a wide array of applications in addition to nanoscale spectroscopy. Today, drug enforcement agencies use Raman scanners to detect more than 100 types of drugs. Raman technology is used to detect trace elements of bacteria, chemical pollutants, and explosives. It is possible that one day Raman scanning could analyze the molecular components in human blood, which could make traditional blood tests obsolete [3].

To target diseased cells, scientists seek to identify biomarkers that are unique to those cells. Biomarkers can identify the diseased cells and can also provide an "address," so a nanocarrier knows where to deliver a drug, protein, therapeutic gene, or other cell-level solution. Biomarkers have been identified for many diseases and offer a promising solution for detecting very tiny early-stage cancer tumors and free-roaming cancer cells that could cause a cancer to metastasize. Some of these biomarkers are being engineered, so they can be identified using Raman technology.

In November 2012, researchers at Washington University announced the development of nanoscale probes that bind to biomarkers at disease sites and light up when hit by an infrared laser to reveal their location. These probes, called BRIGHTs (bilayered Raman-intense gold nanostructures with hidden tags), are made of 20 nm gold nanoparticles covered with "Raman reporter" molecules, and wrapped in a thin gold shell. The BRIGHTs form into polyhedron shapes (with 12 uniform pentagonal faces). The Raman reporter molecules scatter light at specific wavelengths when hit with a laser, and the outer shell and core create an electromagnetic hotspot that concentrates energy in the gap between the gold nanoparticle core and the outer gold shell. This increases the emission from the Raman reporter molecules by a factor of nearly a trillion [4].

4.5 Bringing Biological Nanostructures into Focus

In the past, bioscientists were forced to view only dead biomolecules, immobilized and coated with a material to facilitate viewing. Today we can view and study biomolecules that are still alive and functioning, in their natural environments.

It is easier to treat a disease if we can actually see how a disease is formed, at the nanoscale. Nanomedicine is benefiting from our ability to view and record biological processes in action. Many of the bioscientists interviewed for this book commented on how much has been revealed, including surprise discoveries, as a result of new capabilities to image proteins, viruses, DNA strands, and other structures.

The benefits of this most basic capability are truly profound. For example, researchers on Dr. David Eckmann's team at the University of Pennsylvania Health System use scanning electron microscopes to study how components of the blood accumulate and adhere to the surfaces of materials used in stents and other medical implants. Their goal is to identify better surfaces and coatings for medical implants. One of their goals is to identify materials with the smoothest surfaces, for use in stents where the smooth continuous flow of blood is vital. Nancy Tomczyk, who has been working on Dr. Eckmann's research team since 2008, explained that even the smoothest surfaces used in stents and other medical implants appear to be rough and irregular when viewed under an atomic force microscope. These surfaces have the ability to snag and accumulate plaque and other blood components. "Our goal is to modify the surfaces to make them more biocompatible," she said. "To do this, we need nanoscale images to see what's really happening."

It is impossible to overstate the importance of nanoscale imaging to bionanotechnology and nanomedicine in particular. As we delve deeper into the cell-level causes of a disease, we are discovering the impact of very subtle nuances that can only be fully understood by observing them in motion or over time. Many disease-related factors become clear only when directly observed. In this sense, bionanoimaging is a critical gateway to next-generation treatments and cures.

Until the mid-2000s, most bionanotechnology structures and processes had to be visualized using artist's concepts and computer simulations because scanning probe microscopes were limited in their ability to image "soft" or "morphing" structures in real time. Some samples needed to be hardened with a coating, or imaged in a vacuum. However, most biological processes occur over time (requiring time-lapse images or videos to fully understand). It is difficult to selectively follow the dynamic behaviors and interactions of structures inside a cell, especially when those structures are constantly moving and changing shape. For example, a protein can change shape in a few billionths of a second. Nanoscale structures such as viruses and DNA molecules are also difficult to isolate individually. Thanks to nanoinnovations, scientists are rapidly moving from theory and simulation to direct observation.

In 2012, a group of UK scientists captured the most detailed image to date of the DNA molecule – the iconic double-helix shaped molecule that contains the code that defines us as humans (Figure 4.7). Although scientists had previously theorized that the ladder-shaped molecule could twist clockwise or counterclockwise as a "right-handed" or "left-handed" molecule, this was visually confirmed by the DNA images captured by the research team. To capture the image, the team adjusted the vibration of the cantilever in the scanning probe microscope to extract a greater level of detail. They used a JPG Nanowizard® AFM system, and "homemade" variations that allowed them to couple the system to inverted optical systems that extend the capabilities of the AFM instruments. This allowed the team to study and manipulate single biomolecules at a spatial resolution of ~1 nm, which enabled them to study the function and behavior of biomolecules and processes in their natural (aqueous) environment.

This bioimaging breakthrough was achieved by Dr. Bart Hoogenboom, Dr. Carl Leung, and 10 research colleagues at the London Nanotechnology Centre at University College London [5]. This is exactly the kind of innovative thinking

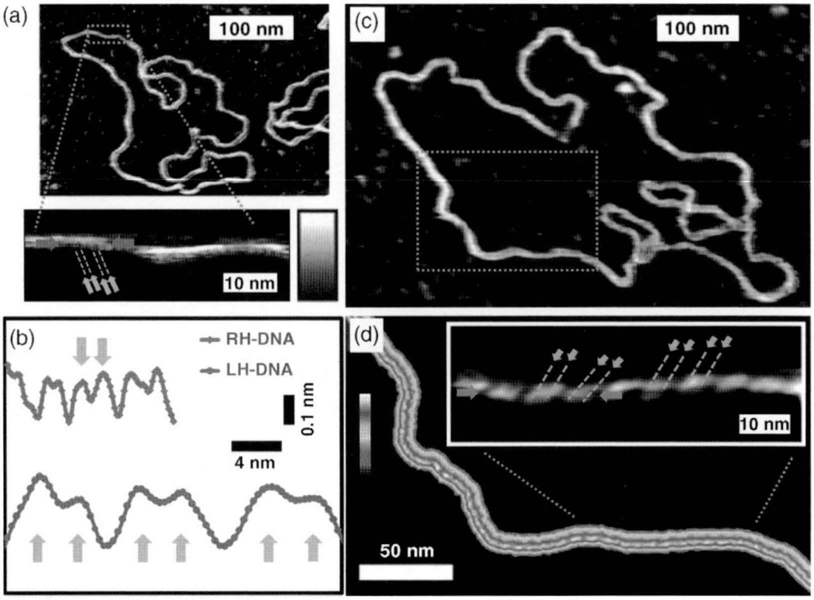

Figure 4.7 These images captured in 2012 are the most detailed images produced to-date of a DNA molecule. The images show two "grooves," which are seen as major and minor grooves. The major groove is the space between the twisting curves of the ladder and the smaller groove is the space between the two strands of DNA that make up the DNA ladder (image courtesy of Carl Leung).

that is needed to open the window to biological processes that were formerly hidden from view. It is a marvelous example of how faculty and student researchers are pushing the boundaries of bioimaging to reveal those parts of the nanoscale landscape that are still hidden from view.

4.6
Using Optical Imaging Systems to View Nanoscale Structures

Most nanoscale microscopes measure electrons instead of light because nanoscale objects are smaller than the bandwidth of light waves. While you may have read that it is impossible to optically image a nanoscale object, scientists are working on methods for producing *suboptical* images of nanoscale objects that are smaller than light waves. Typically, the smallest scale that can be optically imaged is ~200 nm, which is the lower focus limit of visible light waves. However, there are some ingenious techniques being developed that have improved the resolution of optical microscopes to view objects that are 50 nm or less.

A traditional optical microscope (which you probably used in a school biology class) uses a tube with two or more glass lenses to magnify an illuminated sample for viewing or imaging. The sample is fixed on a glass slide or sandwiched between two thin glass plates. The sample may be stained to reveal its features – especially if it is clear or transparent, like a human cell. It may be suspended in liquid, especially if it is a cell or bacterium that lives in an aqueous environment. It may be coated in a material to keep it rigid.

You have never seen a nanoparticle with your naked eye because nanoscale objects are smaller than visible light waves and smaller than the eye can perceive. To view nanoscale samples, we need to be able to see things smaller than 100 nm, while the lowest limit of visible light is about 400 nm and the best resolution of an optical microscope is around 200 nm. That would seem to rule out optical microscopes for nanoscale imaging.

Actually, there are ways to make nanoparticles optically visible, using some very clever tricks. One method is to use a "superlens" or "hyperlens" made of metamaterials to capture images of objects *below the lower limit of visible light waves*. A superlens is a lens that can view features smaller than the diffraction limit of light. Several types of metamaterials including gold have been used to magnify nanoscale objects to capture an optical image. Scientists have used surface plasmons and a superlens to capture a type of light wave called evanescent waves that are normally scattered and lost in conventional optical microscopes.

Research in this field is fairly recent. In 2003, Sir John Pendry, whose seminal writings and research provided much of the foundational work in this field, proposed a design using a curved cylindrical lens to achieve magnification at the nanoscale. The concept was demonstrated by research teams in 2007. In the past few years, various types of superlenses have been developed, including a metamaterial superlens made from an array of nanowires. Several teams have also used an array of nanoholes to concentrate the light so as to improve the resolution.

In January 2013, a team of researchers at UCLA created a method for viewing viruses and other nanoscale objects by using tiny liquid lenses that self-assemble around microscopic objects [6]. The team was led by Aydogan Ozcan, Associate Professor of Electrical Engineering and Bioengineering, who is also on the faculty of the surgery department at UCLAs David Geffen School of Medicine. The team was able to detect and image nanoparticles including viruses and viral particles on a glass substrate. Their system used an LED to illuminate a nanolens object assembly and incorporated a silicon-based sensor array, which is used in cell phone cameras. The system produced holograms of the nanoparticles, which were rapidly interpreted using a personal computer. The prototype was not of as high resolution as an electron microscope, but the field of view was $>20\,mm^2$, which is useful in detecting and imaging nanoparticles that are sparsely distributed. This is only one example of the creative efforts underway to image nanoscale objects using new types of optical microscopy.

Failure to accept the obvious is a hallmark of any innovator. It is a testament to the core values of nanoinnovators that a handful of scientists refused to accept the logic that optical microscopes could not be used to image nanoscale structures because the structures are smaller than conventional light waves. This is the type of ingenuity that makes nanoinnovation such an exciting and promising field.

4.7
Probing the Future

One of the most exciting consequences of being able to view and image the nanoscale world is the ability to *manipulate* atoms, molecules, and other nanoscale structures. Most nanoscale imaging systems today are called SPM because they include the ability to probe and manipulate structures, as well as image them.

The most famous and earliest demonstration of this capability was the creation of the IBM logo using individual atoms on November 11, 1989, when Don Eigler and his colleagues at IBM's Almaden Research Center positioned 35 Xenon atoms to form the IBM logo (Figure 4.8).

Progress in this field has been remarkable, considering that we have only been able to manipulate atoms and molecules for two decades. Today, nanoscientists routinely use a variety of probes for nanoscale manipulation. Most of these are point-to-point probes tipped with a few atoms – as small as one atom at the tip of the probe. To manipulate atoms, a force is applied to the probe, which attracts or repels or otherwise transforms the sample.

Using Light to Manipulate Particles

Another type of probe is called a *nanotweezer* – similar to cosmetic tweezers used to shape eyebrows, although these tweezers are 100 000 times smaller than a human hair. Nanotweezers can be made from carbon nanotubes and nanowires, tungsten, single-crystal silicon, or other materials.

One form, called "optical tweezers," actually uses light as a force to manipulate objects. In 1997, the Nobel Prize in Physics was awarded to Steven Chu, Claude Cohen-Tannoudji, and William D. Phillips for their seminal research on optical tweezers, including the "development of methods to cool and trap atoms with laser light," which derived from earlier work by Bell Labs scientist Arthur Ashkin.[1)]

Dr. Heike Noethe, who edited this book at Wiley-VCH, used optical tweezers earlier in her career to position live cells in certain patterns on surfaces. She recalls, "It felt more like a trap than a tweezers to me because when the laser surface approaches a cell, the cell is pulled into the focus and is 'trapped' there."

Dr. Noethe describes an optical tweezer as "a highly focused beam of light that is able to trap and manipulate individual atoms, molecules, viruses, bacteria and other particles at the micrometer, nanometer and sub-nanometer scales." This

1) "The Nobel Prize in Physics 1997". Nobelprize.org. (September 11, 2011).

(a)

(b)

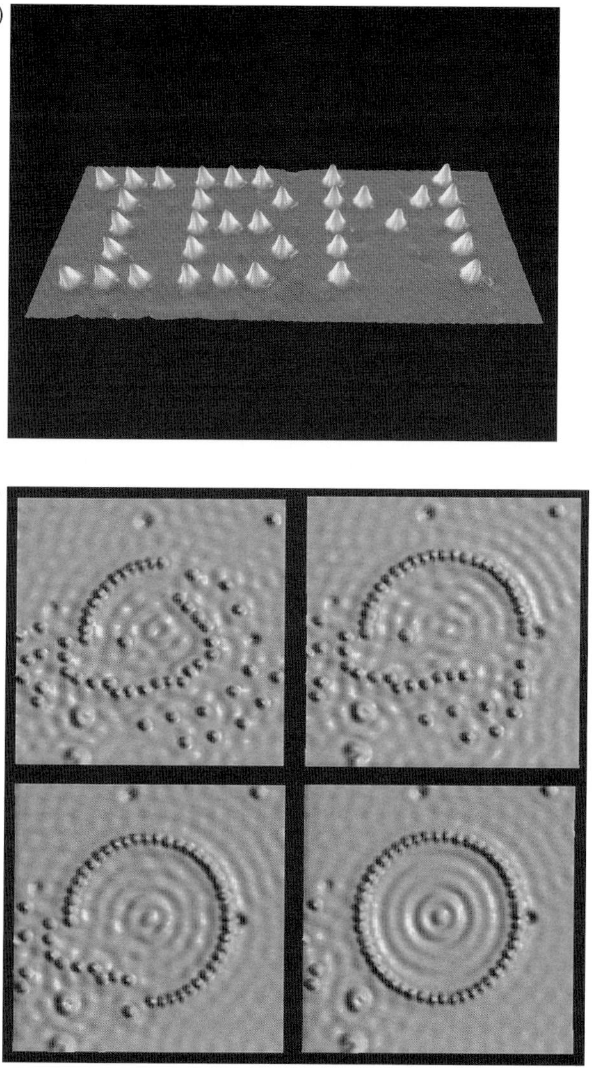

Figure 4.8 (a) This IBM logo was created by Don Eigler in 1989, using a custom-built microscope to precisely place 35 xenon atoms on a nickel substrate. (b) In 1993, Dr. Eigler's group arranged a circle of 48 iron atoms to form a "quantum corral" on a copper substrate – the progression is shown here. Quantum corrals are structures built atom by atom on an atomically clean metallic surface that result in "electron waves" that can be imaged as shown in this figure (images courtesy of IBM).

type of probe is particularly useful for manipulating and imaging biostructures and measuring events such as DNA replication and RNA transcription.

Developing Better Nanoprobes

More robust probes are needed to extend the life and lower the operating costs of scanning probe microscopes. Better methods are also needed to prevent the probes from getting contaminated or coated with atoms from the materials being probed.

Many of these challenges are being actively researched at the Nanoprobe Lab in the Nano/Bio Interface Center (NBIC) directed by Dawn Bonnell at the University of Pennsylvania. Dawn is Trustee Professor of Materials Science and Engineering and Vice Provost-Research at UPENN. The Nano/Bio Interface Center facility is working with industry and academic partners to develop next-generation probes, and makes these innovations available to the research community before they are commercialized.

Professor Bonnell described some of the innovations being researched at the Nano/Bio Interface Center that involve next-generation probes and related technologies. The Bonnell group recently characterized the properties of a *single layer* of protein on an electrode, information that is critical to the development of transformational biochemical sensors. Another project provided new insights into the ferroelectric properties of 50 nm thin-film heterostructures, unique materials that may enable new devices for energy harvesting. Yale Goldman and his team combined total internal reflectance and microfluidics to map how individual proteins move in 3D. Dennis Discher and his colleagues combined atomic force microscopy and fluorescence microscopy to simultaneously probe mechanical properties and chemostructural properties of biomolecules – proteins and ribosomes. Another Nano/Bio Interface Center team including the Savan, DeGrado, and Therien groups developed a new family of artificial proteins that respond to light. Research groups directed by A.T. Charlie Johnson and Dawn Bonnell are contributing to this research. In collaboration with the Johnson and Bonnell groups, these new proteins are being developed into chemical sensors. Along the way, a new mechanism for turning light into electricity (that involved surface plasmons) was discovered. These are only a few examples of research underway at the Nano/Bio Interface Center, which is now benefiting from the completion in 2013 of the University of Pennsylvania's Krishna P. Singh Nanotechnology Center.

"The development of probes that let us 'see' atoms has already transformed our view of the nanoscale world and led to new technologies. Future probes will be multi-function probes capable of detecting and measuring multiple parameters in samples," Bonnell predicted. "Instead of a specialized microscope for each condition we want to measure, we will have multi-function microscopes. Just as nanoscale imaging advanced the field of nanotechnology research, new capabilities in nanoscale probes will result in a similar transformation, taking our scientific understanding and our ability to manipulate matter to a new level. Our research at the Singh Nanotechnology Center will contribute to this transformation."

4.8 Nanoscopes on Mars

If we focus our gaze on the future, one of the most fascinating applications for nanoimaging systems involves space exploration. NASA and its academic and corporate partners have produced some fascinating innovations for conducting nanoscale analysis – on other planets!

In 2007, the Phoenix Mission to Mars carried an Atomic Force Microscope to Mars as part of an instrument package that helped confirm the existence of water on Mars. According to William Boynton, lead scientist for the Thermal and Evolved-Gas Analyzer (TEGA), this was the first time Martian water was "touched and tasted." Water on Mars was previously observed and detected, but this was the first evidence from direct soil sample analysis. This finding was hugely important because one of the goals of Mars exploration was to determine whether surface water ice is available in a form that can be used as a source of power (in fuel cells) and drinking water, which is essential to a human Mars mission.

The AFM used in the Phoenix was contained in an instrument package called MECA (microscopy, electrochemistry, and conductivity analyzer) (Figure 4.9). MECA included four wet chemical reaction cells, an optical microscope, and an atomic force microscope, and was designed to analyze soil samples from a Mars crater thought to contain water ice. In addition to the nanoscale analysis, stereo photos taken by Phoenix revealed an "ice-dominated terrain as far as the eye can see," according to Mark Lemmon, lead scientist for the surface imaging camera [7]. The ability to analyze soil and water on other planets using both optical and atomic force microscopes is amazing, considering that half a century ago, the best we could do is peer at other planets through telescopes – and now we are sending robots to take pictures and analyze the environment down to the nanoscale!

The AFM used in the Phoenix Mars mission was designed by a Swiss Consortium that developed the first AFM for the Mars Surveyor 2001 mission. The AFM had a resolution of 10 nm and an overall image range of 10 μm. Most space probes are equipped with redundant systems. The Mars AFM included an AFM chip with eight adjustable sensors and cantilevers on one chip. This gave the Phoenix team on Earth eight sensors, so if one became broken or dirty it could be broken off to access the next sensor on the chip. The Mars AFM system for analyzing Martian soil used a robot arm to scoop and load soil onto a sample plate on a wheel. The wheel was rotated to the scan range of the AFM, which scanned the sample with the selected cantilever on the AFM chip. The images were stored and transmitted to Earth. This was one of the first attempts to capture AFM images from another planet. Though everything did not work perfectly, the effort should be vigorously applauded for its ingenuity.

When we think of the difficulties and risks involved, it was a miracle that the AFM made it to Mars intact. This groundbreaking – or rather, ground analyzing effort – was an incredibly complex undertaking. An important purpose of this

(a)

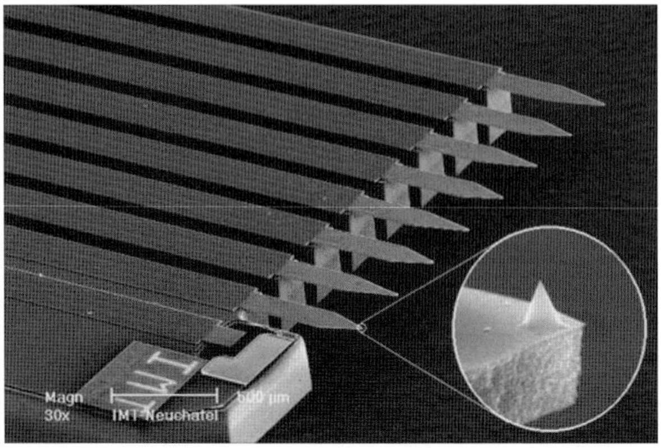

(b)

Figure 4.9 (a) This image showing the eight tips of the AFM on the Phoenix Mars Lander's MECA system was taken before the space probe was launched. These probes mapped soil samples from the surface of Mars in three dimensions. (b) This image shows the scoop filled with dirt preparing to drop it onto the exposed sample wheel directly below – the wheel rotates to the position of the AFM probe that will analyze the sample (images courtesy of the Jet Propulsion Laboratory, NASA).

mission was to prove that an AFM could survive the space trip and deliver an image, which was in fact demonstrated (a fuzzy image of a particle of dust was captured). The AFM was not expected to produce a gallery of nanoimages, but hopefully will do so on a future mission. The primary value of the AFM on the

Phoenix mission was to demonstrate proof of concept. The instruments did not produce a gallery of nanoscale images, but it did produce a test image and imaged a grain of Martian dust, not spectacular but nevertheless remarkable considering the obstacles involved.

Consider what was needed to deliver the nanoscale microscope to Mars. First, the AFM had to survive the vibration and shocks involved in the rocket launch from Earth, and then the impact of the Mars landing. The system had to account for the fact that the Martian atmosphere is mainly carbon dioxide, which can cause direct atmospheric discharges, so all electrical voltages had to be limited to a maximum of 15 V. Furthermore, the Martian temperature ranges between −90 °C (−194 °F) and 10 °C (50 °F). To process the signals, digital-to-analog converters and amplifiers had to be incorporated. During the trip to Mars, solar radiation and bombardment by heavy ions could damage or destroy the CMOS chips used on the spacecraft's circuit boards, including the AFM system. These are just a few of the daunting challenges the Mars Phoenix team had to anticipate and compensate for, in order to deliver the AFM imaging system to the surface of the Red Planet. The excellent writer Andrea Thompson wrote a great series of articles on the Phoenix mission for Space.com, which includes images from the mission. You can find these on space.com. More information on the AFM instrument package can be found online at http://www.famars.unibas.ch/index.htm (*note:* FAMARS stands for first atomic force microscope on MARS).

Of course, the real quest on Mars involves the search for life and nanoimaging is bound to play a role as nanospace explorers look for Martian bacteria, viruses, and possibly alien nano-life forms.

4.9
The Future of Nanoscale Imaging

In the decade ahead, we can expect to see improved techniques for nanoscale imaging, manipulation, and fabrication. The results will include better functionality such as sharper resolution and control, new techniques for imaging across the electromagnetic spectrum, and more affordable systems that are needed to mainstream nanotechnology, especially in schools.

There are many scientific goals and innovations that require better nanoimaging to move forward. For example, one of the areas where we can expect to see some dramatic innovation involves nanovideo.

One promising technology involves a technology called *4D electron microscopy*, which was developed by Ahmed Zewail and his colleagues at Caltech in the mid-2000s (*note*: the "fourth dimension" referred to in 4D is time). Dr. Zewail won the 1999 Nobel Prize in Chemistry for pioneering the science of femtochemistry – the use of ultrashort laser flashes to observe fundamental chemical reactions. These laser bursts can occur at one millionth of a billionth of a second. Capturing a series of femtosecond images provides sequenced visualizations of atomic oscillations that are the equivalent of nanoscale motion pictures. This imaging approach can capture images of atoms forming or breaking apart from molecules, changes over time in physical and biological matter, oscillations of individual atoms and materials, and so on. A patent based on this framework was granted to Caltech in 2006.

Dr. Zewail[2)] and biology professor Grant Jensen [8] are leading research programs that are expanding this research to include biological imaging within cells. There are some key challenges that still need to be overcome such as preserving samples in the vacuum of the electron microscope, extracting three-dimensional information from the images, and recording information before samples are destroyed by the high-energy electron beam.[3)]

As imaging systems are improved and refined, more nanoinnovations will result. There is an indisputable connection between nanoinnovation and our ability to image the nanoscale. It is logical to assume that increased investments in nanoimaging research will pay significant dividends in nanoinnovation across all areas of science.

As much as we've achieved in our ability to image and manipulate the nanoscale world, we're just getting started. There is much more to come – and lots of nanoscopic challenges ahead. The most intriguing question in the field of nanoimaging is: *What haven't we seen yet?*

References

1 He, C., Zhao, N., Shi, C., Du, X., Li, J., and Cui, L. (2006) A practical method for the production of hollow carbon onion nanoparticles. *Journal of Alloys and Compounds*, **425**, 329–333.

2 Oak Ridge National Laboratory (2011) Neutrons provide first sub-nanoscale snapshots of Huntington's disease protein. Press Release No. mr20110518-00, May 18.

3 Aiyar, S.A. (2011) Raman effect: fingerprinting the universe. *India Times*, May 9.

4 Gandra, N. and Singamaneni, S. (2012) Bilayered Raman-intense gold nanostructures with hidden tags (BRIGHTs) for high-resolution bioimaging. *Advanced Materials*, **25**, 1022–1027.

5 Leung, C., Bestembayeva, A., Thorogate, R., Stinson, J., Pyne, A., Marcovich, C., Yang, J., Drechsler, U., Despont, M., Jankowski, T., Tschöpe, M., and Hoogenboom, B.W. (2012) Atomic force microscopy with nanoscale cantilevers resolves different structural conformations of the DNA double helix. *Nano Letters*, **12**, 3846–3850.

6 Chin, M. (2013) UCLA's new nano-lens microscopes can detect viruses, other objects at nanoscale. UCLE Newsroom, January 20.

7 NASA Mars Lander (2008) NASA Spacecraft Confirms Martian Water, Mission Extended. NASA webpage www.nasa.gov/phoenix (July 31, 2008).

8 Caltech Media Relations (2008) Caltech 4D microscope revolutionizes the way we look at the nano world. Caltech News Release, November 20.

2) Physical Biology Center for Ultrafast Science & Technology (UST), www.ust.caltech.edu/index.html, Caltech (April 1, 2012).

3) The Jensen Laboratory, www.jensenlab.caltech.edu/projects.html, Caltech (April 1, 2012).

5
Where Nanoscience Becomes Nanoart

It is good to understand that art, science and technology are not separate. Then we are able to learn from each other.

– *Hugh McGrory, Filmmaker, from a 2008 Interview by Nanotechnology Now*

Many nanoinnovations come from the world of art and design, and this is appropriate considering that much of nanotechnology involves the design of new materials and devices. The connection between nanoinnovation and nanoart is impressive. Nanoart has been used to vividly illustrate how nanostructures can be manipulated with extreme precision. For many researchers, an artistic representation was the first proof of concept of their work. For others, a work of art or artistic pattern provided the inspiration for a scientific achievement.

Don Eigler created an IBM logo from 35 xenon atoms – something familiar and artistic, yet something that had never been done before. Ned Seeman drew his inspiration for a critical breakthrough in DNA nanotechnology from a woodcut by the Dutch artist Mauritz Cornelis Escher. One of Paul Rothemund's first examples of DNA origami was a happy face. The term "DNA origami" took its name from a Japanese art form. A. John Hart used his Nanobliss nanoart Web site to illustrate how a forest of carbon nanotubes could be turned into the first nanoscale portrait of a US president. Neri Oxman, who is both a designer/architect and materials scientist, has shown how elegant nanoforms can be engineered into practical nanofunctions, from construction materials to "carpal skin."

Historically, the link between art and nanoscience is even stronger. The nanoarchaeology examples included in this book are all artistic. The Lycurgus Cup is a work of art. The decorative surface of the Damascus Sword has a distinctive damask pattern that formed naturally on the blade during the sword-making process. Mayan blue pigment was used to decorate buildings, pottery, and paintings.

Nanoscale art – called nanoart – includes artistic renditions of nanoscale structures that are created by manipulating atoms and molecules using physical or chemical processes, as well as artistic images and patterns that occur in Nature. Most of these images are produced in clean rooms by scanning electron microscopes and atomic force microscopes. Often, the black and white

NanoInnovation: What Every Manager Needs to Know, First Edition. Michael Tomczyk.

Figure 5.1 This image of a nanoscale teddy bear has become an iconic example of nanoart since it won a first place prize in the Nanoart exhibition at the 2008 Fall Meeting of the Materials Research Society. The image was created by Helia Jalili, using an aqueous zinc nitrate solution as an electrolyte to prepare these zinc oxide nanostructures. Deb Pradham took the SEM image (image courtesy of Helia Jalili, University of Waterloo).

images are colorized using Photoshop or other software programs to highlight or emphasize various features of interest. Some nanoart may be very important, such as developing new designs and patterns for use in the next generation of semiconductors. Other examples are more whimsical, such as finding an arrangement of zinc oxide nanostructures that resemble a teddy bear (see Figure 5.1).

For more than a century, biological structures such as viruses and proteins have been depicted by artists. Many innovations are conceptualized or visualized by artists before they are able to be engineered, such as the artificial blood cell envisioned and patented by nanomedicine pioneer Robert Freitas.

In the early 2000s, university faculty and students began to discover recognizable images in the nanomaterials they were studying. They observed particles of zinc oxide that looked like nanoflowers. Scientists began to characterize carbon nanotubes and nanowires as thickets or "forests." They digitally fine-tuned their nanoimages and added color to emphasize specific features.

Today, commercial nanolithography products such as MicroAngelo make it possible to take a line drawing of Pablo Picasso's "Don Quixote" or the map of Africa and render it at the nanoscale. MicroAngelo works by converting an image to a list of coordinates and creates a lithography pattern that can be used to manipulate the tip of a probe in an atomic force microscope (AFM).

A Short History of Nanoart

During the 1990s, the first published images of nanomaterials were not visually exciting. Nanoimages are black and white. A single carbon nanotube looks like a smooth tube (or worm) and a cluster looks like a tangle of spaghetti, or a tightly packed mass whose edge resembles the face of a cliff.

By the early 2000s, scientists and students were experimenting with more sophisticated forms and patterns. In Fall 2002, the Materials Research Society began hosting their nanoart competition called "Science as Art." Nanoscale researchers – especially students – began finding and creating identifiable images such as a teddy bear, Santa Claus, or the letter P in images of nanomaterials, similar to how we might see a lion's head in a storm cloud. Since nano-images are typically colorless, many of these images were artificially colored or digitally tweaked using Adobe Photoshop. In some cases, artistic elements were added using nanoprobes.

In Spring 2005, Cris Orfescu pioneered nanoart as a new form of art for which he won the first prize for new media at the "Affaire in the Gardens" art show in Beverly Hills. Cris is a materials scientist and artist, and a nanoart pioneer. He was born in Bucharest, Romania, and since 1991 has lived and worked in Los Angeles. In 2006, he founded NanoArt21^{TM}, which promotes nanotechnology through art and sponsors an international nanoart competition. These images range from STM and TEM images of nanoscale structures and patterns to colorful abstract art.

Cris, who holds a Master's degree in materials science and has studied drawing and architecture, describes nanoart as "the transition from science to art through technology." His philosophy captures the spirit of nanoart: "NanoArt21 is aimed to raise the public awareness of nanotechnology and its impact on our lives. I consider NanoArt to be an appealing and effective way to communicate with the general public and to inform people about the new technologies of the 21st century."

Many nanoart entries illustrate the complex and often exquisite structure found in Nature, such as a grain of pollen or a crystal. One memorable entry in the Fall 2008 nanoart competition at the Materials Research Society in Boston showed a nanoimage of a colored butterfly wing, with one scale tipped over that was grey. This image demonstrated vividly that the color of the butterfly wing did not come from a pigment or dye, but rather from the size, shape, and orientation of the nanosized scales, and the wavelength of light (color) they reflect. Another entry in the same competition depicted a "tower" of nanowires illustrating how nanowires have a tendency to curl and spiral, which can complicate production and control for commercial applications.

Several nanoart images from the Materials Research Society have become iconic, such as the award winning image of the zinc oxide teddy bear. The Nano Teddy Bear was created by Helia Jalili while she was at the University of Waterloo. As one Australian fan commented on Flickr, "awesome shot, love it, what a new way of looking at the world." After winning the MRS award, Helia worked as a postdoc researcher and staff scientist at MIT, and in 2011 joined Panalytical B.V. in Boston as an applications scientist.

Today, hundreds of talented students and research teams participate in nanoart competitions every year. Their contributions have helped push the frontiers of science and demonstrated the proof of concept for many nanoinnovations such as nanolithography and materials science.

In research laboratories around the world, stunning images are being created by nanoscience artists – but you don't need a scanning tunneling microscope to see their images. Nanoart is displayed in digital galleries on the Web, in nanoart exhibitions at prestigious scientific conferences, and at art/design exhibitions.

5.1
Holistic Nano at the Convergence of Nanobliss, Nanoform, and Nanofunction

Nanoart is part of a holistic approach to nanoinnovation, where scientists are using their sense of art and design to develop commercial scientific achievements. One of the best examples of a "holistic" nanoinnovator who personifies this approach is Anastasios John Hart, who leads the Mechanosynthesis Group at the University of Michigan-Ann Arbor.

John Hart is best known as the founder of the Nanobliss Web site (www.nanobliss.com). He was one of the first to demonstrate how to grow "forests of nanotubes" on silicon substrates with finely rendered details. His early Nanobliss examples showed that intricate geometrical shapes and even a presidential portrait could be engineered from nanomaterials.

John's innovation journey began with the use of nanoart to illustrate the scientific principles involved in the formation and manipulation of carbon nanotubes to produce familiar images. His Nanobliss Web site (www.nanobliss.com) has showcased a wide range of nanoscale structures, including the iconic nanoscale portrait of President Barack Obama (see Figure 4.1). The accompanying gallery of nanoimages (see Figure 5.2) illustrates a few of the nanoscale architectures that have been developed as part of this research.

The artistic images showcased on the Nanobliss site are created by using masks and templates as well as more sophisticated techniques that induce these structures to self-organize. Working at the convergence of form and structure has led John Hart and his colleagues to pioneer several next generation innovations such as "nanoskins" for construction and environmental control.

"We believe that nanoskins can be engineered to help control the environment in buildings and rooms, without the use of electronic or mechanical controls," John explains. "Nanoskins can be designed to react to certain conditions, and automatically reconfigure their shape or the size of their pores. Nanoskins can help regulate the flow or moisture, air, light, dust and other environmental factor. We've already demonstrated these capabilities in the lab. Nanomaterials can also increase the strength and durability of existing materials such as cement. Many of the innovations we're working on now came from our work in nano-imaging and fabricating various type of structures. Scanning probe microscopes allow us to see what's possible and experiment with various structures, and are a critical part of our innovation toolkit."

Figure 5.2 The extensive imaging work by Prof. A. Jon Hart at the University of Michigan and his colleagues are giving an entire generation of researchers ideas for innovations based on nanoscale structures, ranging from growing forests of nanotubes to the award-winning development of creating nanoskins to regulate environmental conditions and improve materials used in construction and architectural design (images courtesy of A. John Hart, University of Michigan, Ann Arbor).

In November 2010, John Hart was one of 43 scientists and engineers who shared a total of $16.5 million in competitive grants through the Young Investigator Program of the US Air Force's Office of Scientific Research. This research focuses on morphing carbon nanotube microstructures. John Hart's group has also been collaborating with the Mediated Matter research group led by Neri Oxman at MIT to develop nanoskins and other architectures that have the ability to adapt and respond to the environment.

In 2008, John Hart and Neri Oxman won first place in the newly created "Next Generation" category of the prestigious Holcim Awards, an international prize awarded to innovative architects and designers. Professors Hart and Oxman were honored for their visionary building skin research using carbon nanotubes to develop materials that can be assigned specific structural, functional, and environmental properties. The Holcim Awards are an international competition conducted by the Holcim Foundation for Sustainable Construction, whose primary sponsor is Holcim Ltd., one of the world's leading producers of cement and aggregates.

You're probably wondering, how does cement involve nanoinnovation? Cement was used by the ancient Egyptians and Romans, and it would seem that there wouldn't be much room for innovation in this ancient, traditional industry. Actually, the cement industry has been a major beneficiary of nanoinnovation. The ability to image and manipulate nanomaterials, and to engineer new types of materials with advanced properties, has opened the door to a new era in research for the construction industry. Building materials that incorporate nanomaterials are introducing new properties to cement and other aggregates used in buildings, roads, and bridges – breaking the mold, so to speak, in an industry that has been "rigid" for thousands of years.

In recent years, John's research has led him into the emerging field of nanomanufacturing, with a focus on novel methods for mass-producing nanoscale structures and patterns using readily available materials. In 2010, he introduced his course on Nanomanufacturing at the University of Michigan.

In February 2012, John Hart described his current research interest in nanomanufacturing at an annual event I hosted at the Wharton School called the Emerging Technologies Update Day – the theme was "The Future of Nanoinnovation." John explained that his early research involved growing forests of nanotubes and learning how to turn them into precisely engineered patterns. "I'll probably always be remembered for nanobama," he mused. His early research reflected a keen interest in understanding how nanoscale patterns and processes work, but his ultimate goal is to find new and better ways to manufacture nanoscale structures on a commercial scale. His most recent approach sounds surprisingly simple – and characteristically creative.

"Imagine having a machine like a nano-Xerox copier that allows you to create patterns from nanomaterials on a readily available substrate such as a sheet of paper or plastic, instead of on an expensive substrate like computer-grade silicon," he told the conference attendees. "This would offer exciting possibilities for researchers, students and teachers, as well as companies. It could make nanolithography devices as common as photocopiers and computer laser printers."

This "nanocopier" approach could also enable the creation and manufacture of nanoskins and new forms of nanotubes. At the Wharton conference, John showed an SEM image of small clusters of nanotubes that were standing up like sheaves of corn, and explained how they could be manipulated so that they twist together and arrange themselves in a pattern or matrix. These structures, shown in Figure 5.3, are created using a process called "capillary forming" to create twisting spires. These structures can be up to 10 times stiffer and stronger than polymers currently produced using microfabrication.

John and his team have learned how capillary forces can change the structure of nanostructure geometries. "The starting shape influences how the capillary forces affect the geometry of the structures," he explains. "Some bend, others twist and we can combine them in many different ways. This has important implications for the creation of custom nanostructured surfaces. With our technique, we can design surfaces or 'skins' with many different properties." The process is described in more detail in his article [1].

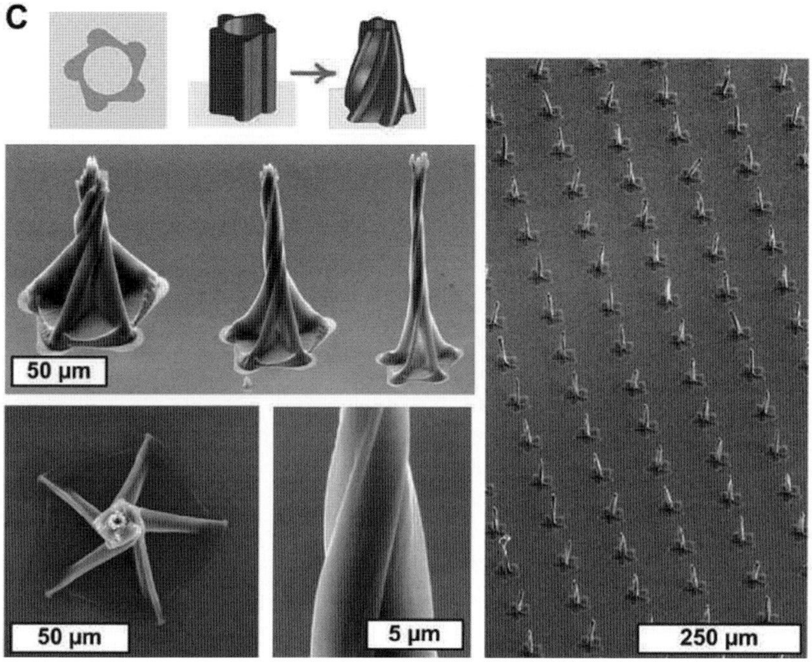

Figure 5.3 The process of creating "nanospires" uses a combination of chemical vapor deposition and capillary forming to create an array of nanotubes that can be made to twist or bend in many different ways. These structures can be engineered into surface materials such as nanoskins that exhibit new properties and commercial opportunities (image courtesy of A. John Hart).

It is striking to see John's progression from experimenting with patterns and processes – using nanoart as a proof of concept – to the development of radical new processes that could offer exciting new solutions for patterning and manufacturing nanomaterials. He is a great example of the winding path that many nanoinnovators take as they sail like ancient navigators into uncharted territory.

5.2
Innovating at the Convergence of Biomimetics, Nanoart, and Nanoscience

There is an obvious link between creativity in science and creativity in art. This convergence is a major driver of nanoinnovation. Neri Oxman's work at MIT illustrates how innovation is evolving at the intersection of biomimetics, design, and nanoscience.

Neri Oxman is without question one of the "trendiest" scientist/designers in the field of nanotechnology (although her work is not limited to nanoscale research). She says that much of her work is inspired by Nature, which she calls a "grand materials engineer" and a "multifunctional designer." She has drawn many lessons from Nature such as studying the matrix structure of an eggshell to determine how material designs can incorporate biomimetic architectures. She is constantly asking intriguing questions such as "What would Nature 2.0 look like?" and "Can we print buildings?"

One of her innovations is a method called "computationally enabled form finding" that she describes as "bringing together material properties and environmental constraints, mixing them together, and then generating form out of them." This approach allows a designer to use these tools to become an "editor of constraints."

A major goal of her research is to use digital design and fabrication technologies to mediate matter and the environment to radically transform the design and construction of objects, buildings, and systems. She is currently exploring how to construct buildings with 50% less material or 50% less energy, using a technology she calls "variable property printing" to allow mapping of structural variations in elasticity, rigidity, and other properties that can be varied by integrating soft, stiff, opaque, and translucent materials and applying or "printing" them where they are needed.

A specific example that illustrates her designer's approach to innovation – developed with Craig Carter from MIT – is a kind of "electronic skin" to help protect against carpal tunnel syndrome. The "carpal skin" is worn like a customized glove and is designed by digitally unfolding the patient's skin to create a pain profile map that determines where hard and soft materials should be placed in the skin-tight glove to provide the best function and least pain. The gloves are "printed" using a 3D printing process.

The Promise of 3D Printing

The concept of printing nanomaterials runs like a theme through nanoinnovation, in nanolithography and nanoelectronic circuits. If we carry the notion of printing to the next level, we need to consider three-dimensional printing – 3D printing – as a technology for designing and manufacturing nanoinnovations. Three-dimensional printing – also called stereolithography – was invented in 1986 by Chuck Hull, the founder of 3D Systems, Inc., the first company formed to commercialize the technology.

In the future, 3D printers will become an indispensable part of the nanoinnovator's toolkit. These nanoprinters could take the form of the "nanocopier" envisioned by John Hart, or could resemble existing 3D printers currently used by product designers to develop and share prototypes for industrial parts.

Looking ahead to the next decade and beyond, 3D printing may be one of the most powerful instruments in the nanoinnovator's toolkit. Three-dimensional printers are devices that use raw materials such as plastic or metal to spray or etch or otherwise create a prototype. Rapid prototyping has been available for a couple of decades; however, using nanomaterials in 3D printing is still a seminal concept.

Most 3D printers look like a glass-walled box with spray nozzles positioned around the center. The nozzles spray a material such as liquid plastic or metal to create a three-dimensional object. The printers take their instructions from a computer, which means that someone in Boston can create the object, and someone in Europe or Asia can "print" it.

Three-dimensional printing enables materials researchers and product designers to quickly render custom prototypes for products they are developing. Three-dimensional printing goes beyond computerized simulations viewed on a two-dimensional screen, to actually render a prototype as an object. Today's 3D printers combine electronics and materials in many ways to create stretchable electronics, customized prosthetics, and bioengineered components used in medical devices and much more.

Cornell University is an academic "home" of 3D printing and emerging robotic technologies. Dr. Hod Lipson, Director of Graduate Studies in Mechanical Engineering at Cornell, was one of the early pioneers in 3D printing. His research group and successive classes of students has created several 3D printing devices and shared those designs online for whoever in the world wants to pursue development of 3D printing. The Cornell Creative Machines Lab includes 3D printing information, blueprints, and applications on its Web site (creativemachines.cornell.edu/Tissue_Eng). One of the most interesting applications is the use of a cell-seeded alginate hydrogel to "print" living tissue in 3D shapes, using CAD data. Also included is extensive (and fascinating) research on a wide variety of robotics, including molecubes, self-replicating and self-organizing robots, microfluidic assembly and much more. Modular self-assembling robots and other structures such as molecubes that are being engineered at the macroscale provide models for similar self-assembly, which is a Holy Grail goal in the field of molecular manufacturing, since many of these self-conforming robots use innovative ways to reconfigure themselves, even if the parts are separated.

It is intriguing and difficult to predict how 3D printers will evolve, but early indications are extremely promising and this is a rich field of study for innovators in fields as diverse as engineering, space technology, and cooking! The first generation of commercial 3D printers is just beginning to enter the consumer marketplace. A 3D printer called the Replicator was demonstrated by New York start-up Makerbot Industries at the 2012 Consumer Electronics Show in Las Vegas. CAD software that is available free online creates an object, which the Replicator creates by extruding a plastic thread from a spindle and moving it through a print head to create the object, layer by layer. A competing system called the Cube was launched at the CES by 3D Systems – one of the Cube's advantages is the use of replaceable cartridges to supply the printer with raw material.

While 3D printing is not a nanotechnology *per se*, this technology can incorporate nanocomposite materials and functions. Nanoinnovators can use 3D printing to customize nanoskins and coatings – not only for research purposes but also for commercial applications. Three-dimensional printing is one of the fastest-growing technologies in the field of innovation. Imaginative scientists and engineers have embraced this technology and are using it to print 3D versions of everything from blood vessels to pizza for space travelers.

Italian inventor Enrico Dini has developed a 3D printing-type device called the "D-Shape" that has the ability to print entire buildings. The D-Shape creates sedimentary stone using a magnesium-based material to bind sand particles for construction applications. Dini has printed structures a few feet high and hopes to use his innovation to construct entire buildings. The inventor founded a company called Monolite to commercialize his technology. Monolite has been collaborating with architects at Foster+Partners and the European Space Agency (ESA) to explore the possibility of using a modified 3D printer to use moon dust to print habitats for moon explorers or colonists. The goal is to use moon dust to "print" cellular structured walls that would provide an effective shield against micrometeroids and radiation, to protect people living or working on the moon. Dini has demonstrated the ability to print building materials at the rate of 2 m/h and says that the next generation of his D-shape will print at a rate of 3.5 m/h, which could be used to produce a building in 1 week.

A Star Trek Nano Replicator?

As previously mentioned, a scanning probe microscope has already traveled to Mars. It is fun to imagine what other nano-enabled devices will find their way into space, to help us "boldly go where no one has gone before."

A few years ago, I was earning a Master's degree in environmental studies at the University of Pennsylvania and took a nanotechnology class from Jody Roberts who is a program director and researcher studying changes in the practice of chemistry caused by technological innovations, social and political pressure, and emerging science.

In one of his classes, Jody mused that one day in the future, we might have a nanotechnology replicator similar to the replicator in the Star Trek TV and film series. This replicator could produce almost anything, from food to materials, tools, and devices. The raw material for this replicator would be a nanoscale form of carbon, which could be manipulated at the molecular level to produce food and other organic materials. This would be an ideal device for future space missions. Astronauts could use carbon molecules as a raw material to "grow" food for a Mars mission. In reality, scientists and engineers are actually taking the first steps to "print" food, using technology that exists today.

Hod Lipson has already taken a first step. His research group at Cornell University has developed a 3D food printer that has demonstrated the ability to "print" food. His Fab@Home project at Cornell is developing versions of 3D printers that are capable of manufacturing custom objects on demand, including food. Cornell scientists have developed substances called hydrocolloids that they can modify with flavoring agents to product different tastes and textures. The device has the ability to "print" new forms of food, such as a pyramid of turkey or a multiflavored dessert.

The French Culinary Institute in Manhattan has been experimenting with this technology for several years. In the hands of some artistic chefs, we could begin to see some new forms and textures of food. Blueprints for the 3D food printer are available for free online, so entrepreneurs and hobbyists can build their own.

The 3D food printer is driven by a computer that requires the "3D food chef" to enter three key parameters: the shape of what you want to create, a description of the materials, and how they should be printed. Several prototypes and early products are in development or poised to enter the market.

It is likely that the first humans who land on Mars will take a 3D printer to "print" a broken screw, a hinge, a seal, or a hamburger. In 2013, the concept of a replicator capable of creating different types of food took one step closer to reality when NASA announced that the space agency is providing $125 000 in seed funding to develop a 3D printer capable of printing food items, for use on long space missions. The research grant was awarded to Anjan Contractor, a mechanical engineer at Systems and Materials Research Corporation (SMRC) in Austin, Texas. Contractor's design uses replaceable powder cartridges as building blocks to create a wide range of foods. The powders used in the cartridges will have a long shelf life of 30 years and could be used to feed astronauts on the first manned missions to Mars, and on longer missions in the future. Contractor has already printed chocolate and the first goal for his prototype is to print a pizza.

Imagine an astronaut using a 3D printer to create a pizza layer by layer, applying the ingredients in layers to build the dough, tomato sauce, cheese, and maybe even sausage or mushrooms, and vegetables. It may look more like a layer cake than the pizza you get from your local Pizza Hut or Dominoes, but it could also include 3D sausages, cheese, and other familiar and tasty ingredients.

The NASA food synthesizer is only one example of science fiction becoming science reality at the convergence of design, art, science, and Nature. NASA has an active open innovation program that encourages inventors, innovators, and commercial partners to work on applications for 3D printers that can be used on NASA missions.

NASA has awarded research grants for the development of 3D printers that can be used to make food in space, especially for deep space missions. Research is being conducted at the Habitability and Environmental Factors Office at NASA's Johnson Space Center – where NASA conducts space food product development and conducts food tasting tests.

An equally important application is the use of 3D printing to manufacture tools and components that may break, or to provide solutions for problems encountered in space. NASA and Made In Space Inc. have announced that they are collaborating on the first 3D microgravity printing experiment scheduled to be conducted on the International Space Station in 2014. The technology was developed under NASA's Flight Opportunities Program and is partly funded under an SBIR grant.

5.3 Using Art to Conceptualize the Future

There is another function of nanoart that has a special relationship with innovation – the ability to conceptualize the future. If you combine something you can

(a)

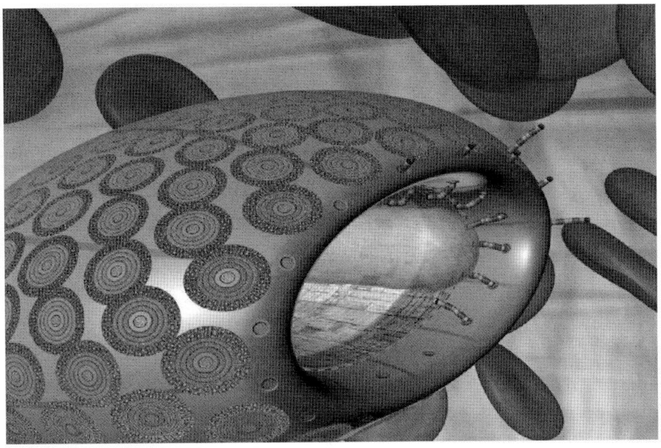

(b)

Figure 5.4 (a) Respirocytes. This illustration shows respirocytes and red blood cells flowing through a blood vessel, including the surface of the respirocytes rendered in fine details. (b) Microbivore. The microbivore is an artificial white blood cell, a form of nanorobot. Both innovations were conceptualized and designed by Robert Freitas (images used with permission of Robert Freitas).

see only using an atomic force microscope with something that might not exist yet, the only way to visualize it is through an artistic representation. Illustrators have been showing us "what's possible" in the realm of fantasy and science fiction, for centuries.

In science, illustrators have illustrated atoms, molecules, proteins, and many other chemical and biological structures long before we had the ability to image them through a microscope. Some complex processes, such as how proteins fold, use sophisticated computer programs to color-code the protein strands, which is something we can't quite do with microscopes, as yet.

Many of the most exciting and radical innovations that we will see in the coming decade exist in the minds of the innovators who are working on them, or in the minds and works of science fiction writers and film directors. However, there are also many innovations that have been granted patents, that haven't been implemented yet, and may not be implemented for another 10, 20, or even 30 years. These innovations are brought to life in illustrations that show us the future.

One of the most striking, well-known, and fascinating examples of this is the work of Robert Freitas, who has shown us the future of nanomedicine in his multiwork series entitled *Nanomedicine*, which was the first book-length technical discussion of potential medical applications using molecular nanotechnology and medical nanorobotics. Much of Dr. Freitas' work is hypothetical or based on inventions and patents that will take decades to functionalize. In the meantime, we can visualize the possibilities through the imagination of the inventors such as Dr. Freitas, as well as illustrators who bring these ideas to life, visually.

Two examples of Dr. Freitas' vision are shown in Figure 5.4, which show artificial red blood cells (respirocytes) and artificial white blood cells (microbivores) conceived and designed by Robert Freitas. The concept of an artificial red blood cell is based on the notion that red blood cells transport oxygen (and some carbon dioxide), and therefore it seems reasonable that an artificial structure could perform the same task. White blood cells destroy pathogens in the bloodstream – Dr. Freitas' artificial white blood cell is conceived as a nanorobot that binds bacteria to the hull, where tiny manipulator arms move the bacterium to the ingestion port (the microbivore's "mouth") where the microbe is digested.

The most fascinating aspect of the relationship between nanoart and nanoinnovation is that, with the right amount of focus, resources, funding, and creativity, we can really make these things happen. We can bring to life the innovations depicted by illustrators. We can turn these artistic representations into functional technologies. We have the tools. We have the talent. We have the creativity. Nanoart is showing us the future, and it is our task, all of us, to make it happen.

Reference

1 De Volder, M., Tawfick, S.H., Park, S.J., Copic, D., Zhao, Z., Lu, W., and Hart, A.J. (2010) Diverse 3D microarchitectures made by capillary forming of carbon nanotubes. *Advanced Materials*, **22**, 4384–4389.

Part III
The Business of Nanoinnovation

NanoInnovation: What Every Manager Needs to Know, First Edition. Michael Tomczyk.

6
Lessons from the First Wave of Nanoventures

> If I know I'm right, I don't need to convince skeptics. I'm proceeding on the basis that I'm right, and if I'm right, I win.
>
> – *James von Ehr, Founder, Zyvex Corporation*

In 2000 while keynoting an industry conference, I coined the term "nanodextrous" to describe the ability to work at the nanoscale as well as the macroscale. Nanoinnovators live in two worlds – nano and macro. They have to be equally proficient with nanoscale materials and properties and the macroscale products they enable. In other words, they need to be "nanodextrous." For an organization involved in nanoinnovation, nanodexterity is a core competence.

The concept of nanodexterity has its roots in the concept of the "ambidextrous organization" developed by Charles O'Reilly and Michael L. Tushman [1]. An ambidextrous organization is an organization that simultaneously (i) *exploits* existing technologies and products and (ii) *explores* emerging technologies.

A *nanodextrous* organization is an organization that works with the nanoscale properties of materials, processes, and technologies while simultaneously developing commercial (macroscale) products based on those properties. Nanodextrous innovators are able to deal with materials and processes they can hold in their hand and see with the naked eye, as well as those that can only be viewed with an atomic force microscope and manipulated with a nanoscale probe. Nanodexterity requires different competencies and skill sets, sophisticated instruments, and a unique understanding of nano–micro–macro properties.

Consider the carbon nanotube (CNT), the iconic symbol of nanotechnology. You don't really see the individual tubes in a baseball bat or golf club infused with carbon nanotubes, or in the hull of a carbon nanotube boat. These products are rigid and hard. However, in bulk form, carbon nanotubes look like a spongy black mass. Incorporating this material in the hull of a boat requires an understanding of how nanotubes are produced, how they can be chemically dispersed in a composite material, and how their unique characteristics can improve the design of a seaworthy watercraft.

Very soon, every organization that manufactures products – cosmetics, pharmaceuticals, cement, solar panels, textiles, cars, boats and more – will be nanodextrous. All of these products will be nano-enabled or nano-enhanced in some way.

NanoInnovation: What Every Manager Needs to Know, First Edition. Michael Tomczyk.

In this nano-enabled world, you won't be able to compete, survive, and succeed in even the most traditional industry unless your organization is nanodextrous.

So, how do you become nanodextrous?

A good starting point is to look at the first wave of nanotechnology business ventures, and the strategies and practices they used to build their enterprises.

6.1
The First Wave of Nanoventures

After more than a decade of commercial development, we are now able to look back at the first wave of nanoventures and draw some intriguing insights. We can also begin to answer questions that can only be answered with the wisdom of hindsight – and from the inside perspectives of the people involved in these ventures. They are packed with insights – successes, failures, survival strategies, home runs, strikeouts, and so on.

This chapter includes profiles of several representative nanoventures, including Zyvex Corporation and its spinouts; Nantero, Inc.; QuantumSphere (QSI); Nanocomp Technologies, Inc.; InsituTec; Nanosys Inc.; and Graphene Frontiers. Most of these ventures survived and prevailed in spite of a challenging economic climate that included two recessions, a tech investment bubble, a liquidity drought, and a multiyear world economic crisis.

Most of these nanoventures were launched by men and women who met in college, were recent graduates, or collaborated with research faculty to develop technologies and ventures. The link between academic research and venture formation is strong and continuing. University and government nanoscience projects have seeded many of the ventures.

All of these ventures have benefited from venture capital (VC) funding. Most early-stage ventures received angel funding from a few large investors or friends and family, and then they sought additional funding from government research grants followed by funding from venture capitalists and corporate partners. The best VCs involved in the first wave of successful nanoventures had a solid understanding of the science involved and the potential for commercialization, and represented "patient money." They understood the distinction between science projects and commercial business projects and structured their funding around a "business context." Those who focused solely on the financial value proposition or the VC exit strategy didn't do as well, given the long lead times required to commercialize even the most promising nanotechnology research. Balancing science innovation, business realities and market needs has required a careful balancing act that is understood by nanotech VCs who include pioneers Jim Von Ehr, Larry Bock, Josh Wolfe, and Bill Tai.

Shakeout and Consolidation

It is noteworthy to mention that there were a couple of investment bubbles that hindered the development of nanoventures. The first occurred in the mid- to late-1990s when nanotechnology and gene therapy were hot and everyone wanted to rush into the field, often blindly. The bursting of the Internet bubble around 2000 (and to some extent the gene therapy and nanotech bubbles) forced venture capitalists to look harder and more critically at nanotechnology investments. In 2008–2009, the global economic crisis created a liquidity crunch that dried up second- and third-tier (mezzanine) funding for ventures already underway. Many ventures were forced to scramble to stay in business.

In addition, longer-than-expected timelines from laboratory to market, combined with scale-up issues and the liquidity drought, made for a toxic cocktail that impacted many nanotech ventures. Analysts have estimated that of the 1200+ nanotech ventures that existed in 2005, only ~10% were selling products in excess of $1 million a year and of these, few if any were profitable. The economic crisis that began in 2008 created a consolidation that shook out some promising as well as failed ventures, and what remains today is a smaller, stronger core of ventures with more realistic commercial goals. Several nanoventures that survived the shakeout are profiled in this chapter; others are described elsewhere in this book.

The lessons from the first wave of nanoventures include some notable failures as well as successes. The most famous example of unfulfilled promise is Carbon Nanotubes Inc. (CNI), which was cofounded by Richard Smalley, who won the Nobel Prize for the discovery of fullerenes. Smalley died from leukemia in 2005 at the age of 62, and in 2007, CNI was absorbed by Unidym. According to insiders, CNI had trouble converting pioneering research and a sizable patent portfolio into commercial products and applications.

Many nanoventures found ways to avert disaster and survived the shakeout. Nanosys Inc. (which is profiled here) seemed destined to follow in CNI's footsteps, but was "resurrected" when their major venture capitalist (Venrock) took action to replace the CEO and set the venture on a commercialization course that undoubtedly saved the company. Another well-known venture had to cut its staff from 50 to 5 people in order to survive. Nantero sold its government business to Lockheed. Zyvex had to develop its Piranha boat prototype using mostly its own resources.

These and other examples confirm that *flexibility* is an important attribute for any nanotech enterprise. Being willing to make changes in leadership, products, and markets, or to tighten the belt and push through a crisis, are attributes that stand any entrepreneurial venture in good stead. The following examples describe achievements and strategies of several US-based nanoventures. I wish to thank the founders and managers of these ventures who generously and candidly shared their insights.

6.2
Zyvex: Divide and Conquer

The first nanotechnology company was Zyvex Labs, founded in 1997 in Richardson, Texas, by serial entrepreneur and computer software pioneer James Von Ehr II. When he founded Zyvex Labs, he adopted a strategy that might be called "divide and conquer." He divided his company into several separate divisions, which allowed him to develop nanoimaging systems and probes, nanomaterials, and medical and marine products – while also conducting riskier long-term research on atomically precise manufacturing. Later, as these business units developed commercial value, he spun them off and shifted his focus to core businesses and research. This strategy not only made Zyvex a pioneer but also gave the company staying power, making it the world's longest surviving nanotechnology company.

Jim Von Ehr is a soft-spoken, earnest, optimistic business pioneer whose business and technology ventures straddle the "real" nanotech world today, and the visionary nanoworlds of the future. He is a serial entrepreneur, Internet pioneer, and philanthropist who began his career as a computer software developer. Jim was born in Grand Rapids, Michigan and earned a computer science degree from Michigan State University and a Master's in Mathematical Sciences from the University of Texas at Dallas.

After college he worked as an engineer and manager at Texas Instruments and in December 1984 he launched his own software firm – Altsys – which developed graphics applications and font editing tools for personal computers. In January 1995, Altsys was acquired by Macromedia, which became a multibillion dollar Web development company that was eventually acquired by Adobe Systems. Von Ehr used the proceeds from these successes to fund the world's first nanotechnology venture.

The Zyvex story is best told in Jim's own words (*and with appreciation for sharing his story in an interview for this book*):

"In 1993, I was developing Altsys and I received an email that Eric Drexler was coming to town to accept an award from Texas Instruments, where I had previously worked," he recalls. "Drexler spent an hour talking about some of the most weird, fantastic stuff I had ever heard. So I special ordered his book (Nanosystems) and as I read it I thought to myself, 'This is a revolution.' I had studied enough physics and engineering to realize that this wasn't antigravity or perpetual motion or time travel or science quackery. This really made sense. So a couple of years later, I sold my company and decided to use some of the proceeds to fund a nanotechnology startup."

"At first, I was thinking of becoming a venture capitalist. I spent about 9 months looking for a startup to fund but I couldn't find anyone I thought was credible so I decided to start my own company. My wife thought I was nuts." He laughs as he recalls her reaction. "After ten years of 80-hour work weeks, I was launching a company that would require another decade of 80-hour work weeks!"

Von Ehr was one of the first nanotech entrepreneurs to embrace Drexler's vision of molecular nanotechnology – which he calls "atomically precise manufacturing." He knew it could take decades to develop nanomanufacturing, which involved finding practical methods to fabricate nanoscale devices on an atomic scale. So he divided his venture into separate companies that could sustain Zyvex and generate profits while pursuing the longer-term vision of using atoms as building blocks. Von Ehr did not get caught up in the scientific debate that was raging at the time (the "Drexler–Smalley" debates), which he characterized as "not constructive."

"You're not going to make it happen by arguing for or against it, but by going to the lab to make it happen," he says. He notes that Richard Feynman, the scientific father of nanotechnology, said that the reason we haven't done it is not because of the physics, but because we're too big.

Surprisingly, Von Ehr says, "I am not a big fan of self assembly or self replication. To me, self-replicating nanorobots are a very tiny subset of nanomanufacturing. It's not uninteresting, but not a critical thing to me. I always envisioned that nanomanufacturing happens in a nanofactory, but it won't come from self-assembly. It will come instead from template assembly."

"Nature uses templated assembly. It uses active biological machines and thermodynamics to transport materials. It uses diffusion to help building blocks find one another. Nature has created machines that move components around and manufacture precise nanoscale parts to create cells and organisms. Biological factories happen to be soft squishy factories rather than hard factories, but we draw encouragement from the fact that we know it's possible because Nature does it. In our nanotechnology world today, we can actively grab and transport parts and assemble them, atom by atom, block by block, subsystem by subsystem – what I call atomically-precise assembly."

"We don't do this one at a time. Nature doesn't build an oak tree from one little molecular machine cranking things out really fast. It uses massively parallel processes. A tree spends a lot of energy replicating itself, surviving droughts, winter weather, famine and so on. We can engineer simpler systems that don't have to deal with these natural problems."

While he is reluctant to make time predictions, Von Ehr believes that the first rudimentary molecular manufacturing systems could be in operation by 2020.

"Some people still say that nanomanufacturing can't be done. Some say it has to be done in solution because so many natural processes occur in solution. How many times have we heard that, in the history of technology and innovation? Before we had airplanes we had birds that flapped their winds and most people thought that airplanes had to have wings that flap. But airplanes don't flap their wings and they still fly!" He smiles and adds, "The same holds true for electricity. What happens with electricity in nature is different from how we transport electricity using copper wire. There are many examples."

"Right now, I don't need to convince skeptics that I'm right. I'm proceeding on the basis that if I'm right, I win. If we can develop methods to do atomically precise manufacturing, we'll make the world a fantastically interesting and wealthier place. We will increase opportunities for humanity all over the world. We will solve the climate, health, energy issues, if we can learn from Nature and obtain precision at the atomic scale."

The venture was developed using a combination of first mover advantage and a multipronged strategy that could be described as "divide and conquer." Some early strategic decisions included forming an instruments company, choosing to process nanotubes instead of producing them, and focusing on solutions for semiconductors, sporting goods, and other large industries and markets.

After self-funding the core venture (Zyvex Labs), Jim Von Ehr divided the venture into several separate companies: Zyvex Labs, Zyvex Instruments, Zyvex Performance Materials (which later became Zyvex Technologies), and Zyvex Asia. Operating as a family of companies, Zyvex has produced a variety of impressive achievements. Zyvex companies have been awarded more than 170 patents, which are used in their own products and licensed to other companies. Zyvex's nanoinnovations range from techniques for dispersing carbon nanotubes in polymers and other high-performance composite materials to nanoscale probes and nano-manipulator systems used in semiconductor fabrication and quality control.

As with all first movers, Zyvex experienced its share of growing pains. In the early days, it was difficult to retain scientists, especially when some technologies failed to work or took longer than expected to commercialize, which is part of the innovation process. Nevertheless, Von Ehr persevered to create a family of companies that has achieved some impressive "firsts" in nano instruments, high-performance materials, and nanomanufacturing.

A key part of Von Ehr's strategy was to partner with academia and government, which greatly extended the small company's knowledge base and resources. Von Ehr has also provided endowments to universities in North Texas and Michigan to provide strong links to the academic community. His contributions to academia included a \$3.5 million endowment to establish the NanoTech Institute and the James Von Ehr Distinguished Chair of Science and Technology at the University of Texas at Dallas, and an endowed scholarship at Michigan State University for 16 engineering undergrads. Jim worked with Zyvex President Dr. John Randall to initiate a consortium of universities and companies called the Atomically Precise Manufacturing Consortium. Zyvex also partnered with and received research funding from several government organizations, including DARPA, the State of Texas, the National Institute of Standards and Technology (NIST), and NASA. Zyvex also partnered with other companies. The Zyvex Partner Program, introduced in 2003, created partnerships with carbon nanotube suppliers, distributors, and manufacturers as well as customers.

Nanoscopes, Boats, and Baseball Bats

Zyvex was very clever in selecting its early target markets and products. An early partner was Easton Sports, which was looking for ways to design stronger sports materials. Zyvex helped create an improved carbon fiber baseball bat called the "Stealth" bat, which incorporated carbon nanotubes. The Easton bat was also one of the first mass market consumer products to incorporate nanotechnology. At the time, it was estimated that as many as half of all the baseball bats (which had lifetime warranties) were getting cracked and returned, which was very costly for baseball bat companies. The carbon nanotube bat eliminated the breakage problem. The original bat was donated to the Smithsonian Institute in May 2006. Zyvex also worked with Easton to create a lightweight bike frame containing carbon nanotubes. Zyvex has also partnered with NASA to develop ways to reduce the size of devices carried on space vehicles and stations, and with race car designers to provide lighter materials for race cars, and adhesives to speed repairs during races.

When Zyvex was formed, the largest market was the semiconductor industry, which was already working at the nanoscale in its computer chips and circuits. Coming from a computer background, Jim Von Ehr was one of the first to recognize that the semiconductor industry was an immediate customer for nanotechnology products – especially instrumentation and nanoscale testing devices. Systems were needed to "fail-test" the architectures of increasingly small semiconductor devices. As semiconductor features dropped from 90 to 65 nm, to 45 nm and smaller, the circuits, channels, and other features became so small that they could only be viewed with an electron microscope and tested only using nanoprobes and manipulators.

Zyvex Instruments created nanoscale imaging systems with multiple probes that enabled precise manipulation and detection of nanoscale structures, which provided a much-needed capability to the semiconductor industry. These imaging systems also benefited Zyvex's own research so in a sense they were also developing instruments for their own use at a time when nanoscale microscopes were expensive and did not have all the features needed by researchers.

Zyvex Instruments pioneered a family of automated multiposition nanoprobes that provided interactive 3D manipulation at the nanoscale, and it was not surprising that these systems found their widest application in the semiconductor industry (see Figure 6.1). The multiprobe systems are used extensively for device characterization; identification and localization of faults at 90 nm and smaller

Figure 6.1 The Zyvex nProber^{TM} shown here is used to measure device quality and conduct failure analysis in semiconductor chips. It is one of the industry's most precise tools for analysis and manipulation (photo courtesy of Zyvex Instruments/DCG Systems).

geometries; and measurement of temperatures, forces, and other factors related to design and production. A typical semiconductor customer uses 1000 probes/year, and a university laboratory typically uses 200 probes/year. Today, Zyvex nanoprobes are installed in virtually every major semiconductor company worldwide.

One of Zyvex's most important lines of business has involved helping customers incorporate nanomaterials into their products, which isn't always an easy fit. For example, carbon nanomaterials are inherently strong and have unique properties, but are not easily incorporated in other materials such as carbon fiber, resins, or plastics. Zyvex's Kentera™ technology (Kentera means "bridge" in Arabic) has provided materials solutions for such client firms as Lockheed Martin, Owens Corning, and PolyOne.

The Zyvex Piranha, made from materials developed at Zyvex, was the world's first boat made with carbon nanotubes. The Piranha is made with Arovex™, a proprietary nanotube-reinforced carbon fiber developed by Zyvex engineers. Zyvex also developed a line of adhesive products called Epovex™, which is used to repair damaged carbon fiber panels in race cars, boats, and other products.

Creating Artificial Retinas to Restore Vision

Zyvex Labs is a founding partner in NanoRetina, Inc., a joint venture formed in 2011 to develop artificial retinas. The company's partner is Israel-based company Rainbow Medical. The founders include Jim Von Ehr, Yossi Gross, Efi Cohen-Arazi, and Ra'anan Gefen. Yossi Gross, one of the cofounders of Rainbow Medical, has filed >500 US patents and founded/cofounded >27 medical device companies.

NanoRetina Inc. is developing an easy-to-implant artificial retina called the "Bio-Retina," which is designed to restore sight. The venture's goal is to provide a bionic retinal implant that can be "glued" to a damaged retina in a 30 min procedure that would "instantaneously" restore sight and enable recipients to watch TV and identify faces. A prototype was created in 2011 to validate the technologies involved. Clinical trials were expected to begin in 2013. The market need for this technology is enormous, given the rate of mascular degeneration especially in older people – the company expects that the addressable retinal prosthesis market will reach 180 000 units annually by 2015.

Company Spinouts

In 2007, the Zyvex venture engine began spinning out its companies as separate entities. The first company spinout was NanoMed, Inc., which was originally launched by several Ph.D.'s: Dr. Rob Burgess, former VP-R&D at Zyvex Instruments; Dr. Gareth Hughes, former Group Leader-Life Sciences at Zyvex; Prof. Rocky Draper from the University of Texas; and Prof. Ellen Vitetta from the University of Texas Southwestern Medical Center. The company subsequently changed its name to Medical Nanotechnologies, Inc. The company's activities include research on nanoparticle-mediated therapies for drug delivery using thermal and photothermal activation, as well as protocols for assessing the purity

Figure 6.2 Zyvex Asia custom designed this ultrahigh vacuum scanning tunneling microscope (shown here under construction) to provide atomically precise manufacturing capabilities – in other words, being able to design 3D structures layer by layer using atoms as building blocks (photo courtesy of Zyvex Asia).

and toxicity of carbon nanotubes in the workplace and in finished products and for studying the interaction of nanoparticle interactions with mammalian cells.

Zyvex Performance Materials was spun out in 2007 and relocated to Columbus, Ohio. The name was changed to Zyvex Technologies and the company is currently focusing on the development and marketing of commercial nanomaterials; nanoengineered composites; nanoengineered adhesives; and molecularly engineered rubber.

In 2010, Zyvex Instruments was acquired by DCG Systems Inc. in Fremont, California and became the company's NanoInstruments Division. Its products include self-cleaning multiprobe nanotechnology instruments with machine vision software. DCG Systems has deployed >1500 systems worldwide. The division continues to partner with other Zyvex ventures.

In November 2011, Zyvex Marine was established as a division of Zyvex Technologies, located near Seattle, Washington. The marine division is focusing on designing and building advanced maritime vessels.

Zyvex Asia Pte. Ltd. was created in 2007 as an independent Asian R&D facility located in Singapore (Figure 6.2). The company is able to draw upon research talent and resources from the Institute of Materials Research and Engineering (where the laboratory is located), Nanyang Technology University, National University of Singapore, and other A*Star Institutes. This strategic move allowed the company to access the very strong science and technology infrastructure available in Singapore and other Asian countries. The company is working closely with Zyvex Labs to develop atomically precise products, engineered and built atom by atom, under computer control. Research goals range from the production of quantum dots to (eventually) massively parallel nanoscale factories.

In 2010, a separate Singapore company, Zycraft, was created to develop and market an unmanned surface vehicle (USV) called the Vigilant. Jim Von Ehr is the managing director of Zycraft and is excited about the prospects for the craft. Like a Predator drone aircraft, the Vigilant can be remotely controlled by pilots using satellite links to guide the vessel, energy harvesting and storage devices, and nanomedicine technologies.

Today, Jim Von Ehr devotes most of his time to Zyvex Labs and Zycraft, working to facilitate his vision of atomically precise manufacturing and commercialize more products enabled by Zyvex nanomaterials.

A Few Recollections from Zyvex Alumni

Over the years, Zyvex has spawned several notable alumni who helped lead and develop the company, and who went on to other significant achievements. Several Zyvex alumni generously shared their recollections and insights for this book to provide additional perspectives on Zyvex.

Tom Celluci was President and Chief Operating Officer at Zyvex Corporation from 2002 to 2006. Tom subsequently became the first Chief Commercialization Officer at the US Department of Homeland Security. Looking back at his Zyvex days, he recalls, "We realized very early that the real value in nanotechnology was not in the raw starting materials such as nanotubes, but in processing them into a variety of nanocomposites. This allowed us to focus our efforts on the processing of nanotubes, which turned out to be a prudent and wise decision. We also spent a lot of time talking to customers. More often than not, unsuccessful technology firms have a solution looking for a problem. We did the reverse. We started with problems and worked to develop solutions, and this included moving Zyvex from a technology push to a market-driven approach."

"Also, there were no standards when we began, so in many areas we had to become the standard-bearer for the industry, creating standards where nothing existed before. The first suppliers of carbon nanotubes couldn't provide quality control and reliability, so we worked with suppliers to provide standards as well as technical approaches to provide reliable, quality nanotubes, which translated into reliable, quality products."

Mark Banash, cofounder and Vice President and Chief Scientist at Nanocomp Technologies, was director of production and quality for nanomaterials products at Zyvex from 2003 to 2007. He managed manufacturing operations, including pilot plant design and construction, and also invented the supply chain certification process that ensured quality in carbon nanotubes at a time when standards did not exist.

Mark recalls the development of the Stealth bat for Easton Sports: "Easton was getting tens of thousands of baseball bats returned every year because they broke. This meant they had to pay for the replacement bat, as well as the costs of shipping, handling, ordering, pulling production off line. We worked with them to provide the solutions for a new type of bat that had a ring of epoxy in the middle that prevents the bat from breaking. This innovation lowered the reject rate by 80 percent, resulting in savings that amounted to hundreds of thousands of dollars. It was one of our major achievements when I was at Zyvex." A profile of Nanocomp with additional comments by Mark is included separately in this chapter.

Taylor Cavanah, President and CEO at mobile software development company Locai, worked at Zyvex from 2004 to 2010, both as a business manager and product manager. He is one of many nanoinnovators who talks about being guided into this field by a genuine epiphany.

"In high school, I saw a double rainbow in the sky and realized that I knew why that was happening, so I decided that instead of being a lawyer which was my original plan, I would be a physicist and try to learn everything there is to know about the universe; so I studied physics and nanotechnology at Rice University," Taylor recalls. "I met Jim Von Ehr at a nanoventures event during spring break, during my senior year in 2002 and suddenly I knew that this stuff is amazing and this is what I wanted to work on."

"Jim suggested that it would be a good idea to go to graduate school, so I attended UT-Dallas, where I studied with Anvar Zakhidov and Ray Baughman, the founders of the Nanotech Institute. This allowed me to simultaneously earn my master's degree in physics, do some teaching and conduct research. My duties included managing the arc discharge carbon nanotubes lab, and included two years working on a NASA research project, studying how to synthesize carbon nanotubes in outer space."

One summer he worked for Zyvex, in the applications group that was developing the nanomanipulator. Taylor had a choice to stay in school and earn a Ph.D. or stay at Zyvex. He chose Zyvex. As an application scientist and product manager, he worked on the Zyvex nanomanipulator, which was developed by Chief Technology Officer Richard Stallcup and his colleagues.

"Our instruments provided tools for failure analysis engineers in the semiconductor industry," he explains. "You may have a billion or more transistors on a single chip. When a chip fails, whether at a customer site or while designing a new process, the chip engineers need to do detective work to find out why it failed. The problem could be a single transistor failure (literally 'one in a billion') or something like an angstrom-sized crack in the gate oxide. Using our four or six or eight positioner system, they can electrically hook up to the transistors, and get the electrical data that points to a problem between the gate and the drain, for example. Once they see the problem, the system electrically tells them what's wrong. But they still need visual proof and more detailed evidence so they may do more work to verify and pinpoint the position and cause of the failed transistor, or the crack."

Taylor piloted the spinout of Zyvex Instruments, which was acquired by DCG in January 2010, and served for a year as general manager of the division at DCG before leaving to establish his own entrepreneurial venture, Locai, Inc., which has developed a location-based marketing application in the emerging field of geosocial marketing.

These personal anecdotes help us understand how Zyvex got into the business of nanotechnology – and retained a leadership role. A better way to understand the company and its strategy is to look at one impressive nanotechnology product created by Zyvex that is large enough for all of us to see – and ride in.

6.3
Zyvex Piranha: The First Carbon Nanotube Boat

Somewhere in the Pacific Ocean, miles off the coast of the state of Washington, a 54 foot slices through the crashing waves. There is something oddly distinctive about this boat. Outside it looks like a large pleasure boat, but that's where the

Figure 6.3 The Zyvex Piranha incorporates carbon nanotubes and patented composite and epoxy technologies to provide a variety of totally new design options for unmanned surface vessels (USVs) The Piranha and future versions could revolutionalize how government agencies conduct surveillance, border patrol, antipiracy operations, and combat support in the very near future (image courtesy of Zyvex Technologies/ Zyvex Marine).

similarity ends. The hull of this boat is infused with carbon nanotubes, which makes it incredibly strong, durable, and lightweight.

The boat is a prototype USV being demonstrated for the US Navy. It's called the Piranha and it's made by Zyvex Technologies' Zyvex Marine division. The craft can be driven by remote control, so the captain of this boat does not need to sit at the helm. He or she can sit at a computer to drive this vessel. In many ways, the Zyvex Piranha is like an ocean-going version of a Predator drone aircraft. This remarkable innovation is the tip of a technological iceberg that could literally sink the current generation of unmanned military seacraft.

The Zyvex Piranha is the world's first boat made with nanomaterials (Figure 6.3). It has been called "the world's first molecularly engineered boat" and one of the world's largest structures built with nano-enhanced materials. It is one of the most fascinating products developed in the realm of nanoinnovation. The story of the Piranha offers lessons for anyone trying to launch a nanoinnovation into a traditional, centuries-old market.

The Piranha can be operated as a manned or unmanned watercraft. The 16 m (54′) vessel can be air dropped. It runs at a top speed exceeding 30 knots and can operate in rough seas. The vessel's 2 tonne payload can include people, weapons, or sensors. Its twin diesel engines can run up to 2000 miles on a tank of fuel. At 5000 kg, the weight of the Piranha is less than one-third the weight of a normal boat the same size. In spite of its performance advantages, the Piranha costs about the same as a comparable boat made with conventional materials. At a cruising speed of 20 knots, the Piranha consumes 12 gallons of fuel/h compared with 50 gallons/h for a traditional boat.

Commercializing the radical new design was a major undertaking that required the company's management to negotiate several early obstacles, according to Mike Nemeth, director of Commercial and Defense Applications for Zyvex Marine and VP-Business Development at the parent company, Zyvex Technologies.

"When we first came up with the idea of a boat made from our nanotube materials, no one would work with us to develop a prototype," Mike explains, "so we had to fund and develop our own prototype, which essentially put us in the boat business. The first Piranha was only outfitted with basic systems to prove the concept. Our second craft started to attract the attention of other systems providers that proved to be important partners. Based on these prototypes, we were able to demonstrate the boat, and this allowed us to make our first sale – to a customer in Singapore in 2011."

Mike explained that the material used to construct the vessel is a proprietary composite called Arovex, a nanotechnology-enhanced carbon fiber composite. In the materials industry, Arovex is known as a "prepreg," which means it is preimpregnated with Zyvex Technologies' Epovex nanotube epoxy resin. The combination of carbon nanotubes, carbon fiber, and resin creates a material that is 40% stronger than conventional carbon fiber composites.

"Just as fiberglass replaced wood in modern boats and ships, we expect that Arovex will displace conventional carbon fiber composites, not only in boats but also in wind turbine blades, helicopters and other high performance applications. The best customers for the Piranha USV are large defense contractors who are searching for faster, better and lighter maritime platforms for their unmanned systems." He added that the Piranha provides an affordable alternative to more expensive unmanned surface vessels that use heavier, costlier technologies. "This is especially important as Defense budgets are being cut and military planners are looking for innovative ways to reduce costs without reducing their mission capabilities."

While applications in the US Navy and Coast Guard offer the most promising near-term applications, the boat's highly reconfigurable design allows the boat to be adapted for a wide range of missions, including antipiracy, surface surveillance, surface action, mine countermeasures, electronic warfare, and antisubmarine warfare. Armament options include stabilized machine guns, Mark 54 torpedoes and over-the-horizon missiles.

"The Piranha could find an important role escorting single ships or convoys," Nemeth suggested, since it can be equipped with advanced sensors and networked satellite or terrestrial communications to detect pirates or other hostiles before they can threaten shipping.

Ironically, the first defense contractor Mike approached was a large defense-aerospace corporation with several large US Navy contracts. To his dismay, the corporation declined his invitation to fund the prototype and copresent the prototype to the US Navy. Why? The reason they gave is that there was no RFP (request for proposal) for a carbon nanotube boat! This corporation seemed to be saying that a totally novel product or system could not be submitted for a development contract if there was no contract request for it. Needless to say, this was a frustrating response.

Figure 6.4 (a) Rear Admiral Nevin P. Carr with Mike Nemeth from Zyvex Technologies, tours the Zyvex Piranha at the 2011 Sea–Air–Space Exposition (US Navy photo by John F. Williams). (b) This photo of the Piranha in the harbor illustrates the size of the vessel (photo courtesy of Zyvex Technologies).

Consequently, the Zyvex team developed and refined their prototype, and approached the Navy on their own. In April 2011, Admiral Nevin P. Carr, Chief of Naval Research, toured the Piranha at the Sea-Air-Space Exposition in National Harbor, Maryland (Figure 6.4a). This was not an easy sell. Designing a USV to military specification would take time. Also, a fleet of carbon nanotube USVs would require an entirely new infrastructure to repair the vessels. The repair process was closer to repairing fiberglass, but different enough to require new technologies for patches and adhesives. Zyvex had already thought of this. Their previously developed line of carbon nanotube adhesives and patching materials was being used to repair carbon fiber parts on high-performance race cars, and this function also worked for repairing a boat made with carbon nanotubes.

Zyvex's experience with the Navy is a good example of how a "white space" – an entirely new market opportunity – can open up unexpectedly. One of the most interesting outcomes of the discussions with the Navy was a request to study how heavy metal hatches and doors on Navy ships and other vessels can be replaced with CNT-based doors and hatches. Zyvex materials with equal or greater strength and durability and decreased weight could make the vessels lighter, and reduce energy and fuel costs. Zyvex is collaborating with military contractors with expertise in this sector to develop this opportunity.

Zyvex Technologies President Lance Criscuolo is currently working with Mike Nemeth and the company's technical staff to turn the Piranha into a full marine product line, including the development of both military and civilian boats in different sizes and designs (Figure 6.4b).

There are lots of lessons embedded in this story. The management team at Zyvex Technologies refused to be discouraged by the initial lack of interest by a potential corporate partner. They grabbed the bull by the horns and developed their own prototype. They went directly to the potential end user, the US Navy. This provided proof of concept and allowed them to make their first sale (to a nonmilitary customer), which got them off and running. In 2014, Zyvex was acquired by Luxembourg firm Ocsial, forming the world's largest nanotech company.

6.4
Nantero: Patent, Fabricate, Outsource

In business, there is a continuing debate over the advantages of being a first mover in a new technology or market, or a fast follower. Some say that a first mover gains an early advantage, which was definitely true in the case of Internet portals (Amazon, Facebook, and Google); while others maintain that a first mover is typically eclipsed and passed by fast followers who enter the market later when the market is better understood (Blackberry eclipsed by smartphones).

In the field of nanotechnology, several first movers such as Zyvex entered the sector early and were able to quickly develop a strong patent portfolio gain in an early foothold based on core technologies.

Nantero Inc. was one of the earliest and best-known first movers in nanotechnology. This was also one of the first companies to figure out how to turn carbon nanotubes into commercial value. Nantero was launched in 2001 in Woburn, Massachusetts, by cofounders Greg Schmergel, Tom Rueckes, and Brent Segal. All three had graduated from Harvard. Tom and Brent were Harvard Ph.D.'s in chemistry and Greg had a graduate business degree, although they did not meet at Harvard.

As Brent Segal recalls: "I had a Ph.D. in chemistry and was working at a pharmaceutical firm, when I met Tom Rueckes. Tom was completing his doctoral degree, specializing in nanotechnology. I remember asking him, 'What's nanotechnology?' and Tom explained to me that it had to do with physical chemistry and being a chemist, this made a lot of sense to me. I found this to be very exciting so I left to start a company with Tom, and we brought in Greg Schmergel as CEO and all started Nantero together."

"The initial venture derived from Tom's graduate work at Harvard, and came from his vision for a new type of memory technology using carbon nanotubes described in his doctoral dissertation. We went out to look for funding at a time when the market was kind of sour, since the Internet bubble had burst in 2001."

Greg Schmergel picks up the story: "We were looking to start a new company in a field with a high entry barrier. Tom, Brent, and I had met through a mutual friend, and they had an idea for a nanotube memory switch. In 2001 when we started, there were only two venture-backed nanotech startups: Nantero and Nanosys. It was a tough time for startups in general. Back then when I met with venture capitalists, almost none of them even knew what nanotech was, and certainly didn't know what carbon nanotubes were."

"We raised $6 million in a Series A round, co-led by two VC firms (Stata Venture Partners and Draper Fisher Jurvetson). Bill Tai and Bruce Sachs from Charles River Ventures led the B round in 2003 and Ullas Naik and Globespan Capital Partners provided funding in 2005, which brought our venture funding to $31.5 million. We have not needed venture capital since 2005, since we were able to sustain ourselves on revenue since then."

Bill Tai, a general partner with Charles River Ventures, led the venture capital firm's investment in Nantero in 2003. He recalls that the venture capital "story" at Nantero seemed solid and well planned. Also, the science at Nantero was very impressive. "Nantero's founders observed that both carbon and silicon sit in the same column in the periodic table – in other words, they both have four valence electrons and thus could be made to electron bond pair in a way that uses very little energy and with structures that are incredibly dense compared to conventional silicon based memory." (*Note: Valence electrons are the electrons of an atom that can participate in the formation of chemical bonds*).

When asked to describe what he thought were Nantero's "best practices," Bill Tai offered his views of three of Nantero's best practices from a venture capitalist's perspective that are applicable to any nanotechnology venture:

1) Innovation is always important, but to the extent a start-up can focus on an area that is massively disruptive, it's more exciting and easier to fund.
2) If you do have important intellectual property, ensure you protect it.
3) Surround yourself with great people. Nantero did a great job of hiring top flight researchers and attracting exciting people to its board of directors who could help get it business. Any business will go through ups and downs and if people don't understand this, the rough patches become really hard.

In addition to venture capital, Nantero drew funding from government agencies that functioned as partners at a time when everyone was still trying to figure out how to leverage the benefits of nanotechnology and specifically, materials and devices that incorporated carbon nanotubes.

"A large part of our revenues came from government programs in the early years," Schmergel explained. "Our government division was started by Brent

Segal. He established some government contacts and realized that they were interested in the radiation resistant properties of our memory technology which we called NRAM. This is a nanoelectromechanical switch using carbon nanotubes, so it is not affected by radiation, which is of obvious importance to the military, and to NASA. You can send these chips into space without radiation errors, and they can be used in combat situations where radiation could also be a problem."

"We at Nantero started using carbon nanotubes in a production semiconductor factory in 2003. At the time, this was considered radical and many people thought this was impossible. Surprisingly, we still run into people who say carbon nanotubes are just a research material, they'll never be used in a production facility, they'll never work, they'll cause safety and contamination problems, and so on. The reality is that carbon nanotubes have been used in semiconductor fabs for many years now. As a nanomaterial, carbon nanotubes are fairly mature and are used in a wide range of applications, including semiconductors. Some carbon nanotube suppliers produce 20 or 30 metric tons a year."

"A key part of our story involves intellectual property. We now have over 200 patents and patent applications pending and more than 125 of those are already granted. Recently Nantero was rated as having the 2nd strongest patent portfolio in the worldwide semiconductor industry in two separate independent studies. Companies like Samsung and Intel were ahead of us and the ones that followed us on the list were companies like Texas Instruments. For a start-up of our size with just under 50 people, to make it onto a list like that tends to validate our IP strategy."

"Some of the broadest and strongest patents were filed from 2001 to 2004. We were able to file broad patents because we were one of the very few and perhaps the only company taking carbon nanotubes seriously as a material that could be used in mass production of electronic devices, at a time when most experts in the field thought it was impossible because of contamination and fabrication issues. People were asking: How do you position a carbon nanotube on a silicon wafer? Do you pick them up one at a time and put them on your wafer – that would take a million years! Do you grow them on the wafer? Students and scientists spent long periods of time trying to position one nanotube on a wafer."

"In 2000, Tom Rueckes published a paper that drew a lot of attention. He showed a carbon nanotube switch with just two nanotubes – one suspended perpendicularly above the other, and in the 'off' state where they are not touching there is essentially infinite resistance. Then he used electrostatic attraction to bend the top nanotube down to create an 'on' state – which was a memory switch with very distinguishable on and off states. It was a fantastic demonstration of the concept of a switch using carbon nanotubes, however, it was completely unmanufacturable."

"So the challenge once we started the company was to figure out how to mass manufacture carbon nanotube switches and get millions or billions of switches all on one chip, since you want to have billions of switches per chip. To provide gigabits of memory we needed tens of billions of switches on one chip. At the time, everyone said that's impossible!"

"One day, Tom said, 'Yes, that's impossible – so what I'm going to do is put the nanotubes everywhere on the wafer and get rid of the ones that are in the wrong place, so what I wind up with is nanotubes in the right place. We put the nanotubes in solution and spin-coated them on the wafer to create a single layer of nanotubes that evenly coats the wafer, just one nanometer thick; then we used lithography and etching to remove the nanotubes that were not in the right place (i.e. where the memory cells are) which gave us a mass manufacturing process, using a combination of spin coating, lithography and etching, which are commonly used fabrication processes. These used no exotic processes or steps not already used in some form at every semiconductor facility in the world. The fact that it's an elegant process with only a few steps means it's also easy to manufacture, and cost effective. If you have a complex process, it's expensive almost by definition. A smaller number of steps is more cost effective. This was one of our first major innovations in the company and the subject of many of our early patents. We trademarked the name for our memory as NRAM."

He went on to explain that in the early years the company worked with single-walled carbon nanotubes. Experimentation led them to find applications that use double- and multiwalled nanotubes.

"Nantero also decided that its business strategy would be to develop as a licensing company, rather than a production company or fabless semiconductor company. Its licensing customers were primarily semiconductor manufacturers. The most urgent need for NRAMs involves replacing a type of flash memory called NAND Flash used to make high density (gigabyte level) data storage such as memory cards used in digital cameras and computers. Another type of flash (NOR Flash) is a higher speed, lower memory nonvolatile storage for solid state drives. Memory storage keeps growing. I recently bought a 16 gigabyte storage card for my 12 megapixel digital camera with HD video—in a couple of years, 16 gigabytes will be small. We're already working on the next generation of solid state drives, which would no longer use flash memory but will use NRAM instead."

Lockheed Acquires Nantero's Government Business Unit

When the liquidity crunch created a tough environment for funding in 2008–2009, Nantero was able to sustain itself from its revenues, but Greg Schmergel observed that for many nanotech companies this was a difficult and in some cases fatal period.

In August 2008, Nantero sold its government division to Lockheed Martin, which generated capital to allow Nantero to sustain its core commercial semiconductor development, while allowing Lockheed to implement a "fast follower" strategy.

The Nantero government business unit formed the basis for the newly created Lockheed Martin Nanosystems group. The acquisition included licenses for technologies and research capabilities for a variety of nano-enabled devices, including memory and logic devices as well as sensors. Approximately 30 Nantero employees joined Lockheed, including Nantero cofounder Brent Segal, who became Chief Technologist in the new Lockheed Martin Nanosystems group.

When the nanotechnology revolution gained traction in the late 1990s and early 2000s, many large corporations resisted the temptation to jump into the nanotech sector. Instead, many corporations opted to take a wait and see or fast follower strategy. It's not unusual for corporations to wait for emerging technologies to develop and define the market, and enter the market later by acquiring companies or divisions after the technology is better defined. This is also a way for companies to acquire portfolios of patents they may need to anchor their own research and it's very telling that the Nantero–Lockheed deal included a broad patent licensing agreement.

Since the acquisition, Lockheed has continued to work with Nantero and the companies have announced some of their joint work on next-generation nonvolatile digital storage solutions and other technologies, and Nantero continues to work with multiple corporate partners and licensees to move NRAM closer to mass production for commercial applications.

6.5 QuantumSphere: Competitive Catalysts

QuantumSphere is a privately owned nanotechnology venture based in Santa Ana, California that engineers and produces advanced catalytic materials called *nanospheres*. Their manufacturing facility is located 10 min from John Wayne Airport in the heart of Orange County.

Every day, the company produces large volumes of nanospheres, which are used as catalysts in a wide variety of industrial processes. The current plant is capable of producing hundreds of kilograms of its proprietary nanospheres each month.

Market applications for QuantumSphere catalysts include multibillion dollar growth sectors such as batteries, fuel cells, emissions reduction systems, and industrial chemicals such as ammonia synthesis for food production. The company is presently focused on metal–air batteries and ammonia synthesis applications of its advanced nanocatalysts.

The venture was established in 2002 by Kevin Maloney (President) and Doug Carpenter (Chief Technology Officer) (Figure 6.5). If you call the company, the phone is likely to be answered by one of the many MBAs, engineers, and Ph.D.'s who are working to commercialize and scale the company's integrated catalytic solutions. The company's scientific advisory team includes a distinguished group of scientists from several universities and companies, including Prof. George Olah, the 1994 Nobel Prize winner in chemistry and Caltech board of trustee member, Jon Faiz Kayyem, Ph.D.

Figure 6.5 Kevin Maloney and Doug Carpenter, cofounders of QuantumSphere.

A Garage Nanoventure

One of the misconceptions about nanotechnology ventures is that the technology is too complicated to develop in a garage like the Apple personal computer. Actually, there are several notable examples of nanoventures that got their start in garage-like atmospheres, and QuantumSphere is a prime example.

Kevin Maloney grew up in Pasadena, California, about 1 mile from CalTech, and was a self-described "beach bum." In 2001, he was a 29-year-old graduate student enrolled in the MBA program at Pepperdine University. He had worked in finance, sales, and marketing and planned to continue his business career, but he also had keen interest in science. While finishing his MBA, he met former rocket scientist Dr. Doug Carpenter, who was seeking a way to use nanoparticles to increase the burn rate of rocket fuel.

"Doug's solution was to create a better, smaller, faster, more efficient catalyst by using nanotechnology to build the tiniest nano metal particles with the largest surface area in existence," Kevin recalls. "So we raised $ 100,000 from two Caltech Ph.D.'s and built a small lab and pilot-scale reactor in the back corner of my brother's neighborhood warehouse, the size of a two car garage. A year later, we had a manufacturing process that became the core technology for QuantumSphere. We succeeded in producing and scaling tiny, several nanometer sized catalysts, and then quickly focused on demonstrating their commercial viability in valuable clean energy and portable power applications (way before these became hot buzzwords in 2007–2009, and quickly shifted away from military applications which can be quite dangerous and take much longer to bring to market).

A nanometer (nm) is one billionth of a meter, or 1000 times smaller than the diameter of a human hair, or roughly the size of a marble when compared to the earth. QSI catalysts typically measure 5–25 nm. A good catalyst is something that facilitates a more efficient chemical reaction, preferably using less energy. The higher the surface area of the catalyst, the more efficient the chemical reaction. A handful of the micron-sized raw materials QuantumSphere starts with have roughly the surface area about the size of an 8.5×11 sheet paper. Once converted using the Company's patented process, a handful of QSI-Nano catalysts have an increased surface area about the size of a soccer field.

QSI's nanocatalysts provide superior functions because of their unique physical properties, including spherical shape, high purity, uniformity, narrow distribution, controlled oxide layer, and extremely large surface area. This can translate into greater efficiency in the generation, storage, and use of energy. The first key patent was issued in October 2007, for a laminar flow gas phase condensation reactor, a closed loop system under vacuum that uses electricity and gases like helium to produce nanoscale particles. The patent was 40 pages long and included about 75 broad claims."

The process they developed involves a ceramic heating element used to liquefy a raw material, in the form of metal wire or micrometer-sized metal particles, and evaporating it in a stainless steel vacuum chamber containing helium. Tiny droplets of molten metals such as nickel, iron, copper, silver, manganese, gold, or palladium are allowed to condense at specific temperatures and pressures, and this creates small uniform spherical particles called nanospheres. A thin layer of oxygen encases the metal particles to keep them from sticking together and bursting into flames when exposed to air. Without the oxide layer, the particles will burn or ignite if they come into contact with air.

Kevin recalls what it was like to be an under-30 CEO: "I was the high risk guy, one of the younger CEOs with a lot of passion and ambition, connected to enough people, with modest capital to get started. In the first year, it was just four of us in a 20 by 20 square foot facility. The culture is different when you start from grass roots and max out your credit cards a few times to keep things alive. We were never overfunded, and always tried to steer away from the hype. We kept our organization lean, focused, and acted as if every dollar we received was the last we would ever receive. We've seen several overhyped, overfunded VC-backed companies lose focus, spend too much on too many things, effectively getting ahead of themselves from a personnel and capacity perspective, and then run out of money. Oftentimes, they were significantly off in their estimations of the time it would take them or their customers to bring products to market; it is the least predictable part of the cycle."

"At QSI, we got a lot of things wrong like most small start-ups, but we found a way to keep the doors open long enough for one or two things to occur and drive some value which allowed us to procure additional capital and time to bring products to market. We did not spend hundreds of millions of dollars using investor capital or government loans trying to build carbon nanotube elevators to the moon, cure cancer using anti-matter, or compete on low efficiency commoditized products. We simply had to do more with less in terms of capital. To this end, we have engaged our customers, tried to understand their biggest pain, their near term market needs, and bring ideas to them that will hopefully solve problems that can be validated and commercialized in a reasonable time frame. Customer driven innovation is a key part of the strategy. Working together to bring the products to market is paramount as the large players in the clean tech industry already have key sales and distribution channels in place and can move quickly with large budgets."

"From the beginning, we were very clear about our status. Like a famous VC once said, 'Saying you're a nanotechnology company is like saying you're an electricity company. It's what you're doing to harness the electricity or nanotechnology that really matters.' We have always viewed nanotechnology as an enabling technology. We happen to be working with very small elemental materials at the nanoscale, but we differentiate by delivering meaningful products and results."

The Advantages of Nanosizing Catalysts

The concept of nanosizing catalysts is a proven "best practice" in the field of nanoinnovation. Nanosizing a catalyst capitalizes on one of the most important nanoscale characteristics that involves increasing the *surface area,* which is a critical feature in chemical catalysts.

Kevin Maloney provided a vivid example to explain how nanosizing works: "Imagine a bulk material the size of a basketball, and we utilize the surface of the basketball for use as a catalyst in a chemical application. The catalyst on the surface of the ball reacts with the chemical to make something happen – that's what a catalyst does. Now, instead of one large basketball, let's fill a basketball-sized bowl with playground marbles and use those marbles as the catalyst. Now, the surface of every marble is available for the chemical reaction, which greatly increases the amount of catalyst available for the process. This is because the ratio of surface to mass increase dramatically the smaller we go, which increases the level of efficiency for the catalyst or imagine lighting a handful of dry kindling to start your fire versus a wet log. Like the marbles, the kindling has a much greater surface area."

Silver is a photocatalyst used in various photographic materials and in nanoparticle inks. It also has bacteria-fighting properties. If you create a solid silver sphere that weighs 10 g, the surface area will be $\sim 5\,cm^2$. But if you create 10 g of silver in the form of nanospheres that each has a diameter of 10 nm, the total surface area of all the nanospheres will be almost $600\,cm^2$ – increasing the total surface area by ~1.2 million times! [2].

Metal–Air Batteries

In 2005, with only five people in the business and very little capital on hand, the company engaged a zinc–air battery expert to help screen its catalysts, design, build, and demonstrate the company's proprietary high-rate gas diffusion electrode (the active layer or "engine" in the battery) in a complete zinc–air battery system. The resulting data and performance led to another external validation, partnership, and subsequent codevelopment and supply agreement with a major global battery manufacturer. This traction gave the company immediate credibility and validation among the scientific and investment community. Several other small and well-known companies are also engaged with QSI, leveraging its zinc–air battery technology for commercial use. QSI plans to launch its own line of portable zinc–air battery systems with key manufacturing partners in late 2014.

After several years of R&D, QSI now considers itself an industry leader in the design, development, and manufacture of high-performance, low-cost nanocatalysts and integrated systems such as metal–air batteries. The company sees this segment of the battery market as a large and emerging opportunity.

QSI's ISO 9001:2008 quality management systems and patented production processes include advanced catalyst materials and high-rate gas diffusion electrodes that enable its line of MetAir™ batteries to deliver the highest energy density of any commercially available battery, at the lowest cost per kilowatt hour (by weight and volume). QSI batteries provide a shelf life up to several years, employ a modular cell construction and customized form factors, utilize recycled zinc (one of the most abundant elements in the United States.), and are lightweight, safe, and environment-friendly. The batteries contain zero lead, mercury, cadmium, or lithium. QSI's portable power systems are designed for remote, outdoor and recreational use, emergency preparedness and response, and are used by manufacturers developing backup power systems for consumer electronics and military and electric vehicles. QSI and its partners offer metal–air battery components (cathodes), replaceable power stack modules, and complete battery systems directly to end user customers, as well as through reseller and distributor channels.

Nanocatalyzing Ammonia

One of the company's largest potential markets involves the production of ammonia, which is a huge 100 billion dollar market. Most of us think about liquid ammonia as a household cleaner, but 85% of ammonia is used to make fertilizer for food production and biofuel feedstocks. Fifty percent of all protein on the planet comes from ammonia. By dollar amount, ammonia is the one of the most massively produced chemicals on the planet, second only to sulfuric acid. Ammonia production consumes 2–3% of the world's annual energy supply and generates 400 million tonnes of greenhouse gases every year. Most of the world's 500 plants that produce ammonia run on coal and natural gas.

Most ammonia is produced in reactors using the Haber–Bosch process, which has been used for more than 100 years. The process was first demonstrated in 1909 by a German chemist named Fritz Haber. The rights to Haber's tabletop system was purchased by the German chemical company BASF, and scaled up by Carl Bosch for industrial level production, which began in 1913. The Nobel Prize was awarded to Haber in 1918 and Bosch in 1931. The process combines nitrogen and hydrogen gases (NH_3), which are run under high-pressure tubes over an ~450 °C heated bed of small iron rocks, which acts as a catalyst. This is how ammonia has been produced for more than a century – until QuantumSphere developed an improved process using high-surface-area nanocatalysts to increase efficiency and output.

"We discovered this about five years ago when we were working on a catalytic converter for automobiles and looked at some ammonia related applications," Kevin explained. "We took our process to the Casale Group in Switzerland, a 90 year old company that designs, engineers, and refurbishes more than 50% of the ammonia production plants in the world, using its patented systems. The company doesn't own or operate the plants. They design and engineer the reactors which use heat, pressure, natural gas or coal, hydrogen, nitrogen and iron catalysts to produce the ammonia. There hasn't been much of a breakthrough by way of new catalysts for these systems for decades."

In 2009, QuantumSphere signed an arrangement with the Casale Group and has been working with them and other major participants in the global ammonia industry, to test and validate its high-surface-area novel catalysts to lower the cost and increase the output of the world's aging ammonia plants as well as next-generation reactors, which may require much less capital to build and operate. QuantumSphere has several patent applications pending on the technology and early on showed >20% increase in ammonia production yield with just a very small loading and coating of nano iron particles onto existing commercial iron support materials and has subsequently improved these results in many instances. The company's current technology has produced as much as a 40% increase in catalyst activity in laboratory validations. The company has been working with a large chemical producer in China, and in November 2013 the company completed a pilot project to validate its high-efficiency ammonia catalyst technology in a real-world scenario and is now working to commercialize its technology for industrial scale.

QuantumSphere continues to work on other large thermochemical opportunities, leveraging the earlier mentioned concepts for hydrogen and methanol production, among others.

Capital Efficiency

Kevin and Doug pride themselves on the company's capital efficiency, noting that QuantumSphere has been able to maintain itself from a combination of revenues generated from operations, private equity funding, and partner funding. Most of the capital secured to date was used to fund R&D, develop and prosecute its intellectual property portfolio, scale manufacturing, and research future growth-related activities.

"To date we have secured more than $23 million in funding and another $1.5 million in government grants. We're still a relatively small company, but we are very excited about the zinc-air battery systems we are introducing with partners in 2014, and longer-term the major opportunities that exist in the chemicals sector," Kevin says.

6.6 InsituTec: From Grad Students to Commercial Venture

Many of the most interesting and successful nanoventures were started by graduate students. You might not think that something as complex as a scanning probe microscope would be developed in a university laboratory, but in the early 2000s, these systems were so expensive that students chose to design and build their own systems.

Developing a scanning probe microscope in a university laboratory is roughly analogous to Steve Jobs and Steve Wozniak developing the Apple computer in a garage – extremely complex and requiring a great deal of ingenuity – but it can be done. That's how InsituTec, Inc. got started.

The North Carolina-based company specializes in designing and producing nanopositioning systems and metrology tools and sensors. The company has been profitable for several years and has a solid research team, growing product line and customized service portfolio, as well as distributors in China and Japan. This is impressive considering that InsituTec was started by Shane Woody and his wife Bethany when they were still graduate students.

In 2001, Shane and Bethany were graduate students in mechanical engineering at the University of North Carolina at Charlotte, studying metrology, the science of measurement.

"UNC had an industrial affiliates program where students and faculty worked on solving problems for industry," Shane recalls. "We were working with the automotive and aerospace industry to measure very small features, about the width of 50 microns, but we wanted to measure surface features that were less than 10 nanometers which is needed in high end precision manufacturing. One application involved measuring the size and shape of nanoscale holes where fuel is sprayed and atomized into the chamber of an engine. The diesel injector industry wanted to understand the dynamics involved, and how improvements could be made. A major bottleneck was having a quality inspection tool to measure these holes and extract information about the process."

"Fuel efficiency in diesel engines is linked to how well the fuel is sprayed into the engine. If the spray holes are worn or distorted over time, too much fuel is injected, resulting in emission of black soot and CO_2. The thinking was, if you could more efficiently atomize the fuel in the chamber you could lower the CO_2 generated, and this is a big deal because the auto industry is hard pressed to lower CO_2 levels. We built a tool called a Standing Wave Sensor that could do these measurements."

"During our research, we learned that being able to accurately measure these processes is directly related to the quality of precision manufacturing. This is especially true in the semiconductor industry and also has applications in biologics and medicine. For the diesel injector project, the sensors had to be long and slender, with a very small sensor tip. If the tip is too large, you can't measure defects such as scratches or pits in the engine part. The smaller the sensor, the more detail you can measure. So the challenge was to scale down the sensor so it could move into the cavities and holes without sticking to the surfaces. At this scale, sensors want to stick to the surface because of the van der Waal's force, or electrostatic or magnetic forces, which is complicated by something called the meniscus effect, a water layer that builds up on the parts you're trying to measure. One of our innovations involves shaking the sensor to create a very pronounced wave pattern which allows us to change the shape and velocity of the wave to measure surface information."

What we did that was actually unique and different – we shook the sensor ~32 000 times per second. Imagine having a jump rope – you hold one end, I hold the other. When you shake it, you produce a wave pattern – to form a wave – when we shake our sensor, we produce a very pronounced wave pattern. My sensor is held at one end where we shake it, it's long and narrow, we can shake it at 32 000 times a second – some really intriguing things happen when we do this. The tip of the probe bends back and forth many times. When that wave is turned on, we can program it so that it can be quite pronounced. The velocity on that is up to 10 m/s, which is one of the highest velocities for a sensing element. In metrology, when we contact surfaces, the wave changes shape and we can use that information to infer what is going on at the surface.

As a result of this research, Shane and his wife designed their first instrument, which they called "Standing Wave Actuator Technology™" (SWAT). In 2001 while still students, they formed InsituTec to commercialize their innovations. Funding was provided by the National Institute of Standards and Technology.

Another discovery involved microfluidics. During a laboratory experiment, Shane activated a wave sensor in a fluid environment and expected the wave to die off, but instead the sensor continued to create a wave pattern. "When we activated the sensor in the liquid, the particles were attracted to the sensor and concentrated there in a matter of seconds. We thought that's really wild – we could turn this on and get particles to concentrate. If you can get certain types of cells to stick to a sensor quickly, you can detect and analyze them. This opened up a new field of investigation."

InsituTec has filed patents on this technology and has been working with Brown University to apply this to advanced fluidics. The company's researchers are also working with Duke University on biointerfaces; and the Carolina Medical Center to study applications for infectious diseases and orthopedic trauma. The company has received a research grant from the National Institutes of Health. A new method for diagnosing disease using this approach would add an important diagnostic tool to the medical science portfolio.

"Currently, when a patient with infection or wound comes into a trauma center, the hospital cultures a sample and sends it to a lab, which takes about 72 hours to identify it," Shane Woody explains. "Technologies such as PCR take less than an hour, but these tests cost hundreds of dollars. We're working on a method for performing the same test in less than a minute, for under $20 per test. Our approach involves chemically attaching an antibody to a sensor, inserting this into the sample, and looking for cells that attach to it. Our first target is a test for staph infection, which is a very large problem in hospitals and nursing homes. This can also be used to diagnose MRSA and other serious infections. We're collaborating with the Carolina Medical Center to develop this technology."

6.7 Nanocomp: Taking Nanotubes to Jupiter and Beyond

Nanocomp is a good example of a nanoventure that focused on commercializing the new forms of nanocarbon. Nanocomp Technologies, Inc. was the first company in the world to master the ability to reliably produce large industrial-size sheets and yarn made with carbon nanotubes. Their production was sold out years in advance and the demand for their sheets was so high that they were able to pay off their investment in their production system in only 3 months. In 2012, the company outgrew its last start-up-sized facility in Concord, New Hampshire and moved into a 100 000 ft^2 manufacturing plant in Merrimack, New Hampshire.

Carbon nanotubes were one of the first nanomaterials. When they were discovered, they provided an intriguing chemical "surprise" that opened the door to a wide range of innovations using one of the most common materials on the planet. In an era when rare earth materials are predicted to become increasingly expensive, scarce, and eventually unavailable, keen attention has been focused on carbon nanotubes and various types of composite materials. While many articles and media reports have extolled the virtues of carbon nanotubes and their inherent strength (lightweight, stronger than aluminum and steel, etc.), the industry faced a very basic challenge: How to turn carbon nanotubes into sheets of material that could be used in commercial products?

Nanocomp Technologies was formed in June, 2004 in West Lebanon, New Hampshire as a spinout of Synergy Innovations, Inc., a technology development company. The venture was initially funded by the founders, along with contracts from the Office of Naval Research, the US Army, and later a substantial SBIR grant from the US Air Force. Partners include some of the nation's largest defense/aerospace companies. A nonexclusive license was obtained to the single-wall carbon nanotube composition of matter patent from IBM. The company has filed >20 patents.

The company's core technology involves the design and production of materials and products that use "long" carbon nanotubes (as long as 1 mm each) that have different properties than "short" nanotubes, which are typically tens of micrometers long. Short nanotubes are typically found in powdered form and have limited industrial uses because of limited macro properties and they are difficult to incorporate into manufacturing processes. Nanocomp's long nanotubes can be produced in sheets in large volumes and exhibit high-performance characteristics needed for manufacturing and commercial use. Nanocomp mostly uses multiwalled nanotubes, although single-walled nanotubes can also be used. Most of their materials are formed with 100% nanotubes.

A decade ago, carbon nanotubes were too expensive to use in many products, partly because the purification process was very costly. Nanocomp developed techniques for growing and producing very pure materials that did not require expensive postgrowth purification. They also developed proprietary methods for fabricating these long nanotubes into structurally strong and electrothermally conductive fibers, yarns, and sheets. Applications include thermal straps and blankets, shielding "skins" and high-strength sheets, and yarns. Yarns are used as core conductors for cables, and in many other applications.

Nanocomp was founded in 2004 by Peter Antoinette, President and CEO, along with its former CTO David Lashmore, Ph.D., who has since returned to academia. They were intrigued by the challenge of finding a way to first produce long carbon nanotubes with compelling macro properties and then turning them into sheets and spun yarns that could be used in commercial products. The challenges of turning carbon nanotubes into useful macro products were daunting.

Peter set the tone for the company by declaring, "This is not a science experiment. This is a company." That sense of pragmatism was essential to the company's development. Until Nanocomp developed its processes for creating nanosheets and yarns, similar form factors had only been achieved by combining nanotubes with a thermoplastic material (polymer), dissolving it in a solvent and casting it in a mold where the solvent was made to evaporate, to form the sheet. "That's fine for a science project," Peter said, "but this won't be effective if we want to make thousands of square feet for industrial use."

The use of a catalyst to "grow" carbon nanotubes was an early trade secret used by the first wave of ventures that produced and used carbon nanotubes. The breakthrough innovation that turned Nanocomp's research from a "science project" to a commercial process was the discovery of a way to stabilize catalyst particles at a desired size and hold it stable to allow the nanotube to grow to millimeter length. This is not as simple as it sounds. Computers control 30 different parameters from temperature levels and gradients to gas flow rates and chemical mix. This allows their engineers to "dial in" settings that create single- or multiwalled nanotubes in the configurations they need.

In 2007, Dr. Mark Banash joined the company as a Vice President and serves as Nanocomp's Chief Scientist. Mark is a Zyvex alumnus who previously served as director of production and quality for nanomaterial products. John Dorr, Vice President, Business Development, and Dave Gallus, Vice President of Engineering, round out the management team.

Mark Banash described some of the engineering challenges they faced: "Carbon nanotubes have a lot of hydrophobic surface area. In bulk form they exist as powders. Short nanotubes are really tough to get into a composite because of the way they aggregate. There are safety issues. They are hard to produce and even harder to incorporate and disperse in a composite material. Making something that can be used in a commercial application – like a sheet of material – requires billions and trillions of nanotubes."

To turn nanotubes into nanosheets, he explained, the nanotubes are synthesized as a cylindrical mesh that is then collapsed onto a metal belt and the sheet is built up layer by layer. The size is only limited by the dimensions of the belt. The longest roll of sheet material made by the company so far is >400′ × 52″ wide. Some sheets sold to industrial customers are produced in a 4′ × 8′ size, which is the standard size of a sheet of plywood, others are cut into custom sizes or even seamed to form tapes. Nanocomp also produces yarns, which are typically shipped in kilometer long spools.

Dr. Banash calls the uses for Nanocomp's products not as "killer apps" but rather "assassin apps."

"This is an exciting time to be working in nanoinnovation," he said. "We are still early enough in the evolution to see some firsts. There are records to be broken. One of the greatest challenges of the last decade was to figure out how to create industrial size sheets incorporating nanotubes or graphene. In 2008, we produced the world's largest sheet made with nanotubes. This clearly demonstrated our ability to manufacture commercial quantities for industrial users. The first nanosheets took several shifts to produce. That time has been reduced to hours with plans to go to minutes over the next two years."

The company has also climbed a very steep learning curve along with its customers. Mark recalls one customer telling him once, "This is too squishy," which he then had to translate into the functionality represented by "squishy." It turned out that squishy meant that when the object was placed in lateral shear, it did not recover its original form fast enough.

Nanocomp has also done extensive testing including air sampling, thus ensuring the safety of its products. Mark indicated they have "shredded, punctured, twisted, pulled, snapped, and ground up their materials and tested the results with air sampling instruments to test for contamination."

Nanocomp is forging ahead with a broad range of exciting applications for its nanosheets. The company is even taking its technology into space. In 2010, Nanocomp's nanotube-based sheet material (Emshield™) was incorporated in the Juno spacecraft, launched on August 5, 2011, to provide protection against electrostatic discharge (ESD) as the spacecraft makes its way through space to Jupiter. Specifically, the Juno development team used Nanocomp material as an ESD protective surface layer on several critical components of the flight system's attitude control motor struts and the main engine housing. Nanocomp worked in partnership with Lockheed Martin, the prime contractor on the project, to integrate Emshield during spacecraft development and construction.

"The Juno spacecraft has many key components throughout the spacecraft that require ESD protection especially as it will be travelling through Jupiter's extremely strong radiation belts," Peter Antoinette explained. "Lockheed was interested in implementing an alternative ESD solution to traditional aluminum foil that is typically bonded to the surface of composites. By adding Emshield

CNT sheet layers during fabrication of the composite, they were able to integrate ESD protection directly onto the structure, making the composite a multifunctional element of the spacecraft. Becoming 'space qualified' against the rigorous standards set by NASA in support of a very important space mission is a major accomplishment for Nanocomp, and another example of the progress being made in the field of nanoinnovation.

A New Application for CNT Sheets

Carbon nanotube sheets have proven applications such as EDS and EMI shielding in data and power cables, according to Peter Antoinette. Data cables of various types have already been developed that are half the weight of their copper-based counterparts. He predicts that carbon nanotubes will produce improvements in the electrical conductivity of low-frequency higher power cables, which today stands at only $\sim 3 \times 10^6$ S/m. Applications for CNTs in battery anodes have been limited to constraining the deformation due to intercalation of lithium in graphite powder, he explains, but so far, CNT sheets have not been used as the current collectors themselves. The advantages of replacing copper and aluminum foils in these batteries seem to suggest a very large commercial market for this material.

"We predict that a significant increase energy density can potentially be achieved while eliminating thermal runaway due to the CNT positive temperature coefficient of thermal conductivity," Dr. Lashmore elaborated. "Non-metallic CNT sheets – comprised of millimeter length CNTs – provide enhanced surface area and properties that manage thermal energy better than metals. For the first time, dissimilar metal current collectors such as copper and aluminum need not be used, saving weight eliminating corrosion couples, and enabling the use of more aggressive and superior high conductive electrolytes. Coating these CNT current anode collectors with silicon or certain metal alloys can potentially result in a high capacity electrode with a comparative reduction of electrode resistance; therefore the Joule heating associated with high current performance is diminished.[1)]"

Carbon Nanotubes and Ballistic Protection

Early work was supported by the Natick Soldier Center and the Office of Naval Research, and duplicated by several prime contractors, confirming that Nanocomp layered CNT sheets can stop bullets.

Near-term benefits have come by combining CNT sheets with aramids and HDPE resulting in dramatic improvement in the overall performance of soft armor systems. The leading manufacturer of soft ballistics has seen a 30% reduction in weight with >20% reduction in thickness, thus utilizing Nanocomp sheets in soft armor. AR500, a commercial supplier of ballistics protection based in

1) This concept is "patent applied for" by Nanocomp and was developed originally under Nanocomp funding.

Phoenix, Arizona, has introduced commercial vests incorporating Nanocomp sheets to enhance performance.

Whereas soft armor protects from handgun rounds and fragments, ceramic hard armor protects soldiers from rifle fire, even those that can pierce metal armor. This level of protection comes at a cost; ceramic vests are heavy, weighing as much as 7 lb each. Nanocomp's sheets used as a surface wrap can reduce this weight by >10%, while improving fragment and multihit resistance. CNT hybrid armor is also showing promise for use in helicopters, boats, and aircraft where it can improve protection over current systems without adding appreciable weight, and reduce fuel costs. In 2012, DuPont Corporation entered into a formal strategic relationship with Nanocomp to further develop ballistics and honeycomb core structures.

6.8 Nanosys: Resurrection and Redemption

Most of the nanoventures in the "first wave" had to reinvent themselves once or twice before they started hitting their commercialization targets. A great example of this is Nanosys, which was launched in Palo Alto, California in 2001 by Larry Bock, Drs. Charles Lieber, Peidong Yang, and Paul Alivisatos.

Nanosys was founded on the premise that strong IP in the emerging nanotechnology field would be the key to building a successful company. Nanosys successfully secured exclusive IP licenses in foundational areas from many leading research institutions, including Harvard, UC Berkeley, UCLA, MIT, and CalTech. Based on this, and the promise of nanotechnology, Nanosys developed a reputation quite early as a high-flying start-up. The company received venture funding from such prestigious VCs as Venrock Associates, Arch Venture Partners, Harris & Harris Group, and Lux Capital.

In April 2004, the company announced that it was going public with an IPO, with the normal amount of media and investor attention that an IPO engenders. However, without any products or customers, the appetite for such an IPO was poor and less than 4 months later, the company announced that they were withdrawing their filing and opted instead for a $40 million private equity financing round. This increased the number of venture capitalists and corporations that had invested a total of ~$100 million in the company to 17. Four years later, Nanosys still did not have a commercial product, and was not profitable. Many in the industry had all but written off the company.

Nanosys is not an unusual case. During the early- to mid-2000s, many nanotechnology companies showed great promise initially and developed impressive research and patent portfolios, but then ran out of funds, failed to scale up their research, or could not commercialize their technology. Some nanotech ventures were simply moving too slow to stay in business and had to cease operations, drastically scale back, or find a stronger company to acquire them.

By 2008, Nanosys had worked on developing its technology in many different fields, ranging from medical devices and implants to printable electronics, advanced semiconductor materials, and even chemical–biological warfare suits. The company still didn't have a commercial product and seemed to be drifting as it struggled to find its identity. That year, a shakeup ousted CEO Calvin Chow and the company's major investors acted to save the company. Venrock partner Steve Goldby became chairman and in October, 2008, Jason Hartlove was recruited to transform Nanosys into a product company.

Hartlove is a Silicon Valley engineer who has been called the "MacGyver" of Silicon Valley. During his 18 year career at Hewlett-Packard and its spinoff Agilent Technology, Hartlove piloted many different technologies out of HP Labs and into products on the marketplace, including optical position sensing (think of the optical mouse), CMOS image sensors (think of camera phones), and others. When he left Agilent, he was a VP and GM responsible for over $500 million in annual revenues. He then moved to South Korea and managed MagnaChip Semiconductor and in 2008 was recruited to become CEO at Nanosys.

"Walking into Nanosys for the first time in 2008 felt a lot like walking into HP Labs," Jason recalls. "The company had great science and technology behind it, a strong team of deeply talented people, but no sense of where the commercial markets were or what their needs were. The *modus operandi* of the company had become to stay alive by pursuing funded research programs; whether or not there was any practical or commercial market or business case for doing so did not matter, as long as the partner with whom the company was engaged was paying for the work."

"Unfortunately," Hartlove continues, "this did not ultimately lead to a recurring revenue stream. Although there were some very well recognized marquee names that Nanosys partnered with, they were always working with the equivalent of a central research lab within those companies, on technologies that were not on the near term product roadmap of any of the businesses within those companies. So when I joined, I began to work with the product divisions of major technology companies, to look for how our technology could solve relevant problems that they had today or that they would have in the near future. From this dialogue, I was able to identify a few opportunities which were high-growth, which were real and happening now, and for which we had a compelling value proposition with a material that we could actually manufacture."

"But that wasn't all. Another key realization was that we had to do the work of nano-material integration as well. While our customers wanted new materials such as high efficiency narrow band emission phosphors, they also did not want to develop new manufacturing processes for integrating them into their products, for doing all of the reliability and quality engineering to optimize them for high volume usage and so on. So Nanosys began focusing not just on doing great materials science but on designing and building ready-to-use component products that manufacturers could deploy easily. We learned a lot about manufacturing things such as LCD display backlights and folded that knowledge into our operations so we could make a product that drops directly into standard manufacturing processes already in use by electronics makers."

With a new commercial market focus and with new management, things began to change. The company began to rise from what looked like a funeral pyre, and started developing products and solutions for major electronics companies.

Within 2 years, the company was back on track and pushing hard into the market with its first commercial products. One of these products is a lithium ion battery anode material called SiNANOde^{TM}, which increases the energy density of batteries, allowing more charge to be stored in smaller batteries, using existing manufacturing processes. The second commercial product is the Quantum Dot Enhancement Film, or QDEF^{TM}, which improves the brightness and efficiency of LED backlights for media-rich devices like tablets, notebook computers, and HD television screens and became the primary focus for the Nanosys display business in 2012.

The first customers for this technology were LG Innotek and Samsung. These sales helped Nanosys achieve its first break-even month (in mid-2010). Samsung, the South Korean electronics giant, has become a major technology partner and investor in the company.

In 2010, Jason Hartlove was able to negotiate a Series E venture round, raising ~$30 million, of which $15 million was provided by Samsung Venture Investment Corporation. Nanosys also negotiated a multimillion dollar licensing agreement with Samsung for use of its nanoengineered structures, what they call "architected materials," in thin-film photovoltaic panels and LCD displays. As Hartlove said, "these are areas which, due to the size of investment required, we cannot fully exploit but which can be brought to commercial products by someone with significant financial and manufacturing scale."

At the end of 2013, Nanosys spun off its SiNANOde energy storage technology into a new company called OneD Material. The split has enabled the two companies to focus on developing technologies in different stages of development for distinctly different end markets. OneD Material continues to refine and scale up the company's SiNANOde materials for consumer electronics and electric car batteries. While Nanosys is now completely focused on scaling up quantum dot manufacturing to meet increasing demand from display customers, it continues to develop a new generation of light-emitting materials.

Today, the company has a staff of more than 120 employees, its patent portfolio includes more than 200 granted and pending patents for light-emitting materials and systems. In 2013, Nanosys achieved a number of significant milestones, including shipping QDEF in their first commercial product—the Kindle Fire HDX 7—opening a new, expanded 60 000 ft^2 manufacturing facility in Milpitas, California and their first ever full year of profitability. With a number of products in the pipeline for 2014, the company is profitable and growing, and continuing to fulfill its destiny as an advanced materials architect.

Paul Alivisatos, cofounder of Nanosys, recalls that Larry Bock once told him that the three most important things in a start-up are the CEO, the CEO, and the CEO. "You could have an absolutely great technology, but that doesn't necessarily mean the company is going to thrive. You could have a lousy science base and a good CEO will find a stronger science base needed to make products. You can put in all kinds of rules about what you want to happen and a lot of that gets thrown out the window when you finally get around to doing it. The people are really important. You have to have the right 'oomf' to push something along."

"It's very important for the CEO to have a deep appreciation for the science, and not a casual one. Almost invariably the inventing scientists are not the best CEOs for these ventures. We need a whole generation of people living in both worlds, science and business. I see a lot of students in physical or electrical engineering programs coming over to the business school to take a few classes, whether their advisor knows it or not. The need has outstripped the availability of people needed to push these ventures along in a business sort of way. There is a need for talent that combines science and management. Companies and universities have to work together to make this happen."

6.9
Graphene Frontiers: Commercializing Graphene

The first wave of nanoventures included several companies that successfully commercialized materials using carbon nanotubes. The next wave of nanoventures will also include some notable innovations that involve carbon, but instead of carbon nanotubes, the next decade's major challenge will be the commercialization of graphene.

Graphene Frontiers, LLC, a nanoventure launched in Fall 2011, is one of a handful of ventures pursuing this opportunity. In this case, the challenge is not just to create products using graphene. The real challenge starts earlier in the supply chain, since there still isn't a cost-efficient method for producing commercial sizes and quantities of graphene. The founders of Graphene Frontiers believe they have the solution.

Graphene Frontiers got its start at the University of Pennsylvania when directors of Penn's Center for Technology Transfer and its UPstart program brought together three talented entrepreneurs: A.T. Charlie Johnson, professor of physics and director of the Johnson Research Group studying experimental nanoscale physics; Zhengtang Luo, a postdoctoral fellow in Johnson's group; and Mike Patterson, a member of the Wharton Executive MBA Program.

Charlie Johnson has authored more than 130 peer-reviewed articles; he holds 2 patents with 18 other patents pending. Dr. Luo has more than a decade of experience in the synthesis of carbon nanomaterials and related product development. Mike Patterson is a proven entrepreneur and business builder with an undergraduate degree in physics.

Producing a sheet of graphene is much different from producing sheets made from carbon nanotubes, Professor Johnson explained. Graphene begins as a flat sheet of carbon atoms arranged in a honeycomb framework, while sheets made from nanotubes are made from millions of nanotubes formed into sheets.

Applications for graphene include replacing silicon in electronics, development of composite materials (especially those based on combining layers of different materials), energy storage (fuel cells), transparent electrodes, chemical and biological sensors, and applications that require high thermal conductivity to control heat, energy storage, and thermal management. Membranes made from graphene can be used as molecular water filters.

"We're still at the very beginning of the graphene revolution," Johnson explained. "Graphene production methods are being developed at several academic research centers, but these approaches use high vacuum chemical vapor deposition, require rare metals and expensive substrates, and provide limited production."

"At Graphene Frontiers, we're commercializing a method to produce large sheets of graphene on an industrial scale, using chemical vapor deposition under room temperature and pressure. The graphene we produce will be developed into a variety of commercial products and also sold to other manufacturers."

The first group of applications include a microelectronic grade conductive film loaded on a silicon wafer, a transparent conductive film loaded on transparent glass slides, and a support structure for samples in electron microscopy.

Mike Patterson described some of the business challenges the venture is facing (Figure 6.6). Mike is a 34-year old business builder with experience in consulting, international business, and manufacturing. He started his career as a physics major at Princeton University and was on the entrepreneurial team that built global supply management pioneer FreeMarkets Online during the dot.com era. After Freemarkets was acquired in 2004, Mike spent 6 years helping Bank of America grow its international operations. In 2010, he enrolled in the Wharton Executive MBA program, planning to be a Finance major but switched to Entrepreneurial Management.

"We believe we have a bold and exciting vision to develop the best company doing graphene research," he enthused. "We are in discussion with several large corporate partners. We're lucky to be based in Philadelphia because there is a lot of carbon know-how here. Pennsylvania companies have been making super thin graphite films for thermal management and other applications. The steel industry has been making graphite anodes for arc steel. These roots go back over 100 years. We are proceeding very thoughtfully to plan how we as a for-profit venture can work with academic and technology centers in the Philadelphia area. We believe our production technology can help Pennsylvania become a hub of a graphene innovation cluster."

"We're also looking very closely at the best ways to fund the venture, and how to position our finance message. We need to have a pragmatic message for VCs. If I tell a venture capitalist that we want to build an organization of 50 Ph.Ds that will create a $100 or $200 million business in 10 years, the VC may say that's too long. If I say we want to build a $5 million business around a graphene niche product, the VC may say that's too small."

Figure 6.6 Can you see the graphene? You will have to look really close to see the 5″ square sheet of graphene being held by Mike Patterson, the CEO of Graphene Frontiers. This sample includes a few layers of graphene applied to one side of a much thicker sheet of polyethylene terephthalate (PET), which is a safe plastic used for beverage containers. As you can see, it is fully transparent, even with several layers (photo by Michael Tomczyk).

"As a new venture, we need to focus and define our goals but we also need to leave some options open. For example, we decided not to focus on touchscreens, which is a very promising market for graphene, but we decided to swallow hard and say, 'We're not going after that market right now.' We want to build an organization that includes entrepreneurial scientists, a core of people who know we've got something special and want to capture value from it, and who are willing to take the risks involved to get us there."

Graphene Frontiers is a terrific example of a nanoventure that is moving into a wide-open market where a lot of applications are still largely undefined. To cross the chasm to this market, the venture team has to negotiate some very tricky technical and financial hurdles. One thing is sure. They have no shortage of talent or enthusiasm.

It would seem that manufacturing solutions such as the graphene production, technology being commercialized by Graphene Frontiers, suggests that "smart production" goes hand in hand with "smart products" and this is a challenge that will continue to face all nanoinnovators.

The same properties that make nanomaterials so promising also complicate the commercialization process. Companies like Zyvex, Nantero, Nanosys, and even Intel took years to figure out how to stabilize carbon nanotubes, nanocomposites, and nanoscale architectures so they can be turned into something with uniform consistency and reliable quality. Thus, it's clear that it isn't enough to simply invent a carbon nanotube boat or a chip that uses 3D architectures or a nanocatalyst or nanosized drug molecule, or a new way of designing a battery or solar cell. You have to also invent the production method, and often the production method is more complex than the original nanoinnovation.

The examples included in these company profiles are representative of the many strategies and different types of challenges faced by the first wave of nanoventures. They offer a variety of lessons and insights, and provide just a glimpse of the commercialization process for nanoinnovations. The good news is that current and future nanoinnovators are able to draw on the lessons of this "first wave" to build the next wave of ventures.

6.10
Carbon Nanotechnologies: Early Promise, Not Fulfilled

Since this chapter draws lessons from the first wave of nanoventures, it is appropriate to mention at least one venture that did not work out as planned – noting that there were many of these. The most famous example of a venture that "lost its way" is Carbon Nanotechnologies, Inc. (CNI). CNI was a nanoventure cofounded in 2000 in Houston, Texas by Nobel Prize laureate Richard Smalley. The venture was based on Dr. Smalley's research, which was licensed exclusively to CNI by Rice University. Professor Smalley's death in 2005 drew some energy from the venture, but the real problem was the inability to commercialize its carbon nanotube technologies. In April 2007, CNI was acquired by Unidym, a subsidiary of Arrowhead Research Corporation.

Daniel Colbert – who was a Rice University chemistry professor and a cofounder of CNI – recalls what happened: "The scale-up problems we had at CNI resulted in the company being acquired, basically at cost. The problem was that we ran out of funds before we could scale up our technology. The scale-up problems we had were a great lesson for me because as I later went into venture investing, it made me personally very sensitive to scale-up risk, and served me very well in a number of investments that I looked at."

References

1 O'Reilly, C., III and Tushman, M.L. (2004) The ambidextrous organization. *Harvard Business Review*, April.

2 Göran, L. *et. al.* (2010) Considerations on a Definition of Nanomaterial for Regulatory Purposes. *JRC Reference Reports*. Page 7.

7
Implementing Your Nanoinnovation Strategy

After decades of nanotech being just around the corner, we've finally reached the corner.

– *Peter Balbus, Managing Director, Pragmaxis*

At the beginning of this book, I pointed out that all of us are touched in some way by nanoinnovation. You may find yourself involved in a nanoinnovation project and you have to contribute ideas, or manage all or part of a nanoinnovation project. You may be involved in planning your organization's nanotechnology strategy. You may have an opportunity to champion a project and contribute ideas and support. As a manager, how do you operationalize nanoinnovation? What are the best practices and strategies? What questions do you need to ask? Is there a menu?

It's easy to say you need to be nanodextrous and deal with the various dichotomies – nano versus macro, research versus commercialization, incumbent versus emerging technologies and applications – but how do you actually *do* this? Where do you begin?

7.1
A Sense-Making Framework for Nanoinnovators

At the Wharton School where I spent much of my career managing Wharton's innovation research center (The Mack Institute for Innovation Management), I learned to think of problem solving in the context of "sense-making frameworks."

The world's simplest (and most elegant) innovation framework I've seen was developed by Larry Huston, Procter & Gamble's longtime innovation guru. I first saw this in 1996 at a Wharton workshop I was helping to organize. Larry asked me to cover the entire front wall of the conference room with flipchart paper. At the workshop, he stood on a stepladder and attached the elements of the framework with examples from Procter & Gamble's product portfolio, which was a dramatic way to present this very simple yet powerful concept. As an innovator, Larry has always been creative and full of surprises.

NanoInnovation: What Every Manager Needs to Know, First Edition. Michael Tomczyk.

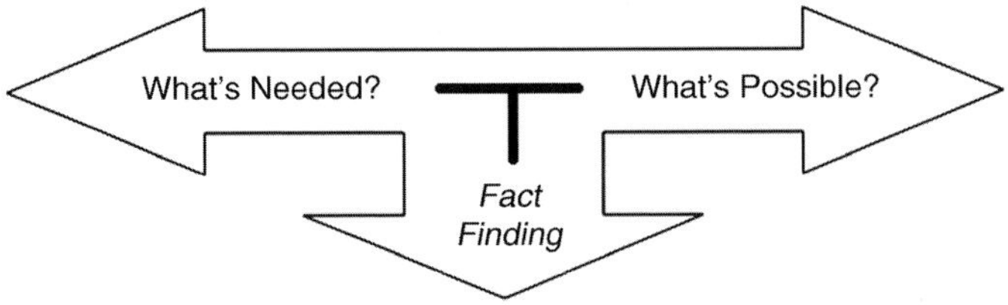

Figure 7.1 Innovation framework. This diagram, developed by Larry Huston at Procter & Gamble in the mid-1990s, is the world's simplest innovation framework. It offers a useful starting point for conceptualizing and describing any type of innovation, including bionanoinnovations.

This powerful concept applies to any area of innovation, including nanotechnology. Using this framework is remarkably simple. You start by asking *What's needed?* – essentially, define the problem. Next, you ask *What's possible?* – identify the technologies and applications that exist today as well as those that need to be developed. Then you conduct *fact finding* in between to find specific resources needed to connect the needs and possibilities (Figure 7.1).

The "needs and possibilities" framework is a conceptual compass you can use to find your way into uncharted technological waters. It can be used to develop any type of innovation. You can use this framework to develop a technology road map, set goals and priorities for corporate and government research initiatives, frame an open innovation challenge, or set the parameters for a technology competition. If you look beyond your organization, across boundaries, and delve deep into other industries and disciplines, you'll be surprised how many solutions and killer apps will be revealed. Often, a problem in one industry can be solved by a solution that already exists in another industry. In nanotechnology, a quantum property of a well-known material such as carbon or gold may provide a solution that was never possible at the macroscale.

7.2
10 Strategic Questions that Nanoinnovators Need to Ask

Most innovations start with a question, or a set of questions. Terry Fadem makes this point in his excellent book, *The Art of Asking: Ask Better Questions, Get Better Answers* (FT Press, 2008). Framing the questions early in any nanoinnovation project – and keeping these questions in mind throughout the process – helps you keep focused on what's important, especially as the answers keep changing. Based on my interviews with nano-insiders who have been working on nanoinnovations for up to 30 years, I can tell you that the answers keep changing, but the questions remain more or less the same.

Here are 10 questions you might consider asking before getting involved in any nanoinnovation project. These apply whether you're planning, developing, managing, approving, supporting, funding, purchasing, or deploying a nanoinnovation. These include some personal observations as well as insights gleaned from the research and interviews conducted for this book.

7.2.1 What's the Value Proposition?

A great starting point for any technology project is to describe it in one sentence. Almost anything can be described in one sentence. Try it. For example, a biopharma project might be described as "nano-sizing our large molecule drug to extend the patent." Graphene Frontiers (described in the previous chapter) is pursuing an innovation that might be described as "development of a cost-effective process to manufacture commercial sizes and quantities of graphene." And here is a nanoinnovation that I personally would like to see: "development of a low-cost scanning probe microscope system that any high school or college can afford."

You should also ask: How does this nanoinnovation improve on existing solutions? This is critical. At this point in the history of nanotechnology, most innovations aren't totally replacing or cannibalizing something that already exists. In most cases, these innovations *improve* something. So what's being improved? Is it functionality, form factor, economics (cost/price), materials, efficacy, strength, durability, weight, and so on? What does this innovation allow us to do now that couldn't be done previously? Is the market willing to adopt this solution? Are the existing solutions acceptable to customers? Are there switching costs? Will this require special handling or infrastructure changes? If the solution is truly radical and disruptive, will you have to develop the market by yourself – which could be very expensive – or is there a smooth path to adoption by customers?

When considering the value proposition, keep in mind that modern product management is holistic. For any product, you need to consider the full product cycle from womb to tomb. This includes by-products that are created during production, energy use, and what happens when it breaks or reaches the end of its useable life span. You also need to make decisions that make things greener, safer, and more cost-efficient – what IBM calls "smart planet" choices.

In a broader sense, you should recognize that the concept of "value" has evolved in the past few decades. Marketing professionals used to talk about the "5 P's" – product, price, place, promotion, and people – but these metrics have given way to a more modern interpretation that is more customer-focused, called "SIVAC" – where we consider *solutions* instead of products, *value* instead of price, *access* instead of place, *information* instead of promotion, and *community* instead of people. In a book chapter I wrote in 2011 [1], I discussed this framework in the context of the marketing challenges facing bionanotechnology and added "community" to the framework that originally included only the first four P's. Last but not least, the value proposition needs to include metrics for measuring success. What defines success in your nanoinnovation project? Is it bringing the technology to market in an acceptable time frame? Achieving a financial target? Are you trying to move from laboratory to market (and get there first)? Sometimes the value proposition involves finding smart ways to survive until the technology or market is ready.

These are a few of the strategic considerations around the concept of "value."

7.2.2
Where Do We Fit in the Supply Chain?

Whether you're in nanomaterials, nano-enabled electronic devices, or nanomedicine, you need to consider the supply chain. For example, if you're doing something that involves nanomaterials, who will manufacture the materials? Will you take on this burden yourself or work with a raw materials supplier?

In the 1990s and early 2000s, most manufacturers did not have the capability to integrate carbon nanotubes and composite materials into their products. Getting consistency from one supplier to another was tricky, especially in composites where the nanotubes needed to be uniformly dispersed. Also, there were safety issues involving the handling of nanotubes that might have required clean rooms and special facilities to minimize contact or inhalation of particles by workers.

The solution for carbon nanotube manufacturers was to move farther down the supply chain and do the integrating and fabricating that their industrial customers would normally do. This gave rise to an entirely new market segment known as "intermediaries." Today, companies that want to incorporate nanotubes or composites can ask suppliers to fabricate the nanomaterials into intermediary products such as chemical coatings, composites, and catalysts. This helped many companies become nanodextrous without having to develop an expensive in-house nanotech production capability.

A good example is the German company Nanogate AG, based in Göttelborn (Saarland), Germany. Nanogate is an integrated systems provider that specializes in incorporating nanomaterials in surfaces and layers – an important intermediary market that bridges the gap between raw materials and end use products. Nanogate provides a variety of nanocomposites and nanoformulations used in nonstick, anticorrosive, and ultralow friction materials and surfaces with a focus on high-performance, optical-quality surfaces. The company's products are used in a wide variety of applications from aircraft windows and utility vehicles to heat exchangers and enhanced headlight lenses.

While "nano-surface chemistry" sounds like a very narrow niche market, it's actually an enormous market since almost any surface can be enhanced by nanocoatings: metals, plastics, wood, stone, ceramics, leather, fabrics, glass, and so on. A material as common and familiar as glass can be coated with nanomaterials to provide coatings that can absorb heat or transport electric impulses without affecting the transparency of the glass. Nanocoatings can control the "surface energy" of a material to attract or repel liquids, oils, or impurities. Nanogate's Nanoplating® technology integrates nanoparticles into custom-made surfaces to provide a barrier against the sun's damaging rays, corrosion, bacteria, and so on. Nanocomposites and nanoformulations are used in nonstick, anticorrosive, and ultralow friction materials and surfaces, including thin materials where optical quality needs to be maintained.

Often, studying the supply chain will reveal a missing link that needs to be filled and this can create a business opportunity. In the field of bionanotechnology, which is explored in the next chapter, the supply chain is constantly evolving as new and better drugs and devices are introduced to treat major diseases. Introducing a smaller pacemaker at Medtronic or a more efficient drug from Merck or Pfizer uses different delivery mechanisms and facilities than administering stem cells, protein therapies, or gene therapy vectors. Who in the supply chain will administer an injection that delivers genes that cure genetic blindness or hemophilia, or stem cells that regrow damaged tissues and organs? Will this be done at any hospital or clinic, at specialized facilities, or at Wal-Mart?

A critical way to think about your project or venture is to understand where you fit in an "innovation ecosystem" – the network of partners, competitors, customers, markets, and applications that are associated with your technology. Where do you fit in the ecosystem? Are you working in a space that is dominated by a large corporation such as Reliance Industries in India, Siemens in Europe, or Samsung in Asia? Will these companies partner with you, or compete with you? How would you describe your niche? Is the barrier to entry by competitors low or high? What is the market opportunity? Where is the low-hanging fruit where the first and easiest revenues will come from? As you answer these questions, be realistic, be conservative, and try to develop a business model that can survive and succeed on the lowest possible market share and slowest pace of adoption.

7.2.3
Can This Be Funded with "Patient Money?"

Virtually all nanoinnovations take longer and require more funding than originally anticipated. This requires access to patient money. In a large organization, you need to lock in (1) the commitment of top management, (2) assurances that your budget won't be misdirected next year to a higher priority short-term project, and (3) the involvement of corporate, government, or academic partners to spread the risk. You need a commitment that your funding will be secure. This means that the budget won't suddenly disappear in 2, 3, or 5 years, or get redirected to a higher priority project.

In a small start-up or venture, you need to ensure the investors are prepared for a long commercialization cycle and reasonably high development and marketing costs. They need to be willing to stay in the course. It is also likely that successive rounds of mezzanine funding will be needed. Your investors may need to accept some dilution or issuance of additional ownership shares to pull in the extra funding. A variation on this theme is the involvement of a corporate partner that can be a good solution, since corporate partners can provide a combination of financial, research resources, and market access, and often they will accept a smaller equity stake because their main goal typically involves access to the technology solutions being developed.

Stable funding may also require a tolerance for failure. You don't want your funding sponsors to pull the plug at the first sign of failure, or even after several failures. In many sectors, including nanotechnology, failing is part of the equation. Most business schools today teach that it's important to "fail fast" to conserve financial resources, although this isn't always possible in nanoinnovation. Some of the most important innovations took 30 years or more – and lots of failure – to achieve. DNA nanotechnology and graphene are two notable examples.

7.2.4
Can the "Lab Solution" Be Translated into a "Commercial Solution?"

As part of your nanoinnovation strategy, you need to describe how you intend to translate the research into a commercial application. Filing a patent is one thing. Commercializing the patent is quite different. Achieving something in a Petri dish – *ex vivo* – is different from getting the same result inside the human body – *in vivo*. Getting a few atoms or molecules to behave a certain way at the nanoscale is much different from retaining those properties when you scale up to a macrosize where those properties are changed or lost.

If you scan the research literature, you'll find hundreds of articles where nanoscientists report "breakthroughs" such as better cheaper solar cells using nanomaterials, or a new way to detect and destroy cancer cells. But have you seen a low-cost flexible solar cell roof you can buy for your house? Have you seen any cancer cures based on delivering drugs inside a nanoshell? In most cases, these solutions are still in the laboratory. They have not yet found a viable commercial path to make them available in the market.

Also, many scientific achievements are really cool, but do not have a "killer application" yet. For example, scientists have trapped a single molecule of water inside a carbon buckyball. That's very cool, right? But how do you trap water in a billion buckyballs – or a billion times a billion, which would be about $1\,m^2$ in size? And once you learn how to do that, what's the killer app?

Technically, these examples are discoveries and inventions, but not yet nanoinnovations under our definition, which is the *implementation* of a nanoscale discovery, idea, or invention.

One of the most interesting and ironic challenges for nanoinnovators is the realization that once you've demonstrated a solution using a small number of atoms or molecules, you need to figure out how to replicate this solution with six or seven orders of magnitude – which often means working with millions or billions of nanoparticles. Scaling up a solution from nanoscale to macroscale or from laboratory to market is a critical factor for any nanoinnovation.

A sheet of material made with nanoparticles that are each 1 nm in size means that a billion or more particles are needed to manufacture that sheet. How do you fabricate something like that? How do you get all the molecules to line up properly? How do you keep the particles from clumping or clustering so you get a uniform dispersion? How do you provide quality control? How many sections of the sheet do you need to scan with a scanning probe microscope to ensure the sheet is uniform (which is both a statistical and technical challenge for quality control technicians)? These are a few of the many important scale issues.

Another scale question involves the properties involved. How do you maintain the properties that worked in a few molecules under a microscope in a laboratory, in a commercial product that attempts to retain those same properties in bulk form? What are the limits and boundaries at different scales? And most importantly, will these limitations delay – temporarily, or perhaps forever – the ability to commercialize the innovation?

Stuart Cantrell, chief editor of Nature Chemistry describes the dilemma like this: "In a process that uses gold in a bulk application, the difference between two million or three million atoms doesn't change the properties. All of the atoms are overlapping and you have an electronic continuum. But when you get to a few atoms, you get discreet states and different properties. Going from 5 to 10 atoms may be a big change. This may not seem like much of a change but effectively changing from 5 to 10 atoms – or from 5 to 10 nanometers – is actually doubling the size and this small change can create a different effect. We're still learning how moving among these lengths and sizes in the nanoscale world can change the properties, to achieve or inhibit a desired effect." The trick is finding the right path and making it happen.

7.2.5 Is the Intellectual Property Secured?

One of the challenges facing anyone in the field of technological innovation is the rapid creation of "patent pyramids" that lock up gateway patents and require you to license the IP for your research. In the field of nanotechnology, many key patents are made available free by universities for academic research and require licenses for commercial applications. Many (if not most) nanoventures began with one or more patents licensed from a university, which then helped the ventures build their own patent portfolios.

A word of caution: A search of the "977" patent category in the USPTO database may still overlook some seminal patents that were filed before the standard was established in 2004, and may also miss some patents that apply to nanotechnologies but do not mention "nano" in the filing, which is especially true for many bioscience patents.

Companies such as IBM have turned their patents into a profit center, by rigorously enforcing claims and requesting technology companies to license IBM patents if they haven't already done so. It should be noted that in some cases, a threatening letter from a company like IBM may make it sound like you're violating one of their patents, but this may in fact not be the case, so don't accept any threatening letter regarding patents on face value. Good patent attorneys can do a lot and are worth their weight in gold. It is not surprising that patent attorneys have been cofounders and are listed on the senior management team at several nanotech ventures.

7.2.6
Do We Have the Right Partners?

There are many reasons for partnering in nanotechnology. Partners share and minimize the risk in an R&D project and provide specialized knowledge, access to patented technologies, and talent. If your innovation involves the formulation and use of nanomaterials such as nanocomposites or metamaterials, you may want to partner with an intermediary supplier to ensure a reliable supply source and consistent quality.

Also, someone else may have already done some of the work. In biopharma, decision-makers often complain about how much duplication there is in drug exploration, which raises development costs and healthcare costs in general. Why does everyone need to work in parallel on the same drug targets? Antitrust considerations aside, it would seem that more collaborative consortia would be a smarter model than the parallel efforts model that exists today. The same may be true in nanotechnology, where so many enterprises are working to do something novel with carbon nanotubes and related materials. We are coming to a point where there should be more pooling of efforts to speed the commercialization process, especially in nanomedicine. This pooling of interests seems to be happening more in Europe than in the United States or Asia.

Often the best partner is a university nanoscience center. University partners provide a stable pool of faculty expertise as well as a constant stream of graduate and undergraduate students who may want to join your organization.

Nanoinnovation is a good sector for "co-opetition," where market competitors join forces to develop a new technology or market. Companies that compete in certain markets may pool their resources to develop hybrid products that occur at the convergence of two or more technologies – where the companies may each have an advantage. Some of the most disruptive and radical innovations in the past century have come from the convergence of seemingly unrelated technology streams.

For example, 10 years ago, who would think that carbon nanotubes would be used to build faster, lighter, stronger boats? Today, Zyvex Technologies is collaborating with boat designers and the US Navy to develop this entirely new product category. It was only a few decades ago that fiberglass replaced wooden boats. It's easy to imagine that in a few years, a Ph.D. student or team of MBAs may do a case study on how carbon nanotube boats completely revolutionized the boat industry!

7.2.7
Is This the Best Team for This Project?

Some people "get" innovation and some don't. Some people "get" nanotechnology, and others don't. The team that actually brings the innovation to market may include a combination of managers who are really well versed in innovation and "nano." Team members may also include people who are not very good innovators or nanotechnologists, but whose special skills and competencies are needed. Everybody on the team doesn't have to be a Steve Jobs clone. There will inevitably be a role for people who know their way around the company and can keep things flowing smoothly. An experienced financial manager is critical. You may need someone who is good at securing research grants from a government agency, or someone to impose budget discipline.

A nanoinnovation team needs balance, more than anything. There should be a mix of older and younger managers. Older managers have a type of insight and wisdom that comes only with experience. They have probably seen their share of failures and know the pitfalls that need to be avoided. On the other hand, younger people are typically closer to emerging technologies, less risk-averse, and more open to new ideas. Younger managers don't have preconceived notions about "what can't be done."

I was once involved in a "future technologies" exercise at a US intelligence agency. We held three meetings to narrow a list of 70 technologies to a smaller list of innovations that might have the most impact and needed to be on the agency's watch list. At the end of the third meeting, the leader of the project asked, "Is there anything we've overlooked that we should consider?" At this point, I raised my hand and said, "Would everyone in the room under the age of thirty please raise your hand?" No one responded. "What's wrong with this picture?" I asked. "Young people are closer to emerging technologies. They have a strong sense of what's coming that's new and different. They have fewer rules and are open to more possibilities. They need to be included." The very next day, the project team gathered up everyone in the building under the age of thirty, including interns and admins. The result was several major new findings that the older group had totally overlooked.

Every innovation team, in any field of technology, should have at least a few young people. Also, there are many occasions where enthusiasm trumps expertise. When I was guiding the launch of the first home computers at Commodore in the 1980s, I was only 32 years old and had very little experience in computing. I was an English major and journalist with an MBA, but I had an insatiable curiosity and fascination for personal computers. I knew where home computing needed to go, and I was a quick study and did a lot of on-the-job learning. Everyone on my product team was under the age of 30 and yet we did the product management and marketing for the Commodore VIC-20, which became the world's first million-seller microcomputer – and a half billion dollar product line for Commodore. My role in that innovation earned me a listing in Wikipedia and a lot of recognition on retrocomputing Web sites in the United States and Europe.

I guess the bottom line is that diversity is needed on any innovation team, especially one as young and open-ended as nanoinnovation. No one knows where this field is taking us but we all know the potential is vast and we have only scratched the surface. To figure out what we can achieve in this uncharted territory, we'll need enthusiastic, dedicated people with fresh ideas who are not bound by the old rules that governed the macro world.

7.2.8 Is Our Strategy Flexible?

In nanoinnovation, it's not unusual to begin a strategy aimed at one target, and along the way you discover another, easier target suddenly popping into view. There are so many discoveries remaining to be made and the science is still so new that it's not unusual to stumble onto something big while researching something else altogether. Remaining open to new possibilities in the technology as well as the marketplace can spell the difference between success, survival, and failure.

Don't leave opportunities lying by the side of the road simply because you're racing to the finish line. The finish line is often quite a long distance away and you need to stop and grab a few trophies along the way if you can – this provides revenues, gives some confidence to your investors, and boosts the morale of the team. Even a small revenue stream from a patent license or providing laboratory samples or nanobiocomponents to the research community can provide revenues. If your goal is to develop a diagnostic test using biomarkers that include a nanoscale component, an interim step may be to license biomarkers to laboratories for use in medical research by other companies, to generate a faster revenue stream.

Sometimes, *adjacent markets* open up that which you normally might not consider because you lack expertise in that market or industry or because it's not part of your core. Moving into adjacent areas can provide an initial revenue stream while the core research is underway, can demonstrate proof of concept for your technology, and can also be used to open markets for your other products and technologies. This strategy is especially important for smaller ventures, but works for large corporations as well.

Almost everything in nanotechnology is still relatively "new," so this sector is a fertile ground for developing adjacencies. The strategic point is to keep your eyes open for a low-hanging fruit that may suddenly appear as you develop a project or venture. These opportunities are not always obvious. Sometimes it seems that the timing is not right and, of course, there are extra risks involved. In this book, I've mentioned the importance of focusing on a core objective and not getting too diluted, which was a problem for many ventures in the early 2000s. However, pursuing well-chosen side opportunities can give you a strategic advantage and a revenue stream.

7.2.9 What Are the Obstacles?

It's very tempting to overlook the obstacles ahead when you begin an innovation project, especially when you're caught up in the moment and invigorated by the exciting prospects ahead. Take a moment to consider the impediments that will hinder or slow your efforts. These should be divided into internal and external obstacles. Technical obstacles can defeat your project or kill your venture if you can't unlock a technological gate. For more than a decade, gene therapy was stalled by an inability to find safe and effective vectors to deliver therapeutic genes. A major complication was the triggering of the body's immune system that reacted to the viruses used to deliver the genes, even when the viruses were rendered "safe." Gene therapy began to show results when scientists found ways to encapsulate or modify the viral vectors to avoid triggering an immune response.

One obstacle you can never anticipate is the human factor. For example, you may find that there are colleagues who resent your involvement in a very exciting project or who do not believe in nanotechnology and think this is a waste of resources, or who are working against you for whatever reason. You need to deal with this. This is a bit of a digression. However, I will mention the several ways I've used to deal with adversaries: (1) turn enemies into friends or "frenemies"– show them the value of the project and involve them in some way, (2) secure the support of top management and ensure your case is strong, which will diffuse lower level resistance in the organization, (3) ignore your adversaries completely, marginalize them, and take them out of the equation, or (4) confront them head-on and push past them. These are tactics that I've used successfully during my career.

7.2.10
Is It Safe?

Holistic innovation requires that you consider safety implications in all aspects of nanoinnovation projects, from safety in the research laboratory to safety in the marketplace, and especially, safety considerations related to recycling/disposal of products and materials.

The overwhelming majority of nano-enabled products and applications are considered safe and do not require special labeling, although this is expected to change. What exactly needs to be labeled is an issue that regulators in many countries have been wrestling with for more than a decade. The first labeling requirements are just beginning to be enacted. For example, cosmetics sold in the European Union need to include the word "nano" in parentheses on the product label, after any nano-ingredients. Many products use nanoparticles today. You should consider the possibility that the products you produce today could require labeling or even new types of testing, in the future.

Going forward, safety issues will play a role in determining the types and forms of materials that you use. There is some evidence that long nanotubes may behave like asbestos molecules that have a similar shape and that are extremely severe carcinogens. This suggests that if your material includes long nanotubes, you may need to ensure that those particles are securely locked in a resin or polymer that will prevent their release now or in the future or you may opt to use short nanotubes that may be safer.

A great deal of academic work has been done in this field and safety studies have not revealed "smoking guns" in nanotechnology. Currently, most nanomaterials appear to be safe in terms of health and exposure, although the impact on the environment is still being studied and questions have been raised about the impact of such materials as silver ions, which can be toxic to aquatic organisms if they find their way into lakes or streams. More details and examples are provided in Chapter 12.

7.3
Where to Learn About Nanoinnovation

So how do you become nanodextrous? How do you educate yourself about the latest developments so that you can make informed decisions? You can read books and articles, surf the Web, and take university courses. Here are a few ideas to help you keep going, after you finish reading this book.

The current generation of nanoinnovators is *nano self-informed.* It's not unusual to have to teach yourself about nanotechnology. Most nanoinnovators and business leaders had to teach themselves about nanotechnology when they got started. They had to inform and educate themselves by reading, studying, experimenting, talking with knowledgeable colleagues and experts, and simply plunging into the field.

The next generation will be *nano-infused.* It will take another decade or so, but eventually, everything you read in this book and more will become second nature to the next generation of students and managers, as nanotechnology moves into mainstream science education and more companies become nanodextrous.

If you prefer not to wait 10 or 20 years for that to happen, you'll need to find ways to learn about nanoinnovations, and to keep learning about them as they are developed. If you want to get involved in creating and managing nanoinnovation, you'll need to be proactive. Contact ventures, companies, academic leaders, and venture capitalists and arrange to meet them to find out how you can get involved, and where the opportunities might exist where you might participate.

An impressive amount of material can be gleaned from online search engines such as Google or Alibaba, and from "push" news services. Announcements concerning nanotechnology research, legal and patent issues, commercialization, safety, industry statistics, and so on can be seen on Web sites such as Nanowerk .com, AZOano.com, nano.gov, nanotechweb.org, nnin.org, nano.org.uk, and nanopaprika.eu; on the sites of government agencies such as the NIH, NSF, NASA, DOE, and DoD in the United States; on Nanothinking.com and SAFE-NANO in the European Union; on Rusnano in Russia; and on STRATNANO (Iran) and NanoForum (Asia). The European Commission's nanotechnology sites are especially good.

An easy way to keep up-to-date with nanoscience is to set up your computer to receive and automatically display nanoscience news, which is constantly being updated. Most major science services have RSS feeds that allow you to channel science news to your Internet portal. Recently, I estimated that I scan about 35 000 technology headlines every year, by using several customized pages to collect and display RSS feeds from a two dozen Web sites and publications that keep a constant flow of headlines coming to my computer home page. Virtually any online portal can be customized to provide science and business news, research updates, and other information. Business and scientific publications such as *New Scientist* and *Scientific American* do a great job of synthesizing the current state of the art. YouTube videos add a visual dimension.

If you look up Nanotechnology in Wikipedia you'll find an extensive series that includes nanomaterials, nanomedicine, molecular self-assembly, nanoelectronics, scanning probe microscopy, molecular nanotechnology, and other topics. Wikipedia's "Nanotechnology Portal" (http://en.wikipedia.org/wiki/Portal:Nanotechnology) offers an excellent introduction if you want to explore nanotech topics.

Where to Study Nanoinnovation

If you have time and resources to take courses in nanotechnology, or earn a degree, there are more approximately 300 university-level nanotechnology programs in more than 25 countries worldwide, with more than 60 nanotechnology degree programs offered by universities in the United States – including 24 PhD programs. More than 30 degree and certification programs are offered in the United Kingdom and Germany. Most of these universities offer educational sites that include tutorials on various aspects of nanotechnology as well as online publications, news releases, profiles of research teams, and project descriptions. There are hundreds of courses at leading universities all over the world that offer nanotechnology programs. You can find a listing of nano-degree programs in the nanoDEGREE database offered online by Nanowerk (http://www.nanowerk.com/nanotechnology/nanotechnology_degrees.php).

There are many excellent "immersion programs" such as the NNIN's Research Experience for Undergraduates (REU) program, which is centered at the Cornell Nanoscale Science and Technology Facility (www.cnf.cornell.edu/). This popular program provides 10 week immersion programs in nanotechnology, as well as internships and travel-abroad opportunities. The REU program graduated 600 students from its program in the first 13 years. REU students conduct actual nanotechnology research projects. Immersion programs like this are helping to build the international nanoinnovation talent pool by involving students in real science and technology projects, giving them a taste for what hopefully will be a career in nanoinnovation.

You can access current research at the Web sites of leading academic nanoscience centers such as the nanotechnology centers at leading universities such as Rice University, MIT, or the University of Pennsylvania. Academic consortia exist in Europe and Asia that describe their projects and achievements on their Web sites. In the United States, there is a consortium of 14 universities that comprise the National Nanotechnology Infrastructure Network (NNIN), which is funded by the National Science Foundation. The NNIN sites include the Cornell Nanoscale Science and Technology Facility/Cornell University; Stanford Nanofabrication Facility/Stanford University; Lurie Nanofabrication Facility/University of Michigan; Nanotechnology Research Center/Georgia Institute of Technology; Center for Nanotechnology/University of Washington; Penn State

Nanofabrication Facility/Pennsylvania State University; Nanotech/University of California at Santa Barbara; Nanofabrication Center/University of Minnesota; Microelectronics Research Center/University of Texas at Austin; Center for Nanoscale Systems/Harvard University; Howard Nanoscale Science and Engineering Facility/Howard University; Colorado Nanofabrication Lab/University of Colorado; Nanofab/Arizona State University; and the Nano Research Facility at Washington University, St. Louis.

Some regional consortia integrate nanotechnology research and commercialization activities, with a strong focus on translational research. An example is the Nanotechnology Institute located in southeastern Pennsylvania. This is a multi-institutional partnership of 13 academic institutions focused on developing "real-world applications" by combining nanotechnology research, commercialization, and company formation.

Keep in mind that most of the innovation surprises will come from the periphery of your vision, in areas that are off your radar screen, so cast a broad net when you are fishing for news that may impact the future.

Reference

1 Tomczyk, M.S. (2011) *Applying the Marketing Mix (5 P's) to Bionanotechnology, Biomedical Nanotechnology*, Springer.

8
International Perspectives

I see nanotech as a great bridge builder, enabling us to make disruptive innovations across conventional boundaries.
– *Dr. Anita Goel, MD, Ph.D., Founder, Chairman and CEO, Nanobiosym*

You may have noticed that the nanoinnovators described throughout this book come from many different countries. Scientists and students engaged in cutting-edge nanotech research come from Australia, Brazil, China, Egypt, Iran, Israel, India, Korea, Russia, Sri Lanka, and many others – representing every corner of the world. If you look at the authors cited in the endnotes of this book, you will see a wide array of ethnic origins. Truly, nanotechnology is an international realm.

In the 1990s, nanotechnology was considered an expensive novelty in many countries, so in 2001 it was considered a bold move when the United States became the first country to launch a national nanotechnology initiative with public funding of $464 million for nanoscale R&D. Within a few years, more than one-third of all countries established their own initiatives. In the decade from 2000 to 2010, national governments invested >US$67 billion in nanotechnology research. Today, >60 countries support national nanotechnology initiatives. The global nanoinnovation ecosystem is well established, and we are beginning to reap the rewards.

It is important to note that the globalization of nanotechnology did not begin with the National Nanotechnology Initiative (NNI) in the United States, although the NNI provided the national model used by most governments. The development of nanotechnology owes its progress to pioneering efforts in many nations. For example, the European Union(EU) funded 80 nanotechnology projects from 1994 to 1998, under the 4th Framework Programme. In the early 2000s, Asian countries, including Japan, South Korea, Taiwan, and China established nanotechnology programs [1]. Countries as diverse as Brazil, India, Indonesia, Iran, Israel, Malaysia, Pakistan, Russia, Singapore, and Spain developed early nanotechnology programs.

From 2000 to 2008, researchers focused on basic discoveries, infrastructure building, imaging, and training. Toward the end of the decade, the focus shifted to commercial products and applications. This transition from discovery

NanoInnovation: What Every Manager Needs to Know, First Edition. Michael Tomczyk.

research to commercial applications has been documented in the superb patent research conducted by Philip Shapira and his colleagues. Dr. Shapira and his colleagues also found that national innovation systems exert a significant influence on the patenting and commercialization of nanoinnovations [2].

By 2015, total global investment in nanotechnology by governments, corporations, and other sources is expected to reach a quarter of a trillion dollars [3]. This makes nanotechnology one of the most universally supported realms of science and technology in history. And why not? Nanotechnology has given the world a toolkit that allows us to innovate by manipulating atoms and molecules. In the world of science and technology, it doesn't get much better than that.

The international development of nanotechnology has been remarkably balanced. A decade ago, there were serious concerns that the United States and other first mover nations would dominate the field of nanotechnology. It was feared that smaller and poorer nations would be excluded or left behind. Critics worried that nanotechnology would evolve like the semiconductor industry, where the emergence of Silicon Valley and the Boston Corridor in the 1970s and 1980s gave the United States a dominant competitive position in digital technologies.

Some analysts suggested that a "nanotechnology divide" would separate the nano and non-nano countries, similar to the "digital divide" that separated strong Internet countries from those with little or no Internet access. In this scenario, wealthy nations would gain an unfair advantage in nanotechnology, with knowledge concentrated in a handful of nanoscience countries. Industrial secrets would be cloistered and hoarded like nuclear technology. Large corporations would lock up gateway patents. Many countries would be frozen out.

Fortunately, these dire predictions did *not* come to pass. In fact, just the opposite occurred. Nanotechnology quickly became a worldwide phenomenon, a source of discovery and wonder, providing a new common language for scientists and engineers.

In many respects, nanotechnology has become one of the most "democratic" technology streams to emerge from the realm of science. In a very short period of time, nanotechnology has found its way into the government, industrial, and academic infrastructures of almost every country and region. The global community of nanoinnovators is vibrant, active, and engaged. Nanotechnology was the first major technology stream to benefit from the relatively new concept of "open innovation" – where anyone in the world can gain access and contribute to the innovation process. Today, scientists, engineers, and business leaders have seamless access to a global network of nanoinnovators. Nanotech patents are easy to search. Articles on the latest research appear every day. Most engineering schools offer nanotechnology courses and degrees. Most corporate R&D programs include nanotech. Nanomedicine is well established. Many of the early risks and failures are behind us. The most promising lines of research are well established, thanks to an unprecedented global effort.

Most large engineering schools have nanotechnology programs. In the United States, more than half of all advanced degrees in science, technology, engineering, and mathematics are earned by citizens of other countries. Cross-border funding and collaboration is common. Most countries – from Brazil to Egypt – host annual nanotechnology conferences. In August 2013, IEEE, the world's largest engineering organization, hosted its 13th International Conference on Nanotechnology in Beijing.

8.1 The Critical Role of Nanoinnovation Ecosystems

Creating a national infrastructure is essential for nations that want to leverage nanoinnovations to help grow their economy, and to keep from falling behind. In most cases, this involves cultivating a national nanoinnovation *ecosystem*.

In the natural world, an ecosystem is a network of interactive relationships among people, organizations, and their environment. In the world of science and business, a nanoinnovation ecosystem is *an interactive network of people, organizations, and environments designed to implement nanotechnology ideas, discoveries, and inventions*.

Ecosystems can be created by governments, corporations, industries, and even customers. Many different types of organizations have evolved their own innovation ecosystems. Facebook, Google, IBM, Intel, Lockheed-Martin, Microsoft, NASA, and Samsung are a few examples. Google is not just a search engine or advertising platform. It is an innovation engine whose culture includes a legion of users who participate and contribute to Google's innovation engine. These organizations are no longer just entities that design and provide products and services. They have evolved into open innovation communities where everyone generates and implements ideas.

For example, the nanotechnology interest group on Facebook includes >20 000 people representing countries from all over the world. Many nanotechnology researchers are linked by online networks. Many of these are sponsored by government initiatives that fund research, encourage collaboration, and incentivize commercialization.

Based on my research for this book, I've found that countries with the best-developed nanoinnovation ecosystems include a framework that includes several core elements that are depicted in Figure 8.1. The real value of this framework is unlocked when the infrastructure is linked to all the other nanoinnovation ecosystems in the world to form a knowledge-sharing network. These networks often resemble a massive, complex spider web. In the context of a global network, a country can be located at the *hub* of the global network; it can exist as a *node* in the network; or it can be off the grid altogether.

The most complete national or regional ecosystems include a National Infrastructure that consists of essential institutions and facilities, a research core to fund and promote R&D, programs and incentives to stimulate commercial output, and a focus on solutions and benefits. Typically, these elements are linked to the country's national science and technology priorities.

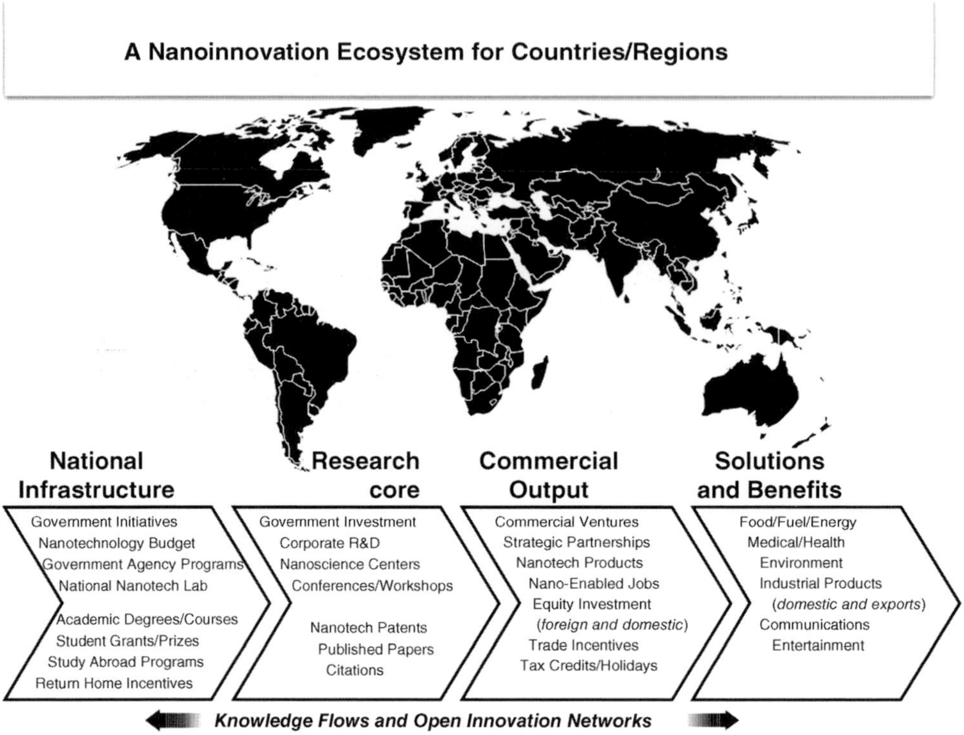

Figure 8.1 A nanoinnovation ecosystem for countries and regions. Countries that have benefited most from the nanotechnology revolution are those that created their own national nanoinnovation ecosystem. This is a framework that can be created by even a small country with a poor economy. The key is to create and support a proactive network that can connect with the rest of the world (diagram by Michael Tomczyk).

Everything starts with a government initiative. Typically, the initiative is led by a small group of government and industry sponsors who promote, frame, and guide the initiative, with public leadership provided by well-known champions who keep waving the flag for nanotechnology. Most nanotechnology initiatives are anchored by a national government laboratory and university-based nanoscience center(s).

Every nanoinitiative is anchored by a flagship nanoscience laboratory, which may involve sponsorship by a cluster of government agencies and corporations with an interest in the research outcomes. Building a nanoscience center can be expensive and complex. Although atomic force microscopes have gotten smaller and less expensive, facility costs remain high and the design and location of a nanotechnology laboratory can pose a variety of challenges.

A good example of a nanoscience laboratory is the Krishna P. Singh Center for Nanotechnology, a 78 000 ft^2 research and education complex, which cost $92 million and opened in October 2013 (Figures 8.2 and 8.3). The facility is located in the heart of Philadelphia, Pennsylvania, on a busy street where a

Figures 8.2 University of Pennsylvania Researcher Erie Morales sets up an atomic force microscope (Figure 8.2) at the new Singh Nanoscience Center (Figure 8.3) (Photo by Steve Minicola, University Communications, University of Pennsylvania).

stream of cars pass the gleaming silver and glass building. Through the glass walls, you can see students sitting at study carrels on the upper floor. Located in the basement are two advanced state of the art nanotechnology laboratories. The laboratories possess capabilities for nanocharacterization, nanofabrication, and measurement.

Locating an ultrasensitive nanotechnology research facility in the heart of a busy urban environment posed a variety of technical challenges. As you stand

Figures 8.3 The Singh Nanoscience Center is architecturally striking and was designed to be extremely functional for nanoscience research (Photo by Michael Tomczyk).

on the street, you can feel the vibration of the cars and trucks passing by – this posed special challenges for the designers, since vibration and even airflow can interfere with the stability of ultrasensitive nanoscale instruments and samples.

I had the privilege to tour the Singh Nanoscience Center while the center was under construction. I was impressed by the thought that went into the planning. Dr. Karen Winey, director of Penn's Nanotechnology Institute, explained that a great deal of preplanning went into the architecture and construction. The center's state-of-the-art laboratories are encased in concrete and acoustic materials. To minimize vibration, nearby access to upper and lower floors is provided by staircases instead of elevators, and the laboratories are located as far as possible from the street traffic. Special airflow systems provide heating, cooling, and ventilation, so that delicate samples aren't affected by changes in temperature or air pressure. These are only a few of the many considerations that are required to design a modern nanoscience facility.

Most of the world's nanoscience centers are connected through a global network of collaborative programs and projects. This enormous network includes scientists and engineers from many countries and is a major source of nanoinnovation.

How these initiatives are designed and operated varies from country to country, but one of the most important threads that run through the most successful nanotechnology initiatives is *collaboration*. Shapira suggests that fostering a collaborative environment for nanotechnology research is one of the best ways to foster an effective nanotechnology initiative.

In every region of the world, nanoinnovation has been enabled by cross-border funding of nanotechnology projects, which includes sharing of facilities, pooling of resources, and sharing of findings, although most of the research results seem to be concentrated in about a dozen countries.

In their groundbreaking research on cross-border collaboration, Philip Shapira and Jue Wang documented the impact of cross-border collaboration on nanotechnology research [4]. They analyzed >91 000 nanotechnology papers published between August 2008 and July 2009 and found that these papers were authored by researchers from >150 nations. Almost a quarter of the papers were authored by researchers from more than one country. However, >90% of the papers came from only 15 countries. The top four countries producing nanotechnology research papers during this year were the United States (23%), China (22%), Germany (8%), and Japan (8%). Key findings from their research are depicted in the accompanying infographic that was originally published in *Nature* (Figure 8.4).

The nodes in this infographic graphically depict cross-border funding of nanotechnology projects and confirm the value and impact of international collaboration (Figure 8.5). The size of the nodes are proportional to the total number of publications funded, and the thickness of lines between nodes is proportional to number of articles acknowledging funding from sources located in both countries. Only lines representing 50 or more coacknowledged publications are shown. Research programs of the EU are shown as a separate funding source.

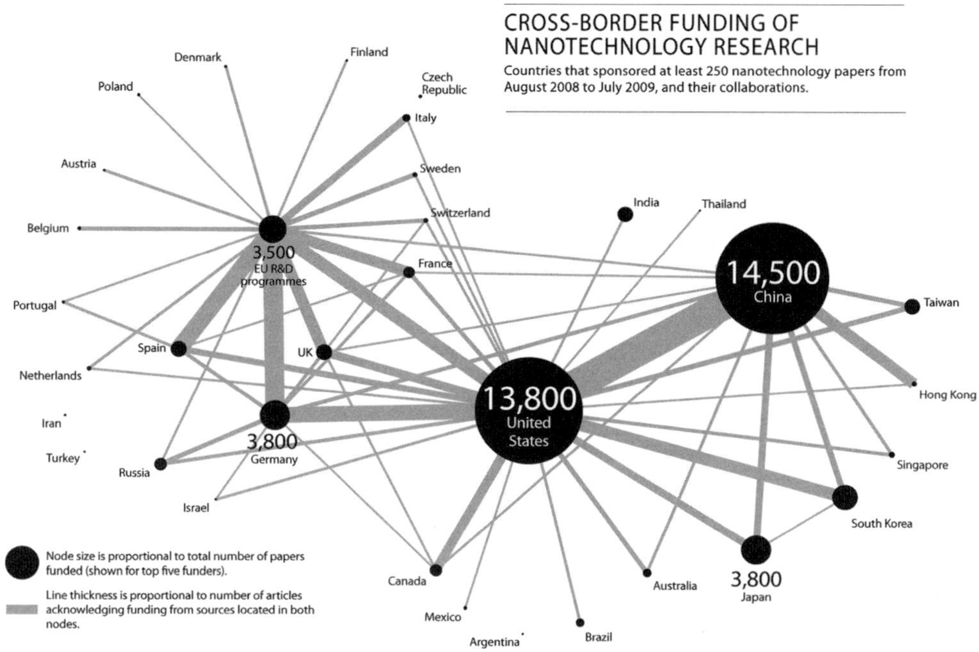

Figure 8.4 Transnational grant cofunding of nanotechnology research. (*Source:* Reprinted with permission from Ref. [5]).

Figure 8.5 Dr. Anita Goel is an award-winning nanoinnovator who believes in crossing boundaries to bring nano-enabled innovations to emerging markets as well as developing countries.

8.2 Nanoinnovation in the Asia–Pacific Region

To gain a better understanding of the global nanoinnovation ecosystem, it helps to look at individual countries and regions, and how they have implemented various national initiatives and programs. As you will see in the country examples presented in this chapter, there are many ways to configure a national nanoinnovation initiative. Following are some notable examples of achievements, best practices, and strategies in key countries and regions.

China, Korea, and Japan are the "three tigers" of nanoinnovation in the Asia–Pacific region. In 2011, Japan, China, and Korea respectively ranked second, third, and fourth (behind the United States) in number of nanotechnology patents filed. Asian–Pacific nations account for about half of all nanoscience publications. Following is a summary of some notable achievements and best practices in this group.

China

China's national commitment to science and technology, and the modernization of Chinese institutions during the past decade have made China a world leader in nanoinnovation. China's impact and influence on nanotechnology is indisputable. China has surpassed Europe as America's leading research partner in nanotechnology, as evidenced by citations in research publications (China produced more nanotechnology research publications than any other country in 2013).

Nanoinnovation has helped China improve its manufacturing processes, and helped to speed the modernization of key industrial sectors ranging from advanced materials and biomedicine to solar technology (China is the world leader in the production of solar products, and the top manufacturing nation, overall).

China's national support for nanotechnology follows the "ecosystem" framework adopted in most countries. China has also cultivated economic and technological development zones that involve focusing an entire city or region on a specific technology or industrial theme. For example, the Dezhou Solar Valley was designed to develop solar energy and other clean-energy technologies.

The China Nanotech International Innovation Cluster is located in the historic city of Suzhou, which hosts >100 nanotechnology-related companies and >2500 nanotechnology-related experts. City planners are developing a nanotechnology ecosystem called "Nanopolis Suzhou," which is envisioned to become a global nanotechnology hub. Completion is scheduled for 2015.

In August 2013, China's leadership role was recognized when the 13th IEEE International Conference on Nanotechnology – one of the world's largest nanotechnology events – was held in Beijing. The conference was held jointly with the 4th International Conference of the Chinese Society of Micro/Nano Technology. Two dozen technology tracks featured sessions on nano/biomedicine, nanoelectronics, nanopackaging, nanomanufacturing, and other topics.

South Korea

Korea's relentless commitment to nanotechnology and the country's aggressive development of a Korean nanotechnology ecosystem is impressive. The Korea National Nanotechnology Initiative was launched in December 2000, providing a strong first mover advantage. Korea moved quickly and early to develop a nanotechnology infrastructure to groom students and research faculty, and business leaders. Iconic Korean companies such as Samsung have incorporated nanomaterials and nanoelectronics in a wide variety of products, from mobile phones and tablet computers to high-definition TV.

Korea's National Program for Tera-Level Nanodevices was established in 2000. The Nanotechnology Industrialization Support Center was created in 2001. This was followed by the Nano Particles TIC (2002), Nanotechnology Information Support System (2003), Korean Nanotechnology Research Society (2003), Nano Particles Application Center (2004), Nano Convergence Practical Application Center (2004), National NanoFab Center (2005) and Advanced NanoFab Center (2006), the National Center for Nanoprocess and Equipment (2007), Nanotechnology Roadmap (2008), Korea National Nanotechnology Policy Center (2010), and the Nano Convergence Foundation (2012) [6]. Today, 24 South Korean universities have nanotechnology research programs.

As a result of this early and strong commitment to nanotechnology, South Korea has arguably produced the greatest commercial return from nanotechnology activity of any nation other than the United States. Examples of Korean nanoinnovations are Samsung's use of nanosilver in appliances (Silver Nano), nano-enabled electronics such as flat-panel TVs, smartphones, and smart watches. Korean nanotechnology companies manufacture carbon nanotubes, chemical nanomaterials, electronic displays, graphene, nanocatalysts, nanocomposites, nanometal powders, nanofibers, and a variety of bionanotechnology products.

So how did South Korea create its nanotechnology powerhouse? Korean policymakers linked policies and investments to economic growth and created a process to periodically review and adjust their industrial development strategy, nanotechnology milestones, and economic goals. If there is anything that can be attributed to Korea's success in nanotechnology – beyond commitment and superb execution – it is Korea's decision to move very early to cultivate its national nanotechnology ecosystem. This early commitment earned it a spot in the top tier of countries that have leveraged nanoinnovations for competitive advantage.

Korea's nanotechnology strategy also benefited from an earlier decision to enter the semiconductor market. Korea's Samsung Group began producing memory chips in the early 1980s. In 2013, Korean semiconductor sales exceeded $50 billion, giving them a market share of almost 16% and moving them into second place, globally [7]. When nanotechnology began to catch fire in the late 1990s, countries with a strong position in semiconductors had a running head start in nanotechnology.

Linking technology and innovation to economic development has helped make South Korea the world's 15th largest economy. South Korea's current science and technology goal is to become one of the top three leading countries in nanotechnology, by 2015.

Japan

Japan's achievements in nanoinnovation are well documented and include several innovations described in this book. Japan has ranked consistently in the top five countries in nanotechnology-related inventions as measured by patent filings and publications.

Japan has been a consistent world leader in nanoinnovation, thanks to a multi-billion dollar investment in nanotechnology research that began in the early 2000s. Like Korea, Japan also benefited from their early leadership position in semiconductors. By 2006, US- and Japan-based companies accounted for as much as 75% of global corporate investment in nanotechnology R&D.

Most government funding is provided by the Ministry of Education, Culture, Sports, Science and Technology; and the Ministry of Economy, Trade and Industry; with funding coordinated by the Japan Science and Technology Agency. In 2009, nanotechnology represented 5.2% of Japan's 3rd Science and Technology Basic Plan.

Japan's 4th Science and Technology Basic Plan, which extends from 2011 to 2015, has identified life innovation and green innovation as key priorities, and nanotechnology research is being targeted toward these national priorities. The 4th S&T Plan also includes a special focus on *social needs* such as safety, recovery and reconstruction, green innovation, and life innovation. Japan is leveraging its nanoinnovation efforts to address a variety of national issues such as the need for alternative energy solutions, waste management and pollution control, green technologies, and a variety of technical challenges related to recovery from the Tōhoku earthquake and tsunami, which caused the Fukushima Daiichi nuclear disaster.

Including specific social needs in national nanotechnology priorities is a somewhat recent and growing trend, and is a welcome development. More nations need to refocus their nanotechnology research from "discovery research" to "application development" to address critical-need solutions that will directly benefit their societies.

Japanese nanoinnovations are extensive. Of special note is the use of nanotechnology to develop alternatives to precious metals and rare earth materials. As previously discussed, nanomaterials are providing replacements for rare earth elements such as rhodium and palladium, which are becoming increasingly scarce and expensive. The depletion of rare earth elements is a ticking time bomb. There are no acceptable substitutes for most rare earth elements used in consumer electronics, solar panels, and other products. Fortunately, scientists are demonstrating that earth-abundant materials exhibit properties at the nanoscale that can replace rare earth materials.

For example, palladium is a rare earth that has the ability to absorb hydrogen gas, which is useful in catalytic converters used in automobiles. Demand and prices for palladium have skyrocketed, while the supply remains low ~200 metric tons per year. In 2010 and 2011, Japanese researchers, led by Hiroshi Kitagawa, announced that they had used nanotechnology to create an alloy of palladium by mixing silver and rhodium. Silver and rhodium are two separate elements that normally do not mix. The Japanese researchers mixed the elements in solution, and treated them with heated alcohol to create a mix that resulted in particles of the new alloy that are about 10 nm in diameter. Dr. Kitagawa has indicated that the new alloy would be difficult to produce commercially, but believes this technique can be used to produce other alloys for use as alternative rare metals [8]. Dr. Kitagawa directs the Kitagawa Laboratory at Kyoto University's Graduate School of Science. This is only one example of a wide array of research projects currently underway to find substitutes for rare earth elements.

Some of the most important breakthrough research is being sponsored by Japanese auto companies, working with university scientists. In 2012, Osaka University Professor Hideaki Kasai and researchers at Daihatsu Motor announced that he and his colleagues had developed an emission purification catalyst that does not require rhodium, a rare earth used in automaking for which no replacement existed. The catalyst is made from an oxidative form of copper. In 2011, researchers at Kyoto University and Sumitomo Electric announced that they had developed a new battery that uses sodium in place of lithium ion, which could cost 1/10 as much as lithium batteries and enable a vehicle to travel twice the distance compared with a lithium battery. Researchers said they hope the new battery will be available in 2015.

8.3 Nanoinnovation in Latin America

Latin America has a well-developed nanoinnovation ecosystem, anchored by strong networks and state-of-the-art research facilities in Brazil and Mexico and strong initiatives in Argentina and Chile. Colombia, Venezuela, and Uruguay are also active in nanotechnology research.

Nanotechnology centers at major universities were created as early as 1999 under the World Bank's Millennium Science Initiatives, which sought to give developing nations infrastructures and resources that exist in developed countries [9]. Brazil, Chile, Mexico, and Venezuela have received World Bank funding for nanotechnology centers and projects. Two of the earliest centers created with World Bank funding were the Technical University Federico Santa Maria and the Andrés Bello University in Chile.

Some of the best and most insightful research on nanotechnology in Latin America has been conducted by Guillermo Foladori and Noela Invernizzi [10], whose research provided some of the statistics cited in this section. Professor Foladori is a Uruguayan-born anthropologist with a Ph.D. in Economics and teaches in the doctoral program at the National Autonomous University of Mexico. Both Guillermo Foladori and Noela Invernizzi are Co-Coordinators of the Latin American Nanotechnology and Society Network (ReLANS). Noela Invernizzi is with the Federal University of Paraná, Curitiba, Brazil.

Professors Foladori and Invernizzi developed the accompanying chart from their research, which shows the years when nanotechnology was incorporated into the public agendas of a dozen Latin American countries [9]. Foladori notes that the list shows when each country declared nanotechnology as a priority or strategic area for science and technology development; the dates do not necessarily indicate when the initiatives were implemented or when funded projects began.

Incorporation of Nanotechnology in the Public Agendas of Selected Latin American Countries

Year	Country	Institution
2000	Brazil	Ministry of Science and Technology
2001	Mexico	National Council for Science and Technology
2003	Argentina	Science and Technology Secretariat
2004	Colombia	Administrative Department for Science, Technology and Innovation
2004	Costa Rica	National Council for Scientific and Technological Research
2005	Guatemala	National Council for Science and Technology
2005	Ecuador	National Secretariat for Science and Technology
2006	El Salvador	National Council for Science and Technology
2006	Peru	National Council for Science, Technology and Innovation Technology
2008	Dominican Rep.	State Secretariat for Higher Education, Science and Technology
2009	Uruguay	Ministerial Advisory Office for Innovation
2010	Panama	National Secretariat for Science, Technology and Innovation

Source: Ref. [9].

Some notable achievements in Brazil and Mexico are described in the following summaries. Both countries have well-developed nanotechnology infrastructures and collaborative programs with other countries.

Nanoinnovation in Brazil

With the largest economy in Latin America and one of the world's fastest growing economies, Brazil has been able to invest substantial resources in nanotechnology research. From 2001 to 2007, Brazil's investment in nanotechnology programs exceeded $100 million. More than 160 Brazilian companies conduct research in nanotechnology.

Brazil was the first Latin American country to provide public funding and support for nanotechnology development. Starting in 2001 with a few institutions and laboratories and a nucleus of a few hundred researchers, Brazil has developed the most advanced nanotechnology initiative in Latin America. The first networks were originally supported and expanded by the *Redes Nacionais de Nanotecnologia* (*Rede BrasilNano*) program and consolidated under the National Program of Nanotechnology. Today, >2000 researchers are involved in nanotechnology projects including collaborations with colleagues in other countries. Brazilian researchers have authored >1000 publications and secured >100 patents.

As the hub of South America's nanotech ecosystem, Brazil has established several cooperative initiatives with other Latin countries, including the Brazilian–Argentinean Center for Nanotechnology (2005), Brazilian–Mexican Virtual Center for Nanotechnology (2009), the Brazilian–Chilean Center for Nanotechnology, and many other institutions and programs.

In August 2013, the Ministry of Science, Technology and Innovation officially launched a national nanotechnology initiative known as IBN, funded by an initial investment of $440 million Brazilian reals (US$186 million). This initiative is part of Brazil's ongoing National Strategy for Science, Technology and Innovation (ENCTI), which focuses on developing innovations in information technology, biotechnology, and nanotechnology. Brazilian nanoinnovations are showcased on the Web site of the Brazilian Nanotechnology National Laboratory (lnnano.cnpem.br/).

Brazil's approach to nanotechnology and research in general has benefited from a unique national research registry called Lattes, which was originally developed to manage public funding of science and technology projects. Researchers and institutions conducting research in Brazil are required to provide detailed information in the Lattes database. The registry includes past and present projects, credentials, collaborations, and other information. Each scientist is given a unique ID number. The Lattes database has been expanded to include collaborators in other countries and now includes >3 million Ph.D.s, doctors, engineers, and other researchers, making this the world's most comprehensive research registry.

Lattes was named for César Lattes, an Italian Jewish scientist born in Brazil who became one of the nation's most distinguished and honored physicists, best known for his discovery of the pi meson (pion), a subatomic particle made from a quark and antiquark.

Lattes is available in Portuguese, Spanish, and English, and is being adopted by other countries. The database incorporates an innovation/collaboration platform that makes it convenient for international collaborators, media representatives, and others to identify experts working on areas of interest. This very interesting and ambitious strategy has helped position Brazil as an active Latin American hub in the international nanoinnovation ecosystem.

Brazil's government and industry priorities include evaluation of safety issues involving nanotechnology, however, the country has taken a *laissez-faire* approach to regulation. In Fall 2013, the Brazilian Congress rejected a bill that would have required labeling on food, drugs, and cosmetics that contain nanostructures – with politicians agreeing that there is no scientific basis for warning people about nanotechnology products, and such labeling could be misinterpreted as a "warning" that would be unduly alarmist.

Nanoinnovation in Mexico

With >50 Mexican universities and institutes involved in nanotechnology research, it is not surprising that Mexican researchers have produced the second largest number of nanotechnology research papers in Latin America. One of the leading institutions in Mexico is the National Autonomous University of Mexico, which sponsors nanotechnology activities at more than a dozen departments and institutes.

The Mexican nanotechnology effort has benefited from free trade agreements with >50 countries. The best developed nanotechnology sectors include semiconductors, electronics, and biomedicine, thanks to the country's close association with the United States and proximity to the US market.

While Mexico does not have a nationally coordinated nanotechnology initiative, the government identified nanotechnology as a strategic priority as early as 2002. The Silicon Border Development Park is a 10 000 acre science park established in 2004–2005 on the border between Mexicali and San Diego, which claims to be the first high-technology park in America specializing in nanocomponents. In 2008, a specialized nanotechnology cluster was established in a high-tech business park in the city of Monterrey in the state of Nuevo León. This 175 acre science park brings together research centers, universities, government agencies, and private businesses on a variety of collaborative projects. Two incubators specialize in emerging technologies, one for biotechnology and the other for nanotechnology.

When the Nanotechnology Cluster in Nuevo León was opened in 2008, it had 16 members, which has since grown to ~30 members. Approximately 50 publicly funded linkage projects are sponsored, which are designed to promote collaborations between scientists, research centers, and private businesses, to commercialize new technologies and create high-value jobs. Starting with a core of 17 businesses, the Cluster hopes to host ~100 nanotechnology businesses by 2015 [11].

Mexico continues to make significant progress in nanotechnology, thanks to the courage and determination of nanoscientists who often work under a bizarre threat of violence. In 2011, several scientists and laboratories were targeted for violent attacks by Luddite-style eco-anarchists who are specifically opposed to nanotechnology. One biotechnologist was murdered, several people were injured by bombs, and letter bombs have been sent to universities. To date, these attacks represent the strongest and most violent opposition, worldwide, to nanotechnology. As a result, Mexican universities have had to institute stringent security measures. The attacks have scarred the science community, but stiffened their determination. It is a tribute to the Mexican nanoscience community that they continue to publish research papers, file patents, and keep developing nanoinnovations that can help their nation and the world.

8.4
Nanoinnovation in the European Community

Europe's approach to nanotechnology research and technology in general has been to adopt "challenge goals" that relate public funding to specific outcomes. Faced with a complex task of diffusing nanotechnology across a network of multilingual, multicultural European countries, the European Union started with a series of "framework programs" to build and strengthen Europe's scientific institutions, networks, and facilities.

The European Community's Framework Programme 7 (FP7), which ran from 2007 to 2013, invested millions of euros in several specific technology streams, of which nanotechnology was the largest. Nanotechnology accounted for more than €900 million of the more than €50 million in FP7 research funding, augmented by substantial project funding by the European Research Council. The Marie Curie Fellowships provided grants to students, young scientists, and host universities and facilitated a variety of skill-enhancement activities. A Marie Curie training network called ManETEI (Management of Emerging Technologies for Economic Impact) provided skill-building activities to business schools, business students, and industry partners.

While I was researching this book, I had some firsthand experience with the ManETEI network. In June 2011, I spent a week at the University of Leeds in Manchester, England at the invitation of Krsto Pandza, a Slovenian-born academic who is Professor in Strategy and Innovation and Terry Wilkins, CEO and founder of the Nanomanufacturing Institute at Leeds. Terry is also Professor of Nanomanufacturing Innovation, and Chair of the FP7 Expert Advisory Group that advises the EC on nanomaterials and production technologies. Pandza and Wilkins are two of the most ardent innovation champions in the United Kingdom.

At Leeds, I had an opportunity to work with a group of doctoral students from more than a dozen European countries who were attending a week-long innovation workshop. The students were required to select an innovation-related topic for their doctoral thesis, and were encouraged to choose projects that would benefit their home countries in some tangible way. I recall one young lady who was investigating the best systems and protocols for improving wireless communications in her home country of Romania. Another student, from Spain, planned to do a thesis on nanoinnovation.

While Europe is a world leader in government funding of nanotechnology research, commercialization of nanoinnovations has been slower to develop. For example, Europe's funding of commercial nanotechnology ventures is only about 20% of venture capital funding in the United States. Translating technology investments into commercial value has become a critical issue in the wake of the economic crises that have affected many European countries.

When the FP7 initiatives ended in 2013, policymakers in the European Community realized that the next round of research and technology needed to address new challenge goals geared to economic needs. Consequently, the Horizon 2020 program, which runs from 2014 to 2020, folds technology research and development into broader priorities that emphasize economic development and job creation. Support continues for emerging technologies such as nanotechnology as well as information and communication technologies, materials, biotechnologies, and space technologies – however, the new round of funding focuses on technology projects that will bridge the gap between research and market.

Nanotechnology does retain a prominent role in the Horizon 2020 program. The Future and Emerging Technologies initiative features a €1 billion, 10 year commitment to investigate and exploit the unique properties of graphene. This flagship effort involves >100 research groups and 136 principle investigators including four Nobel laureates.

Online nanotechnology resources can be accessed through the EC's CORDA (COmmon Research DAta Warehouse) database, as well as through nanotechnology Web sites in each country. For example, the annual Italian Nanotechnology Census includes a detailed survey of Italian nanotechnology companies, institutions, and resources. Some Web sites contain infographics that allow you to search or link to a specific area of interest such as nanomaterials.

Insights from European Nano-Insiders

According to many of the nano-insiders interviewed for this book, European nanoinnovators tend to collaborate more than other global communities. Patent sharing is easier in Europe than in the United States.

"Some of this collaboration is by necessity," one European nanoscientist observed. "Imagine a dozen or more countries all hoarding patents and trying to gain competitive advantage. It works better for us if we collaborate and pool our resources."

Another view was expressed by Paolo Samori, an award-winning Italian-born author and a nanochemistry pioneer: "I'm Italian. I've worked in Germany. Now I'm in France. I've seen the Japanese, American and Dutch systems, and I can tell you that nanotechnology is like a car race. Some countries are conducting nanoresearch like racing a Ferrari and in other countries. Nanoresearch is like driving a Cinquecento (a small Italian city car). To stay in the race, some countries have to strive for creativity to make up for what they may lack in resources or funding. I often challenge my students to show me how to develop a science project with 10 euros instead of 100,000 euros. Unfortunately a cutting edge field like nanotech requires strong investment which does limit competition in some areas of research and technology."

Samori knows what he's talking about. He has demonstrated a high degree of creativity and productivity in his work and the projects he leads. For example, he recalls leading a group of young scientist–students who wanted to experiment but didn't have access to silicon wafers, so they improvised by using a CD-ROM as a substrate.

Paolo Samori is Distinguished Professor and Director of the Institut de Science et d'Ingénierie Supramoléculaires and head of the Nanochemistry Laboratory at the Université de Strasbourg. He has published >160 papers on a variety of topics including hierarchical self-assembly of hybrid architectures on surfaces, supramolecular electronics, and the fabrication of organic-based Nanodevices. He is the author of the book, *Scanning Probe Microscopies Beyond Imaging: Manipulation of Molecules and Nanostructures* (Wiley 2006).

Additional insights were offered by Mark Morrison, CEO of the Institute of Nanotechnology, a charitable organization based in Glasgow, Scotland that works with government, academic, and industrial organizations on nanotechnology policy, technology and market analysis, project management, and funding.

Dr. Morrison has seen many of the nanoinnovations that are coming out of Europe, including nanoinnovations that are helping to address many of Europe's "challenge goals" such as helping to prevent and remediate pollution. One example he described is a photocatalytic cement that cleans itself and also cleans the air by helping to oxidize pollutants. The cement is called "TX Active" and is made by Ital Cementi, a 150-year-old Italian company, using cement and concrete that incorporates nanoparticles of titanium dioxide, which helps the cement stay clean and resistant to nitrous oxides in the atmosphere. "There are paints and coatings and other materials that can reduce pollution just by using a simple photo catalyst," he said. This is only one example of how nanoinnovation is being applied to address society's needs for smarter materials, even a material like cement that is thousands of years old. Increasingly, Morrison observed, nanoinnovations are being developed to address environmental remediation, clean contaminated air and water, and to produce more efficient renewable energy.

"We're not going to change our lifestyles immediately," Morrison cautioned. "We won't suddenly become less reliant on cars and other energy hungry devices. This will take 20 years or longer, so anything that reduces the power hunger of these devices such as nanocoatings that reduce the consumption of energy or resources, or better rechargeable batteries for electric vehicles and other alternatives, these are some of the commercial innovations that nanoscientists are developing now in Europe and other regions." He noted that the EU's FP7 framework program covered up to 100% funding for nanotechnology researchers, and covered some costs for industry partners as well.

Dr. Morrison noted that some of the best nanoinnovation in Europe has come from collaborative programs that bring together researchers from different countries. He cited as an example the Interuniversity MicroElectronics Centre (IMEC) in Belgium, which has developed worldwide connections by sending faculty and students on missions to form collaborative relationships with researchers in other countries such as India.

Best Practices and Strategies

As a leader in nanotechnology, not only in Europe but worldwide, Germany is a good choice to look at more closely, to gain insights into best practices and strategies. Nanotechnology has played an important role in the growth of Germany's export markets.

More than 800 publicly funded research institutions in Germany are engaged in nanotechnology research. More than 1000 German companies are currently working on the development, application, and marketing of nanotechnology products – representing >60 000 jobs. Approximately one-third of these companies describe nanotechnology as a core competence, representing >30% of their business activity.

Germany's R&D spending reached a record 79.5 billion euros in 2012, under the Federal Government's 2015 Nanotechnology Action Plan. Priorities include using nanotechnology to contribute to growth and innovation in Germany; making nanotechnology safe and sustainable; tapping the potential of nanotechnology in education and research; and tapping the potential of nanotechnology to meet such global challenges as environmental protection, energy, health, sustainable agriculture, and food safety. The Plan also includes a "National Electromobility Development Plan" to support one million electric vehicles in Germany by 2010 [12].

International projects led by Germany that involve nanotechnology education and innovation include joint projects with Canada, China, Egypt, India, Ireland, Jordan, Russia, and South Africa. To navigate Germany's extensive network of nanotechnology centers and resources, the Federal Ministry of Education and Research provides an online nanotechnology "competency map" that lists >2200 institutions, showing the type of institution and their geographic location in Germany. This map is cross-referenced to more than a dozen industry sectors such as automotive/traffic, chemistry/materials, and optics; and can also be sorted by 10 technologies (nanoanalytics, nanobiotechnology, nanocoatings, nanooptics, nanosystems/sensors, etc.) [13].

There are many examples of German nanoinnovations: In November 2013, researchers at the Technische Universität Dresden and the Fraunhofer Institute for Electron Beam and Plasma Technology announced the development of a miniature spectrometer that used nanowire antennas and was small enough to fit in a mobile phone. The device used the nano antennas to collect, amplify, and redirect light to a CCD/CMOS chip [14]. Applications ranged from noninvasive medical monitoring to monitoring the ripeness of fruit. The spectrometer can be mass produced using thin-film manufacturing.

In contrast to Germany, other European countries have pursued a "start small and expand" strategy. Management experts call this a "fast follower" strategy. For example, Spain began with relatively small scale projects during 2004–2007 under Spain's National Plan. This was followed by larger scale infrastructure building under the 2008–2011 National Plan. Spanish scientists self-organized the Nano Spain Network, which has expanded into a community of ~300 R&D groups and >2000 nanoscientists. NANODYF (Iberoamerican Network for Dissemination and Education in Nanotechnology) promotes education in Iberoamerican countries where people communicate in Spanish. More than 60 Spanish companies are involved in the production of nanomaterials, nanocomposites, nanobiotechnology, and nanoparticles.

Spain is also leveraging its resources through regional cross-border initiatives like KEEN-REGIONS. KEEN stands for Knowledge and Excellence in European Nanotechnology. This EU-funded initiative linked three established nanotechnology clusters located in three different countries/regions – Veneto Nanotech (Italy), Minatec (France/Rhône-Alpes), and Gaia (Spain/Basque Country). Together, these regions represented >4000 researchers, 250 laboratories, and hundreds of companies that were provided an opportunity to participate in nanotechnology infrastructure building, knowledge sharing, research and commercialization, and joint problem solving. The participants were encouraged to identify and map their nanotechnology resources, develop innovation projects, increase utilization of nanotechnology by local industry, and share best practices across regions and specialties.

8.5 Insights from Selected Nanoinnovation Ecosystems

Every national nanoinnovation ecosystem has its own unique approach to nanoinnovation. Looking at how other countries have cultivated their initiatives can offer some ideas and best practices that may be of value. Following are a few examples of nanotechnology programs from Israel, Kenya, Russia, South Africa, and Sri Lanka. These countries were chosen because they are representative of several different strategies and practices.

Nanoinnovation in Israel

In 2006, Israel's Forum for Research and Infrastructure established a national 5-year Nanoscience and Technology Program to develop the nation's nanotechnology infrastructure and in 2007, Israel designated nanotechnology a national priority, guided by the Israel National Nanotechnology Initiative (INNI).

Israel's nanotechnology community has been mapped in an online infographic (knowledge portal map) that provides direct online links to 120 companies involved in nanotechnology as well as >600 nanotechnology researchers in 7 industry sectors (http://usasearch.dainfo.com/INNI_2011/). This searchable database makes it especially easy to search and locate nanotechnology resources. Israel has also used its nanotechnology programs to lure top scientists and engineers from other countries to further strengthen their initiative. Providing a searchable database is one of the best practices that countries are using to foster collaboration.

Israel's nanoinnovation program has already produced an impressive array of commercial technologies and products. An example of Israeli nanoinnovation is the aluminum–air battery developed by Professor Arie Zaban and his research group at the Bar Ilan Institute of Nanotechnology. The metal–air battery, which uses novel nanomaterials, can drive a fully electric vehicle a distance of 330 km on a single battery charge. The technology is being commercialized by Phinergy, an Israel-based technology firm. A prototype of the electric vehicle using the aluminum–air battery was shown to President Barack Obama and Prime Minister Netanyahu during President Obama's visit to Israel in March 2013.

As a nation, Israel works hard to cultivate and recognize its most talented scientists and engineers. A question that is often asked by international policymakers is "How do we turn nanoscience and nanotechnology research into commercial products?" One way is to support problem-focused research groups that are committed to developing real-world solutions, as Israel has done. This goes hand-in-hand with recognizing and showcasing researchers and groups that are "serial innovators." Honoring and publicizing the achievements of star innovators is a critical part of any national nanotechnology initiative.

A good example is Ze'ev Zalevsky, a 41-year old engineering professor who heads the Electro-Optics study Program at Bar-Ilan University, Israel's fastest-growing university. In 2012, Zalevsky was recognized as Israel's outstanding Young Nanotechnology Researcher by the INNI. Zalevsky and his colleague Javier Garcia at the University of Valencia in Spain developed a device called the "Opto-Phone," which uses the principles of nano-photonics to read heartbeat, blood pressure, and blood glucose levels from up to 100 m away. The device integrates a laser beam, advanced camera, and software to analyze heartbeats, blood flow, and other metrics that normally require physical devices or blood tests. Zalesky has also collaborated with Dr. Hamutal Slovin to develop a "nano-probe" for use in monitoring activity centers in the brain. Professor Zalesky, his colleagues, and students are working on more than a dozen other nanoinnovations for applications ranging from sensors to satellites.

Nanoinnovation in Kenya

With the exception of South Africa and Egypt, we very seldom hear about nanotechnology in Africa, although the continent has benefited greatly from nanotechnology, especially in wireless communications. There was a time not too long ago when someone in Nairobi who wanted to call someone in another African capital had to route the call through Europe! Thanks to pioneers like Mo Ibrahim and other visionaries, Africa's archaic landline system was bypassed and revolutionized by a wireless phone network that allows people in most countries to participate in the digital phone and Internet revolution. The next step for Africa will be to create technology centers where African scientists and student-researchers can develop nanoinnovations to address African needs for better healthcare, safe drinking water, energy solutions, and security systems.

In 2013, Kenya broke ground for the $14.5 billion Konza Technology City, which is being developed under the country's "Vision 2030" economic plan. Konza has been called "Silicon Savannah" because it is being created 60 km south of Nairobi on the African savannah, where it will exist in sight of grazing wildebeests, elephants, and other wildlife. The city will include research facilities, a university, hotels, schools, hospital, and a large residential development. The first phase is slated for completion in 2017. By locating the city in an undeveloped area of the country, the government hopes to avoid sectarian and regional conflicts as well as corruption [15].

The Kenyan government hopes the new city will eventually create >200 000 new jobs and attract investment from entrepreneurs and technology partners. Nanoinnovation will play a role in the new technology city, especially in semiconductor and electronics facilities that incorporate nanocircuits in their products.

Kenya is already on the road to becoming a major African technology center. Hundreds of technology ventures have been created in Nairobi in recent years, enabling Kenya to export $360 million in technology products in 2010, compared with $16 million in 2002. IBM established its first African research laboratory in Nairobi, in 2012. Google, Microsoft, Intel, and other companies also have regional offices in Kenya. Kenya has also experimented with several state-of-the-art technologies and applications such as M-pesa, a mobile banking system that allows people to use their mobile phones to make purchases and payments [16].

Kenya is not alone in its ambitions to develop a high-tech infrastructure. Other African countries such as Egypt, Nigeria, and Rwanda have made notable progress in improving their technology infrastructures, despite periods of armed conflict and political turmoil.

Nanoinnovation in the Russian Federation

Russia ranked 8th in nanotechnology publications between 1991 and 2007 and 16th in the number of nanotechnology patents during this period. However, this low patent ranking is misleading, since Russia's strategy of investing in overseas ventures has provided additional access to a global pipeline of nanoinventions and nanoinnovations. These investments include licenses for the use of commercial products including materials and therapeutic drugs. In many ways, Russia has done more than any country to take advantage of the globalization of nanotechnology, through scientific collaboration and through commercial investments designed to bring nano-enabled solutions from other countries to Russia.

The Russian experience has shown that the road to a national nanoinnovation ecosystem is not always smooth. Most national initiatives start with basic research and discovery and expand to commercialization. Russia's national nanotechnology initiative – known as Rusnano – was created as a joint stock company owned by the Russian government and designed to commercialize nanoinnovations. This approach – somewhat unorthodox and unique at the time – created a commercial entity instead of a government initiative, in hopes of hastening the path from research to commercial results.

Rusnano was capitalized with assets worth 130 billion rubles (US$4.39 billion) from the government. The original company was 100% owned by the Russian Federation, with a long-term objective to sell 40% of the company to private investors. Since 2008, Rusnano has been led by general manager Anatoly Chubais, a guiding force and leader of Russia's economic reforms.

In 2009, Rusnano launched an annual international award called the Rusnanoprize, to incentivize nanotech research and innovation and identify investment candidates. The prize fund is over 3 million rubles (US$90 000).

The original goal was to establish by 2015 a nanotechnology industry capable of producing commercial products worth 900 billion rubles (US$29 billion). Priority was given to projects that would help modernize Russia's industries. Target sectors included solar energy and energy saving, nanostructured materials, medicine and biotechnology, mechanical engineering and metalworking, optoelectronics and nanoelectronics, and infrastructure projects. In the first 5 years, >100 investment agreements were signed.

Several subsidiaries were also established including Rusnano Capital, which facilitates investment in foreign projects and tech transfer for use of emerging technologies in Russia. Rusnano Inform promotes information technology. The Rusnano Metrological Center promotes imaging and instrumentation. Rusnano USA Inc. represents the parent company's interests in the United States and Canada and promotes Russian nanotechnology products overseas. Rusnano Israel Ltd. promotes cofinancing and development of nanotechnology projects with Israel.

By acting as a global venture capitalist and investor/licensee for international nanotechnology start-ups, Russia has moved far upstream in the product development cycle to access nanoinnovations at their source. This is accompanied by a horizontal investment strategy that allows Rusnano to buy equity and partner with nanotech (and biotech) ventures in many other countries.

Rusnano's strategy of investing in innovations developed in other countries as well as in Russia has allowed the enterprise to tap into some of the world's best innovation streams. In addition to the >US$4 billion invested in Russian ventures, Rusnano has invested >$1.5 billion in US-based technology ventures, including BIND Biosciences, Selecta Biosciences, and other US biotech ventures; and sizable venture investments in the United Kingdom, China, Israel, Italy, and Saudi Arabia. Anatoly Chubais has indicated that ~20% of Rusnano's holdings are in projects designed to spur the development of advanced medical technologies, and to make these innovations available to the Russian people.

In 2012, Rusnano partnered with US-venture capital firm Domain Associates in a $760 million joint investment that included over $300 million to build manufacturing facilities in Russia to produce products developed by ventures in Domain's portfolio. As part of this arrangement, Rusnano plans to work with life science firms to conduct clinical trials in Russia and to license rights to drugs and treatments to make them available in Russia. A new pharmaceutical company called NovaMedica was established in 2012 to develop, market, and commercialize next-generation products from the Rusnano-Domain Associates collaboration.

While many countries have sought investments from the United States and other countries to develop their infrastructures, Russia has reversed this model to invest in those other countries in order to secure access to emerging technologies. As one Russian scientist observed, "We need nanotechnology expertise more than we need financial investment, especially in bionanotechnology. We have a strong scientific base, large markets, and strong demand for innovations that come from nanotechnology and nanobiotechnology. We want to speed dissemination of these technologies in Russia. We're investing in joint projects that will provide shared value to Russia and to our partners."

In 2013, Rusnano was at the center of a scandal related to its failed ventures, lost investments, misdirected funds, and possible misappropriation. For a few months it appeared that the scandal could derail the flagship initiative. It was reported that $450 million was invested in a silicon factory that was never operational and US$1.37 billion in loss reserves were depleted. After considerable publicity and public discussion that resulted in tighter financial controls, Rusnano survived these controversies. The company has announced a plan to sell the parent company to investors by 2020.

Nanoinnovation in South Africa

South Africa's national nanotech strategy focuses on (i) "social cluster" technologies involving health, water, and energy technologies; and (ii) "industrial clusters" that support competitive/strategic industries including advanced materials, chemicals and bioprocessing, mining and minerals, and manufacturing. More than US$77 million has been invested in nanotechnology research since the national research plan was launched in 2006. Sixteen South African universities have significant research programs in nanoscience and nanotechnology, according to Joseph Molapisi, Manager for Emerging Research Areas, Department of Science and Technology [17].

South Africa is the leader in nanotechnology research on the continent of Africa, although several other countries are working to establish technology parks.

An example of a nanoinnovation from South Africa is a "tea bag" developed at Stellenbosch University that contains nanoscale fibers and grains of carbon to clean hazardous contaminants from drinking water. According to the developers, one bag can be used to filter up to 1 l of water and costs pennies to produce [18].

Nanoinnovation in Sri Lanka

Sri Lanka, the island nation formerly known as Ceylon, is a good example of a developing country that is building a modern nanoinnovation ecosystem following the framework described earlier in this chapter. Like the NNI in the United States, Sri Lanka's National Nanotechnology Initiative was championed by a nucleus of scientific and industry leaders who worked with Sri Lanka's National Science Foundation and political leaders to create a nucleus of institutions, facilities, and talent needed to provide a nanoinnovation ecosystem. It is admirable that they were able to launch this ecosystem in 2008 despite the global economic recession that hit all countries including Sri Lanka.

With constrained government resources, Sri Lanka was able to involve the country's largest industries and companies, which helped provide funding and management expertise. This active involvement by the nation's industrial community helped focus nanotechnology R&D on areas that can benefit the country's major industries, including food production and processing, and the design and production of fabrics and garments.

The country's Ministry of Technology and Research launched its first nanotechnology initiatives in 2008. That year, the Sri Lanka Institute of Nanotechnology (Pvt.) Ltd. was established as a private company to develop nanotechnology applications, working with a group of private sector partners. Two institutions – NANCO and Sri Lanka Institute of Nanotechnology (SLINTEC) – were created to oversee the activities under the NNI at the NSF. SLINTEC was established in a building complex in the Biyagama Export Promotion Zone.

Deshamanya Mahesh Amalean has been an inspiring and prominent nanotechnology champion who has played a leadership role in the country's nanotechnology initiatives. Amalean is chairman and cofounder (with his brothers Ajay and Sharad) of MAS Holdings, one of the world's largest and most successful apparel companies.[1)]

In 2012, I had an opportunity to spend a day at the Wharton School, hosting a series of meetings with Amalean and a dozen members of his management team from MAS Holdings, one of the world's largest textile groups. Amalean is also one of the guiding spirits who helped lead the development of Sri Lanka's national nanotechnology ecosystem. He served as founding chairman of the Sri Lanka Institute of Nanotechnology and helped build a government–science–business coalition of nanotech business leaders, scientists, engineers, and educators led by the National Science Foundation of Sri Lanka. This type of partnering is especially effective in smaller countries where resources and funding may be limited.

As Chairman of the SLINTEC, Amalean leads a public–private partnership between the government of Sri Lanka and six leading private sector companies, including MAS Holdings, Brandix Lanka, Dialog Axiata, Hayleys, Loadstar, and Lankem. SLINTEC was established in 2008 with 50% of the seed funding provided by Sri Lanka's National Science Foundation and half invested by the five partner firms. The government also provided tax incentives for research investments.

Amalean's degree in chemical engineering gave him an early understanding of the value of nanotechnology. His interest in innovation was enhanced by visits to companies such as Google and Facebook in Silicon Valley, and his company's involvement as a sponsor of academic technology programs such as the Mack Institute for Innovation Management at the Wharton School.

"Innovation requires conviction," he says. "You have to build a culture of innovation so everyone involved shares this conviction. That's how we achieve creativity. Nanotechnology has the potential to enhance the industries where Sri Lanka is already a competitive leader, and will help us develop new export markets." He has been involved in the nation's nanotechnology initiative since its inception, and has played a guiding role in bringing together government and industry sponsors.

SLINTEC began conducting research in 2009 and produced more than a dozen nanotechnology patents in its first 2 years. SLINTEC's research is focused

1) The honorific title "Deshamanya" means "pride of the nation" and is Sri Lanka's second-highest national honor. Amalean received the honor in 2005.

on problems and opportunities of importance to emerging countries such as Sri Lanka. In 2010, SLINTEC and NANCO were merged.

In March 2012, SLINTEC formed its first international venture, with Nagarjuna Fertilisers and Chemicals Ltd. (NFCL) an integrated agrifertilizer firm based in India. SLINTEC is researching next-generation nanotechnology-based plant nutrition solutions, which Nagarjuna will commercialize. Researchers will focus on developing solutions that will enhance nutritional efficiency as well as environmental solutions.

Also, in March 2012, SLINTEC signed an agreement with the Sri Lankan company Laugfs Holdings Ltd., to produce titanium dioxide (TiO_2) from sand. SLINTEC has developed a proprietary process to produce TiO_2 and nano TiO_2 from ilmenite obtained from pulmoddai sand. Sri Lanka has ~18 million metric tons of ilmenite reserves, the world's ninth largest deposit. Pulmoddai ilmenite contains ~50–60% extractable titanium dioxide.

Laugfs Holdings Chairman W.K.H. Wegapitiya observed that Sri Lanka imports processed sand, which exerts a negative impact on the economy; he said this initiative could reverse the trend, and believes that scientific research can unlock other ways to take advantage of Sri Lanka's natural resources [19].

In October 2013, President Mahinda Rajapaksa opened Sri Lanka's first nanotechnology research center and nanoscience park, a 50-acre technology campus designated as a "Nanotechnology Centre of Excellence." The facilities include state-of-the-art nanotechnology microscopes and instruments. The research park will serve as a knowledge hub for nanotechnology research in Sri Lanka and South Asia, and will work with academic and commercial partners to develop commercial nanotechnology-based solutions to enhance Sri Lanka's major industries, including agriculture, apparel, rubber, and energy, as well as solutions for health security, water purification, and other social needs.

Several research projects are currently underway. One example is a joint research agreement with Stretchline Private Ltd., to apply nanotechnology solutions to improve the efficiency of energy, water, and chemicals in textile processing. The first research target will investigate processes involving the dyeing and finishing of Nylon. Stretchline is a three-way joint venture established in 1996 between Stretchline (the United Kingdom), MAS Holdings (Sri Lanka), and Brandot International Ltd. (the United States). Stretchline has manufacturing plants in seven countries and is the world's largest producer of crochet, woven, jacquard, and printed elastic.

SLINTEC CEO Harin De Silva Wijeyeratne said, "Supporting the garment export sector is of particular significance to SLINTEC as the industry is a key contributor to the national economy. Providing local industry with cost effective solutions to meet global price competition is one element of SLINTEC's mandate" [20].

Sri Lanka is doing an excellent job of leveraging limited economic resources to develop a world-class nanoinnovation ecosystem. The country has actively engaged industrial leaders in the planning of programs, infrastructures, and research facilities. Nanotechnology institutions such as SLINTEC were established and managed as private companies. Large companies with an interest in emerging technologies were involved in the development of the country's first nanotechnology research park. Scientists and engineers include local experts as well as experts from other countries. Research projects are collaborative and focused in industry goals. Nanotechnology research is designed to develop enhancements that will help make Sri Lanka's major export industries more competitive. Partnerships were also formed with companies in other countries that were collaborating with Sri Lankan firms to identify and fund targeted research that will reduce costs, improve processing, minimize pollution and energy use, and enhance the features and functions of products. It is also noteworthy that nanotechnology, which involves cutting-edge science, is being applied in novel ways to centuries-old traditional industries such as agriculture and textiles.

Turkey

Most people would not list Turkey as a nanotechnology leader. However, after Iran, Turkey is the second-largest producer of nanoscience papers in the Middle East and ranks 22nd in the world. The country's nanotechnology community published >1000 nanotechnology articles in 2012. The number of nanopapers published annually by Turkish scientists has grown almost 500% since 2005 when the Turkish government made a $120 million grant to create the state-funded National Nanotechnology Research Center (UNAM). Since then, dedicated nanotechnology research divisions attracting large private donations have opened in universities across Turkey, including Bilkent University, Ankara University, Sabanci University, and Middle East Technical University.

Turkish scientists have been collaborating across borders for many years. Cross-border collaborations during 2001–2009 included collaboration projects with the United States, Germany, the United Kingdom, France, Italy, Japan, Russia, Azerbaijan, Romania, Sweden, Spain, and other countries.

There are several nanotechnology companies in Turkey. Examples include Nanotechnology-NNT, which produces an antifriction additive for automobile engines, made from suspended boron nanoparticles; DYO-Nano, which produces dyes and wall paint; Yeşim Tekstil (Textile) Company, which produces nanostructured smart fabrics; and Grafen Chemical Industries (GCI), a producer of carbon nanomaterials and metal/oxide nanoparticles. Turkish companies that specialize in nanofibers include Nano109 and INOVENSO Ltd. Nano109 produces nanofiber-based air filtration products for clean rooms and other applications [21].

In addition to Iran and Turkey, other Middle East countries that are actively engaged in nanotechnology include Saudi Arabia, Egypt, and Iraq.

8.6
Critical Issues for International Nanoinnovation

A lot of insights can be gained from looking at strategies and best practices in the international community, but we also need to address critical issues and challenges. There are many blind spots and asymmetries in nanoinnovation.

These factors can provide competitive advantages – or disadvantages – for countries and companies. As a manager, you need to be aware of these issues.

Issue 1: Economic Value

In most countries, the *economic value* of nanotechnology is measured in terms of employment, education, research activity, and commercialization of products and processes. These metrics can vary widely.

It is difficult to define the metrics of nanotechnology. We can count patents, but not all patents are commercialized. We can evaluate research by counting published articles and papers, but research varies widely in its importance and value. Nanotechnology degrees earned by university students provide a useful metric, along with research projects and outputs, to gauge the value of academic activity. However, economic metrics are more complex.

If you're in a company or entrepreneurial venture, commercial activity is easily measured by calculating revenues and profits, although tying a nanoinnovation to a specific product or process can be tricky since many innovations such as nanomaterials or nanocircuits are embedded in systems, and nanotechnologies occur throughout the supply chain.

Evaluating the return on investment from nanotechnology investments is much more difficult for government agencies and policymakers. Often, a government agency is asked to justify the value derived from millions or billions of dollars in nanotechnology investments. These investments can be evaluated in terms of job creation, reduction of manufacturing costs, new company formation, contribution to export industries, and creation of new products or services – but rarely is there a straightline metric that translates public funding of nanotechnology initiatives directly into commercial value.

As you can see from the descriptions of the examples from various countries in this chapter, each nanoinnovation ecosystem has its own characteristics. Some are extremely open and allow the activities to flow in many directions, while others are tightly focused on helping to generate commercial revenues or support key export industries.

In March 2012, an international symposium entitled "Assessing the Economic Value of Nanotechnology" was hosted by the US NNI and the Organization for Economic Cooperation and Development (OECD). The event was hosted in Washington, D.C. by the American Association for the Advancement of Science (AAAS). The organizers declared that there is enough commercial activity to justify a serious examination of the "economic value" that is being derived from nanotechnology. Presentations featured insights from nanoscientists, business, and government leaders who presented viewpoints and examples from many different countries. *This event was arguably one of the most important nanotechnology conferences since the dawn of nanotechnology.* The conference report is a "must" reading for anyone involved in global nanotechnology. The full conference report and four background reports are available online as downloadable documents (http://www.oecd.org/sti/nano/internationalsymposiumonassessing theeconomicimpactofnanotechnology.htm).

Issue 2: Asymmetries in Nanoinnovation

Asymmetry is an imbalance where one party has an advantage and another party has a disadvantage. An *asymmetry* in technology, national wealth, or market access can create a competitive advantage for one country or company and create a competitive disadvantage for other countries/companies. While nanotechnology is generally available to all nations and nanotechnology researchers come from all nations, the focus of nanoinnovation is not equally distributed. Most of the nanoinnovations being worked on address challenges and opportunities of interest to wealthier nations.

In nanotechnology, countries with the largest and best financed nanoinnovation ecosystems have the privilege of selecting their innovation targets and market opportunities. Unfortunately, this means that problems that exist in smaller countries and remote or rural regions may not be addressed by larger industrialized nations that have the best-funded nanotechnology programs.

Large multinational companies tend to focus their nanotechnology initiatives on products that will yield high profits, rather than low-cost solutions that could be marketed in developing or underdeveloped economies. For example, a large drug company may be more interested in developing an expensive blockbuster drug to fight cancer than finding a pennies-per-dose solution that cures a disease that exists in rural tropical rainforests. Computer companies build increasingly powerful – and expensive – computers and smartphones.

In remote regions of the world and in poorer nations there is a need for better, cheaper technologies – in healthcare, energy, pollution control, food processing, and other areas. Nano-enabled methods are needed to prevent, diagnose, and treat diseases such as malaria and waterborne diseases, and portable diagnostic devices for use in remote villages are also needed.

As the late Dr. C.K. Prahalad observed, large markets in poorer economies should not be overlooked, since there is a "fortune at the bottom of the pyramid." My mentor Jack Tramiel, the founder of Commodore, believed that computers and other technologies need to be developed "for the masses, not the classes." As an innovator and author, I share these convictions.

So how does a smaller, poorer nation access the cutting-edge solutions offered by nanotechnology? In many cases, neglected countries and regions need to develop their own solutions, or find creative ways to obtain solutions from other countries and companies that are willing to work with them. Countries that are not in the mainstream of nanoinnovation need to somehow get connected to nanoinnovation ecosystems. This does not mean that every country needs to develop its own nanotechnology infrastructure. Smaller countries need to clearly identify high-priority problems that can be addressed by nanotechnology, and find ways to join innovation networks and ecosystems. Where possible, they need to create their own innovation centers and encourage entrepreneurs in their home countries.

A good example is Vérone Mankou, a 26-year-old entrepreneur from the Republic of the Congo. Mankou has designed an inexpensive smartphone he calls "Elikia," which means "hope" in his native Lingala dialect. The phone is made in China by Mankou's company VMK, and is marketed by Airtel and other telecoms in Africa.

Of course, not everything is based on technology. Better business models are needed to transfer bionanotechnology solutions from advanced nations to lesser developed nations. Matching the needs of people living in poorer and remote regions with the possibilities that industrial nations can provide is a goal that we all need to pursue.

Issue 3: Invention versus Production

Most nanoinnovations are invented in industrialized countries. The irony is that most of the products that use nanoinnovations are manufactured in lesser developed nations where labor costs are lower. This asymmetry has existed since the 1970s and 1980s when semiconductors and consumer electronics were designed in the United States or Europe but manufactured in Asia. A similar imbalance is occurring with nanotechnology.

The United States, Europe, Germany, Japan, and other nanotechnology centers pump massive investments into nanotechnology research and innovation, while most of the products that use nanotechnology are manufactured in other countries. Wealthier nations produce more patents, but much of the value from those patents is captured by lesser developed countries that do the manufacturing and assembly. For example, it has been estimated that Europe produces 37% of patents that combine nanotechnology and energy – while most energy products that incorporate nanotechnology are produced in China [22].

Issue 4: Using Nanotechnology to Benefit the World

One of the greatest challenges – and also most gratifying achievements – is using nanotechnology to help people in remote areas of the world. Thanks to nanoscale circuits, miniaturization of computing devices, and wireless networks, millions of children are gaining access to computers, classroom education, and medical care that was previously not available or affordable.

There are many examples. The Telesecundaria program in Mexico and Central America enables students to receive postelementary classes via satellite at small classrooms located in remote areas of the country that do not have electricity. Nicholas Negroponte's One Laptop Per Child initiative is bringing affordable, rugged laptops to children in developing countries – under this initiative, >200 000 children have received laptops in the strife-torn nation of Rwanda. Millions of Intel's Classmate PCs have been placed in countries as diverse as Argentina and Nigeria. In India, Minister Kapil Sibal has been pursuing a goal to provide low-cost India-produced laptop computers to Indian schoolchildren, with a long-term cost target of $10 per device.

Issue 5: Participation – or Not

Despite the large number of countries that are funding and supporting national initiatives, many nations are still not actively participating in the nanotechnology revolution. All countries – especially smaller countries – need to support and build a nanotechnology infrastructure or risk becoming a nanotechnology "laggard."

Nano-enabled products and processes are creating new applications, expanding markets, improving productivity, and providing competitive advantages at every stage in the value chain. Staying on the sidelines will be costly for those countries that opt out of the nanotechnology revolution or who remain on the fringe of international networks. These countries will be forced to deal with bulk materials instead of nanomaterials, to work with obsolete technologies, and will be forced to buy nano-enabled products at higher prices from other countries.

In the manufacturing centers of the world – which now include many lesser developed economies – nanoinnovation plays a special role. Nanocoatings are used to reduce corrosion and maintenance costs in systems used to process fruit preserves and other food products. Nanofertilizers are helping to improve the quality of poor soils in harsh environments. Quality control engineers are monitoring tolerances at molecular and atomic scales, and use nanoimaging to detect the most minute amounts of contaminants. Chemical manufacturing costs are being reduced by the availability of more efficient nanocatalysts. Nanoinnovations have enhanced new generations of semiconductors, wireless communications, and solar energy. All of these benefits extend across national borders.

Many problems that plague Third World nations – especially those with large jungle, island, and desert environments – can be solved only by nanoinnovations. Nanotechnology is critical to making solar energy cheaper and more efficient. Nanomedicine is providing new treatments for tropical diseases that don't exist in industrialized nations located in temperate zones. Nanofilters are helping to filter contaminated water and remediate pollution.

With international markets and technology ecosystems so closely intertwined, it's important to go beyond science and technology to consider how nanoinnovations will bring economic, environmental, social, health, and other benefits to all countries.

8.7
Nanoinnovation for the Bottom of the Pyramid

It is appropriate to end this international chapter with the story of an American-born scientist of Indian heritage who is working to apply nanotechnology to solutions that will benefit India and other countries with large rural populations.

Dr. Anita Goel, MD, Ph.D. is a Harvard–MIT physicist, physician, inventor, and internationally renowned pioneer in the new science of nanobiophysics – a nanoinnovator based in the United States who is using nanotechnology to bring disruptive solutions to the world's greatest problems in healthcare, energy, and the environment. Dr. Goel has a particular focus on leveraging nanotechnology to build desperately needed affordable solutions that democratize access to her breakthrough inventions. Dr. Goel was awarded the 2013 XPrize, winning the \$525 000 grand prize in the first competition of the Nokia Sensing XCHALLENGE, which is designed to advance innovative sensing technologies that capture data about a consumer's health and surrounding environment. The Sensing challenge is cosponsored by the XPrize Foundation and Nokia, and is a feeder into the foundation's \$10 million Tricorder competition sponsored by Qualcomm.

She described her prize-winning invention – called Gene-RADAR® – that won her the grand prize in the XCHALLENGE: "The basic idea is, you put a drop of blood or saliva on a chip, place it in a portable mobile device that provides a real-time readout based on a DNA/RNA nanosensor. This process can detect any kind of DNA/RNA signal, such as an infectious disease, food or waterborne pathogen, bio-weapon, swine or avian flu, and so on. We program the target DNA sequence we want to detect and our mobile device will tell us if it's there or not. Our approach works at the nanoscale, enabling better signal-to-noise ratios, and orders of magnitude improvement in accuracy and precision. For example, if we're fishing out a target from a contaminated background we can achieve much higher sensitivity than traditional methods."

The Gene-RADAR platform uses a nanochip to analyze a drop of blood, saliva, or other bodily fluid. The chip is inserted into a mobile device that can detect the presence (or absence) of a disease pathogen in real time. Gene-RADAR is being commercialized by Nanobiosym, a Cambridge-based research incubator and institute led by Dr. Goel and its commercial partner Nanobiosym Diagnostics.

Dr. Goel is an American of Indian descent who grew up in rural Mississippi where she became fascinated by physics, biology, and medicine. "I developed a deep belief that there is an underlying unity in Nature. This belief fueled a lifelong quest – a dream, if you will – to bring fundamental physics to understand life and living systems in a unified theory. Long before it became fashionable, I found nanotechnology is a great bridge to help me bring biomedicine and physics closer together."

In 2004, Dr. Goel established the Nanobiosym Research Institute and Incubator to pioneer disruptive innovations at the nexus of physics, nanotech, ICT, and biomedicine fields traditionally disparate disciplines. DARPA and various prestigious US government agencies tapped Dr. Goel to help them prepare for the next generation of biological threats from bioterrorist agents and pandemics like SARS or avian flu. Nanobiosym has the potential to disrupt how healthcare is practiced, similar to how mobile computing and wireless communication devices disrupted the telecom industry. Dr. Goel wants to bring medical diagnostics outside of a centralized hospital or pathology laboratory to the patient's bedside, doctor's offices, people's homes, and even rural, remote villages in the developing world.

"It's not enough to have a disruptive technology platform. You need a testbed ecosystem, almost like a sandbox where the technologies can be brought into various levels of society and integrated into the economic fabric of countries and regions. We have to reprogram supply chain alliances, an ecosystem that helps us launch and integrate technologies into healthcare and other areas. We see this as an evolving global ecosystem that transcends boundaries, and includes a broader humanitarian mission."

To provide an ecosystem for testing and delivering the Gene-RADAR platform in the developing world, Dr. Goel has begun forming innovative public–private partnerships with governments in emerging markets to develop multipurpose Nanobiosym Technology Parks. Her goal was to bring emerging technologies into emerging markets and especially resource-poor areas where cutting-edge scientific technology can provide new solutions to old problems.

"I realize that our strategy is not traditional," she says. "The conventional approach in our industry is to get your product to a U.S. or European market, try to charge as much as you can to maximize the profit, then flip the company by selling to a larger company to extract value. Our approach is different. We are consciously *cutting* our profit margins so we can reach a larger volume market at the bottom of the pyramid as well as in emerging markets in the industrialized world. This is our way of democratizing access to healthcare, the way Google democratized access to information."

References

1 Roco, M.C. (2005) International perspective on government nanotechnology funding in 2005. *Journal of Nanoparticle Research*, **7**, 707–712.

2 Shapira, P., Youtie, J., and Kay, L. (2011) National innovation systems and the globalization of nanotechnology innovation. *Journal of Technology Transfer*, **36**, 587–604.

3 Centifica (2012) Global Funding of Nanotechnologies & Its Impact, July.

4 Kay, L. and Shapira, P. (2009) Developing nanotechnology in Latin America. *Journal of Nanoparticle Research*, **11**, 259–278.

5 Shapira, P. and Wang, W. (2010) Follow the money: what was the impact of the nanotechnology funding boom of the past ten years? *Nature*, **468**, 627–628.

6 Iranian Nanotechnology Initiative Council (2013) Milestones in Korea Nanotechnology Policy between 2000 and 2012; South Korea Plans to Stand among Top 3 World States in Nanotechnology, December 28.

7 The Chosunilbo (2014) Korea overtakes Japan in semiconductor market, January 25.

8 Grant Banks (2011) Japanese scientists produce artificial palladium. *Gizmag*, January 3.

9 Foladori, G. and Invernizzi, N. (2013) Inequality gaps in nanotechnology development in Latin America. *Journal of Arts and Humanities*, **2**, 37–40.

10 Foladori, G. and Invernizzi, N. (2007) Nanotechnologies in Latin America. ReLans, December.

11 Lau, E.Z. (2011) Nanotech cluster in Nuevo León, Mexico: reflections on its social significance. *Nanotechnology Law and Business*, **8**, 49–59.

12 Federal Ministry of Education and Research (2011) Action Plan Nanotechnology 2015.

13 Federal Ministry of Education and Research (2014) Competency Map Nanotechnology in Germany. www.nano-map.de (accessed January 11, 2014).

14 Johnson, D. (2013) Tricorder-like mobile phones enabled by nanotechnology. IEEE Spectrum, November 11.

15 Steadman, I. (2013) Kenya breaks ground on its 'Silicon Savannah' city project. Technology, January 25.

16 Ghosh, P. (2013) Kenya starts building 'Silicon Valley' city: as Quiet Tech Revolution transforms country. *International Business Times*, January 23.

17 Molapisi, J. (2012) Nanotechnology development in South Africa,

International Symposium on Assessing the Economic Impact of Nanotechnology, March 27–28, 2012.

18 (2010) Latest invention: tea bag that uses nanotechnology to clean drinking water. InfoNIA.com (accessed August 17, 2010).

19 Jayasuriya, S. (2012) Laugfs, SLINTEC tie up for natural mineral sand processing. *Daily News*, March 2.

20 (2013) SLINTEC and Stretchline kick-start joint research. Daily FT, December 20.

21 Nano Statistics (2013) Nanotechnology in Turkey, Iranian Nanotechnology Initiative Council, August 27.

22 OECD and NNI (2012) Synthesis Report: Symposium on Assessing the Economic Impact of Nanotechnology.

Part IV
Where Bio Meets Nano

9
Innovation at the Frontiers of Nanomedicine

> Interpreting, replicating and modulating our biology in a bid to make our lives healthier and happier is one of the aims of the modern nanoscientist.
> – *Alok Jha,* The Guardian, *September 5, 2011*

Bionanotechnology is the place where bio meets nano. This is a space where we focus on the biological impact of nanoinnovations on plants, animals, and humans.

Whether you manage a small or a large organization in industry, government, or academia, you need to know what's happening in bionano – and especially, nanomedicine – because medical nanoinnovations could save or prolong your life, and the lives of family members, friends, and colleagues. Nanomedical innovations are not just game-changing innovations. They are life-changing innovations.

Bionanoinnovation can be defined as *the controlled use of biological materials, organisms, by-products, and processes with nanoscale dimensions to achieve a desired result.*

Nanomedicine is the application of nanotechnology to medicine and healthcare. Nanomedicine is the subset of bionanotechnology where nanoinnovation is creating new healthcare solutions. Bionanoimaging is used to view and manipulate biological processes that cause diseases. Nanofluidic chips are used in the emerging field of nanodiagnostics. Most labs-on-a-chip have nanoscale features and properties. Nanomaterials are used in stents, pacemakers, and other medical implants. Much of what we call "molecular medicine" involves nanoscale structures.

As humans, our lives are enabled and sustained by biological nanomachines. Most of the cells, proteins, bacteria, and viruses that comprise the human ecosystem are nanoscale in size or have nanoscale features. Being able to view, manipulate, and control these structures is critical to preventing and treating disease, keeping us healthy and helping us live longer.

In this chapter and the next, we'll explore the implications of bionanoinnovations for human healthcare and take a closer look at some of the most promising emerging technologies and discoveries that are driving the future of medicine.

NanoInnovation: What Every Manager Needs to Know, First Edition. Michael Tomczyk.

9.1
Medical Miracles and the Nanomedicine Landscape

Many of the breakthroughs coming out of nanomedical research sound like medical miracles. Curing a potentially fatal disease with a nanosized drug is a miracle. Replacing a defective or missing gene to cure inherited blindness – restoring the victim's sight – is truly miraculous. Engineering nanoparticles to find and destroy a cancer cell, that's a miracle. How about putting a tiny computerized "brain" in a drug tablet that sends information to the Internet after you swallow the pill? Or nanosizing a cooking spice that fights the cellular processes that cause disease and aging?

Some people would say these things are just "clever science" but for the people who would die without these solutions, and who dodge a medical bullet that could kill or cripple them and live longer healthier lives, these are miracles. In this sense, the research teams that are working on these innovations are miracle workers.

Thanks to a legion of hardworking scientists, researchers, educators, and students – and research communities enabled by the Internet – the world's collective genius is slowly but diligently unlocking the nanoscale structures and processes that cause disease. They are finding ways to make drugs more effective by nanosizing them to make them soluble, more bioavailable, and safer.

Just as nanotechnology is enabling semiconductors, thin and flexible display screens, and smaller electronics overall, nanoinnovations are helping to create the next generation of medical devices. The entire concept of mobile medicine is being fueled by nanoinnovations in semiconductors, smartphones, robotics, and other technologies. Smartphones are rapidly morphing into mobile diagnostic platforms. For example, smartphones are being equipped with microscopes that can magnify and identify pathogens. Medical implants such as pacemakers and stents are becoming smaller, more efficient, and durable. Nanotechnology is providing scaffolds for growing stem cells into organs. Nanomaterials are improving stents used to repair heart vessels. Nanotechnology is reducing the processing time and cost of analyzing medical samples. Cancer cells are infused with nanoparticles, so they show up more vividly on MRI scanners. Nanoparticles can also be heated to kill tumor cells without damaging healthy cells. Nanofluidic biochips are analyzing DNA molecules. Research teams have engineered biodegradable nanoparticles that can actually punch holes in harmful bacteria. And for the first time in medical history, scientists are getting close to an antiviral therapy that can treat a broad spectrum of nanosized viruses that cause everything from the common flu to AIDS and tuberculosis.

Mapping Bionanotechnology

The accompanying bionanotechnology map (Figure 9.1) shows some of the key areas where bionanotechnology is impacting human healthcare. This map is not intended to be comprehensive but rather representative of the functions, applications, and innovations encompassed by bionanotechnology. This dynamic

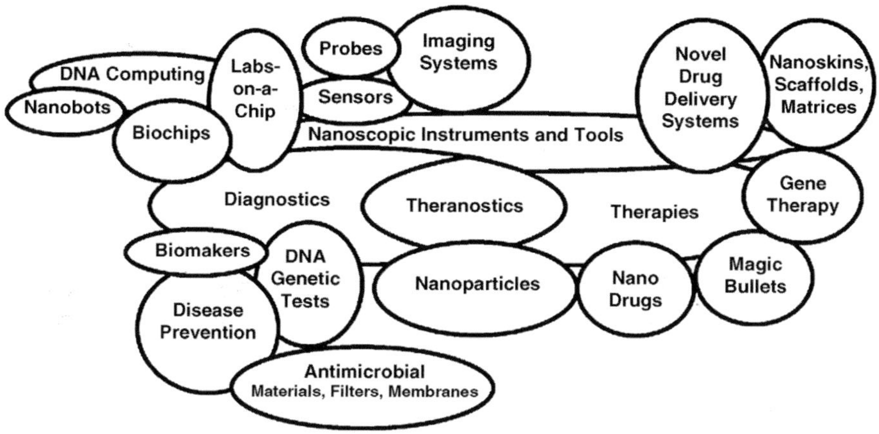

Figure 9.1 Bionanotechnology map (copyright 2012, Michael Tomczyk).

map, which is constantly changing, offers a starting point to help understand the broad scope and potential of bionanotechnology.

Typically, a nanomedical innovation gets its start at a university science center where a discovery is made by a research group, usually with government grants and corporate funding. If the innovation is successful, it may be spun out into a business venture or licensed to a biopharmaceutical company. In this sense, the business of nanomedicine follows the model of biotechnology boutiques that have helped maintain the drug pipelines of big pharma companies during the past three decades.

Nanomedical ventures and projects have to negotiate the same maze of challenges as any innovation, as well as some daunting obstacles that are unique to health-related technologies. Bionanoinnovations have long development cycles. Solutions that work in the laboratory need to be converted to commercial products. A nanomedical drug that works *ex vivo* – outside the body – may not work *in vivo* – inside the body. Most medical products need to survive expensive animal studies and human clinical trials before they are approved. Novel therapies, no matter how effective, still need to prove that they are substantially better than existing technologies that are familiar to the medical community. And with nanoparticles especially, there are safety concerns that need to be addressed.

Funding for nanomedical research is provided by government grants, corporate partners, venture capitalists, and angel investors, although funding has been tight in recent years. Overinvestment in gene therapy, early nanotechnology ventures, and ecommerce ventures led to “bubbles” that burst in the late 1990s and early 2000s. These notorious investment bubbles, combined with the economic crisis triggered by the subprime mortgage meltdown in 2008–2009, have tended to make venture capitalists and corporate R&D centers more risk-averse and cautious when it comes to bionanotechnology.

Fortunately, a solid core of nanomedical ventures survived the venture capital drought and this sector has been growing steadily.

The Business of Nanomedicine

Like any medical innovation, bionanotechnology needs to justify its existence, by saving and extending the lives of human patients, generating reasonable profits for its providers, and ideally reducing healthcare costs.

As a business sector, nanomedicine is tricky to analyze because the metrics are not yet well defined. However, there are ways to quantify nanomedicine. In the United States, we can count how many nanomedicine drugs and therapies are in clinical trials at the FDA. We can also quantify the value of the sector by calculating revenues, research grants, and milestone payments. One way to measure the progress of any medical technology is to track the number of projects that are in clinical trials. Currently, there are >250 nanomedicine projects in clinical trials at the FDA. Those nanomedical innovations that survive the rigorous clinical trial process will have an opportunity to generate value in both human and economic terms.

As a manager, you understand the critical need for better, cost-effective medical and healthcare solutions, especially in an era of spiraling healthcare costs. Economically, healthcare represents more than 17% of the US economy and is increasing as a percent of GDP in most countries in the world. Nanomedicine (and nanotechnology in general) has a unique opportunity to contribute innovations that reduce costs, replace expensive biological compounds with less expensive substitutes, and prevent/treat/cure medical conditions that keep inflating healthcare costs in all countries.

We can also think about nanomedicine as a sector of innovation that creates value in the form of entrepreneurial ventures, corporate and government research projects, educational programs, and jobs associated with these activities. The most important value proposition of bionanotechnology is that it is opening a portal to an entirely new set of medical solutions that were previously unavailable.

The Nanomedicine Market

Nanomedicine can be grouped in four basic categories: (1) **novel platforms for improving medical treatment and delivery of secondary products** (e.g., *finding new and better ways to deliver chemotherapy molecules to tumors or coating stents with nanoparticles*); (2) **nanomaterials used in medical applications** (e.g., *antibacterial nanocoatings used to sterilize work surfaces in hospitals and scaffolds to grow human tissue*); (3) **pharmaceutical products** (e.g., *nanosized drugs and biomarkers*); and (4) **nano-enabled medical devices** (*biochips, DNA analyzers, and portable diagnostic devices*).

A report from the Business Communications Company (BCC) projects that the market value of the global nanomedicine industry will grow from \$63.8 billion in 2010 to \$130.9 billion by 2016, with most of the current value concentrated in six therapeutic categories: cancer, infection, inflammation, cardiovascular problems, nervous system disorders, and other medical conditions [1].

During my research for this book, I came across an insightful article in *The Journal of Drug Delivery* [2]. Valentina Morigi and her colleagues calculated the economic value of the nanomedicine market by measuring sales revenues, licensing fees, and milestone payments. The authors determined that the US nanomedicine market has been growing steadily despite the impact of the 2008–2009 recession and the global nanomedicine market is growing at a compound annual growth rate (CAGR) of 13.5%, increasing from $53 billion in 2009 to a projected figure of $100 billion in 2014.

Dr. Morigi and her colleagues explored strategic and funding issues involved in nanomedicine projects, and conducted a detailed analysis of the relationship of nanomedicine markets, competitors, and investors. They found that the United States accounts for 53% of nanomedicine patent applications followed by Europe (25%) and Asia (12%). Drug delivery is the largest segment in nanomedicine, accounting for 76% of publications and 59% of nanomedicine patents. The second largest segment is *in vitro* diagnostics (11% publications and 14% of patent filings).

They analyzed the growth of nanotechnology investments from 2007 to 2009, which showed that the largest increase was in the healthcare sector (Figure 9.2a). They also found that the growth of the global nanomedicine market has been comparable to the growth of the anticancer products market (from 2006 to 2014 projections) (Figure 9.2b).

Based on their research, they observed what they call an "asymmetry" between investors and scientists involved in nanomedicine: "While most members of the investment community are able to grasp the meaning of nanotechnology and can expertly launch and manage a viable product into the market, they are limited in their conceptual understanding of this scientific discipline and the intricate inner workings behind the product's functionality. On the contrary, those involved in the scientific research recognize that nanomedicine is an expansion of nanotechnology but they have very little understanding of the business expertise required to develop their technologies into a commercial product" [4].

This asymmetry in mind-set between investors and scientists was mentioned by many of the nano-insiders interviewed for this book. It is essential for the investors in a nanomedicine project to understand that there are inevitably long lead times involved, high costs associated with animal studies and human clinical trials, and high uncertainties and risks.

Funding Issues for Bionano R&D

The high levels of risk, long-term commitments, and uncertainties in nanomedicine (and biotechnology in general) have made it difficult for large traditional investors to support bionano research. Consequently, most start-up ventures and pioneer technologies are started with the inventors' own money and funding from angel investors, government grants, and corporate partners.

The biopharmaceutical industry, which you would think would be the major player in this sector, has been unusually reluctant to invest in bionanotechnology. This bias was discussed in a 2006 Lux Research report entitled "Why Big

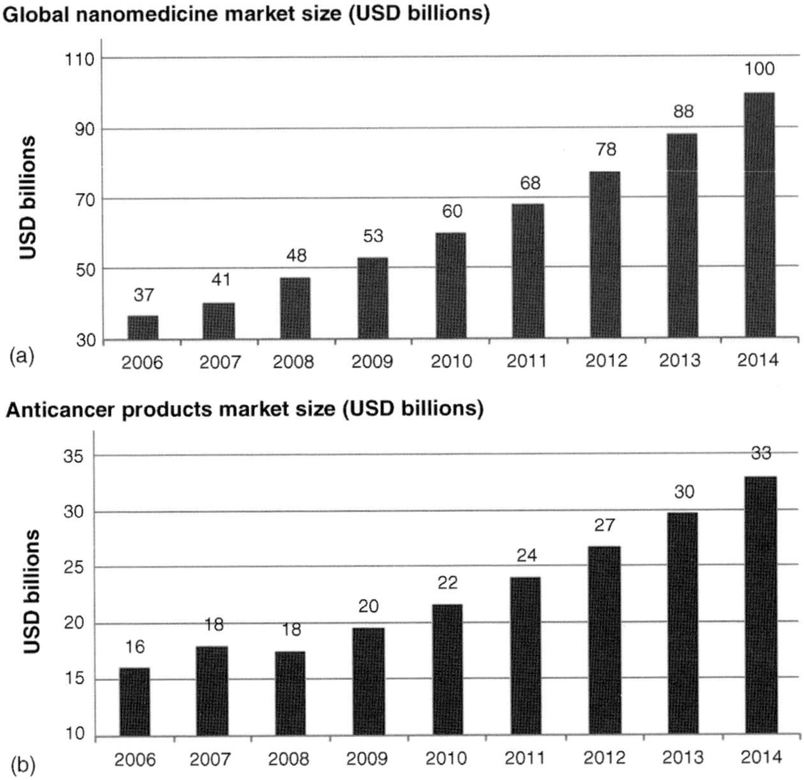

Figure 9.2 (a) Global nanomedicine: market size. (b) Market size for anticancer products (*Source:* BCC Research, Nanotechnology in Medical Applications: The Global Market, and "Nanotechnology in Medicine from Inception to Market Domination" [3]).

Pharma Is Missing the Nanotech Opportunity" [5]. The research firm based its opinion on interviews with 33 global corporations that yielded some fascinating findings:

- No life sciences interviewee rated nanotech as a high corporate priority, as opposed to 78% of interviewees in electronics and materials.
- Only one out of six life sciences respondents claimed to have an explicit strategy for nanotechnology, compared with two-thirds of those in other electronics and materials.
- Big pharma companies on average committed 16 people and less than half of 1% of R&D spending to nanotechnology research, whereas like-sized electronics and materials firms committed more than 100 people and more than 8% of R&D.

The Lux Research report went on to suggest that large drug manufacturers were paying little attention to nanotechnology for three reasons: organization, history, and hubris. The analysts observed that a big pharma focuses on drug discovery, whereas nanotechnology focuses on drug delivery. Many pharmaceutical firms felt they had nanotechnology experience because they developed small-molecule drugs, but this does not necessarily translate into bionanotechnology expertise.

The report concluded that big pharma's *laissez-faire* attitude toward nanotechnology could have serious consequences as generic drug makers create nanosized versions of branded drugs. The Lux Report recommended that big pharmaceutical firms should consider adding "nanoscale reformulation specialists" to their technology teams.

Anecdotally, I have spoken with executives at large corporations and pharmaceutical firms who confirm that they argued against or abolished in-house nanotechnology R&D projects, citing the high costs and risks and long development cycles. One former VP at a large multinational corporation recalls that he killed a billion dollar nanotechnology research commitment based on his belief that nanotech projects would be too expensive, the lead times were too long, and the therapies that would result did not seem to synch with the company's drug-oriented research focus. This confirms the notion that biopharmaceutical firms favor drug discovery over drug delivery and seem to have a bias against therapies that do not involve drugs. This bias exposes the industry to risks from disruptive technologies that cure diseases using nondrug therapies.

Nondrug Therapies: A Paradigm Shift?

Many of the most exciting innovations in nanomedicine involve therapies, not drugs. The question is: Will large pharmaceutical firms embrace a portfolio of therapies that are not necessarily drugs or will device companies such as Medtronic, AbbVie, or smaller biotechs provide nondrug therapies enabled by nanotechnology?

Nanomedicine is helping to create an entirely new portfolio of medical therapies to combat disease and allow people to live longer, healthier lives. Many of these innovations have the potential to *replace* drugs, surgery, and other traditional medical treatments. This could represent a major paradigm shift in the healthcare industry. Drugs and surgery have been the dominant medical treatments for more than a century. Nanomedicine not only scales down the treatment of medical conditions to individual cells and molecules but it also includes solutions that are already changing the concept of how we diagnose, prevent, treat, and cure disease.

For example, nanoparticles of gold or iron, or carbon nanotubes have been shown to deliver cancer-killing heat and radiation to tumor sites. Near-infrared radiation is a promising cancer-fighting innovation that involves injecting gold or iron nanoparticles at the site of a tumor and administering radiation to heat the cells, which kills the tumor cells but leaves the healthy cells undamaged. The heated nanoparticles are technically not a "drug." They are a *therapy.*

Experimentally, nanoparticles are being used to focus and intensify heat at tumor sites. But heat is not a drug. It is not a surgical procedure. It is a treatment or a therapy. Heat has been shown to kill temperature-sensitive cancer cells. In Europe, cancer specialist Dr. Carlo Pastore has been designing "temperature therapy centers" in Italy, Spain, and other countries, where the primary cancer-fighting therapy is the administration of heat to destroy tumors, especially tumors that can't be treated with conventional drugs or surgery. In the United States, Dr. Lance Becker and his colleagues at the University of Pennsylvania pioneered the use of therapeutic hypothermia to stabilize critical care stroke and cardiac patient during treatment and recovery. Temperature therapy goes far beyond simply bringing down the temperature of a patient. Special cooling systems and instruments are needed to administer cooling evenly without damaging the body. For heat treatments, the opposite is often true, as heat needs to be directed and focused not only to small tissues but also, in the case of cancer, to tumor cells.

A more established medical technology that is benefiting from nanoinnovation is the cardiac stent, essentially a tube used to sustain blood flow through constricted or clogged blood vessels. Stents are typically smooth or mesh tubes implanted in blood vessels to increase blood flow. Drug-eluting stents are infused with drugs that prevent cells, platelets, and other cell fragments from proliferating on the stent surface. Magnetic nanoparticles are being used to extend or renew the drug coatings, and biodegradable nanoparticles are being developed for applications where the stents are intended to dissolve over time. Nanomaterials are also used to smooth the surface of metallic stents to improve blood flow.

Gene therapy is one of the best-publicized "new vistas" of medical research that involves nanotechnology if because the DNA molecules that carry genetic codes are only 2 nm wide. Gene therapy involves administering a single injection of genetic material to cure a protein deficiency with one shot. In some cases, a single treatment can last a decade or a lifetime. It has taken a decade and a half to achieve, but gene therapy is finally showing signs of success. Scientists are beginning to cure several rare diseases using gene therapy – including hemophilia B and genetic blindness. There are signs that gene therapy will also provide new methods for treating cancer, heart disease, and even the flu.

Nanomedicine includes preventive medicine as well as medical treatments and cures. This includes the development of nanosized versions of natural foods and chemicals that have disease-fighting and even age-extending properties. Natural compounds with cancer-fighting properties such as curcumin (turmeric) are being nanoengineered to make them more soluble, safer, and able to penetrate cell membranes. One day, your mother or wife may ask, "Would you like some nano-cumin in your soup?"

New Therapies, New Issues

These new therapies raise some interesting questions and issues. For example, a drug or medicine is prescribed and administered by a doctor, often in a hospital or clinic. However, if a therapy delivers heat or radiation or genes to a patient, does this have to be done in a hospital or clinic? Can it be done at a new type of facility, such as a nanotherapy or gene therapy center?

Another issue involves pricing. How do you price a therapy that is not a drug or a surgical procedure? It's relatively easy to price a pill but how do you price a *treatment?* How does a biotech or pharmaceutical company recover its R&D investment if the medical solution cures a disease with a single injection of genes? If it takes a billion dollars to develop that single shot therapy and the therapy works with a single treatment, does the company have to charge $1 million for each injection to recover its R&D costs?

If a worker gets sick because he or she was exposed to toxic nanoparticles, how does a hospital detect the nanoparticles? How does a diagnostician know if a symptom is caused by a toxic that is known, or a reaction to nanoparticles, which are not yet well understood? Will medical labs need to have atomic force microscopes?

These are a few of the critical issues involved in the emerging field of bionanomedicine.

Nanomedicine Is a Solution, Not Just a "Product"

Thinking of nanomedicine as a "product" is an outdated concept. Nanomedical innovations are more appropriately categorized as *solutions*. To deal with these and other nanoinnovations, you need to adjust your concept of marketing.

If you studied marketing in college, you probably learned about the "five P's" of marketing – also called the "marketing mix." This includes five market factors that decision-makers can control and incorporate into their business strategy. Traditionally, the 5 P's include Product, Price, Place, Promotion, and People. That's the traditional framework previously used to promote and sell products and services.

The new concept of marketing is more customer-centered, and thinks of innovations as solutions rather than mere products and services. In 2005, Chekitan Dev and Don Schulz presented an expanded next-gen marketing mix called SIVA [6] that stands for **S**olution, **I**nformation, **V**alue, and **A**ccess – where Product becomes Solution, Price becomes Value, Place becomes Access, and Promotion becomes Information. You might notice that SIVA only replaces four of the "5 P's." In a book chapter I wrote in 2011, I suggested adding a fifth element to SIVA, by replacing "People" in the traditional marketing mix with "Community," so SIVA becomes SIVAC (Solution, Information, Value, Access, and Community) [7]. These five elements can be used to develop and analyze markets for almost any innovation. The SIVAC framework works especially well for entirely new innovations and markets that previously did not exist. This is a "nextgen" way of thinking about technologies, products, services, and markets. To paraphrase Tom Friedman, we live in a flat world, where solutions rule. We need to be constantly updating our frameworks for dealing with this fast-changing world.

Nanomedical Solutions Need to Be Affordable

It is becoming clear that most economies, even in the industrialized world, can't afford to keep increasing their healthcare costs. Either we have to learn to prevent a disease or detect and cure it earlier, or we need to develop a portfolio of affordable solutions that are effective and available in any country in the world.

In the near future, it will be necessary to have healthcare solutions that are more affordable. We can't keep developing increasingly expensive technologies and solutions that most countries and socioeconomic groups can't afford – creating a world of medical haves and have-nots (which exists to some extent, already). We can't allow the burgeoning healthcare budget to consume a disproportionate share of national economies. Fortunately, there are remedies looming on the near horizon and many of these involve nanotechnology and nanomedicine.

For example, instead of using customized drugs to treat hundreds of different viral diseases, nanomedicine may be able to deliver one therapy that can treat 80 or 90% of all viruses. Instead of waiting for an inherited genetic disease to manifest itself, we may analyze an individual's genome and see the disease coming before it strikes, allowing us to preempt and cure the disease. Scientists are already developing binding agents that identify, target, and deliver cancer killing drugs to tumor cells, and at some point we may be able to draw from a mix-and-match menu of cancer fighting systems that target specific types of cancers and target individual cells. Also, we are close to being able to "print" human organs using our own cells, to replace damaged or diseased organs, using scaffolds made with nanomaterials. This would not only provide organ transplants without worrying about rejection of donated organs but could also provide a means to replace failing organs in older people, to extend their lives and provide a better quality of life.

These are only a few examples. In the next chapter, we'll examine bionanoinnovations that have the potential to radically transform the practice of medicine and healthcare – with solutions that are both effective and affordable.

References

1 Chai, Cameron (2012) Global Nanomedicine Market to Reach $130.9 Billion by 2016, Azonano.com (accessed January 23, 2012).

2 Morigi, V., Tocchio, A., Pellegrini, C. B., Sakamoto, J.H., Arnone, M., and Tasciotti, E. (2012) Nanotechnology in medicine: from inception to market domination. *Journal of Drug Delivery*, **2012**, http://dx.doi.org/10.1155/2012/389485.

3 Morigi, V. *et. al.* (2012) Nanotechnology in Medicine: From Inception to Market Domination, *Journal of Drug Delivery*, p. 3.

4 Flynn, T. and Wei, C. (2005) The pathway to commercialization for nanomedicine. *Nanomedicine: Nanotechnology, Biology, and Medicine*, **1** (1), 47–51.

5 PR Newswire (2006) Big Pharma Is Missing the Nanotechnology Opportunity, February 15.

6 Dev, C.S. and Schultz, D.E. (2005) A customer focused approach can bring the current marketing mix into the 21st century, *Marketing Management*, **14** (1), 18–24.

7 Tomczyk, M.S. (2011) Applying the Marketing Mix (5 Ps) to Bionanotechnology. *Biomedical Nanotechnology*. Humana Press, 393–411.

10
Areas Where Nanoinnovations Are Creating Medical Miracles

The role of the infinitely small in nature is infinitely great.

– *Louis Pasteur*

If you've read the news about nanotechnology and especially nanomedicine, you would have probably noticed that every discovery sounds like it's going to produce a cure for cancer, help us live past 100, or cure the common flu. Given the flood of medical breakthrough articles, it is hard to figure out what is really happening.

The truth is nanomedical research is already changing how diseases are prevented, detected, diagnosed, treated, and cured. Nanoinnovations are helping to enable medical breakthroughs in every area of medicine. Thanks to nanotechnology, medical science can now diagnose virtually any disease, from new forms of influenza to rare forms of cancer. Medical laboratories use automated labs-on-a-chip. DNA analysis uses biochips that incorporate nanofluidics. Molecular biomarkers are used to target and deliver chemical bullets to individual tumor cells. The first gene therapies are working. New antiviral approaches could prevent or treat up to 90% of all viral diseases. Scientists are using stem cells to grow human tissues and organs. Foods with health and medical benefits are being nanosized and formulated in a new generation of nutraceuticals. Therapies and diagnostics are converging in a variety of innovations called theranostics.

In this chapter, you'll learn about some of the most creative and ingenious achievements occurring at the convergence of biology, nanotechnology, and medicine. You'll also learn about some very remarkable scientists who are inspiring and guiding this research. Many of these innovators did not start out in bionanotechnology, but came to the field from other disciplines. All of these pioneers have experienced frustrations and failures – some tragic – in the course of their research.

One of the sad realities of medical research is that many patients die during research experiments and clinical trials, or before a cure can be perfected. Gene therapy pioneer Jim Wilson drew inspiration from the tragic death of an 18-year-old gene therapy patient who died due to a fatal immune response. This tragedy provided insights and motivation needed to develop a family of gene

NanoInnovation: What Every Manager Needs to Know, First Edition. Michael Tomczyk.

therapy vectors that are being used in the first wave of gene therapies that are actually curing diseases.

The stories of how researchers discovered their ideas and turned them into innovations are fascinating and full of insights. One of the interesting things I learned from my interviews with nanoscience pioneers is that they are not only pragmatic in their research but also creative. For example, many nanoscience pioneers create a theme or catchphrase to help define and guide their work. For example, Proteus CEO Andrew Thompson talks about his vision of "intelligent medicine." Don Ingber at Harvard describes a concept he calls "collaboratories." Todd Rider at MIT – a rocket scientist-turned-bioengineer – coined the terms PANACEA (pharmacological augmentation of nonspecific anti-pathogen cellular enzymes and activities), CANARY (cellular analysis and notification of antigen risks and yields), and DRACO (double-stranded RNA activated caspase oligomerizer) to describe his work on pathogen detection and treatment.

Another insight is how much work is involved. It is important to acknowledge the legions of unsung heroes – students, faculty, postdocs, networks of collaborators sharing information as well as entrepreneurs, investors, government and corporate champions, and so many others who contribute to the innovation process but may not receive recognition or accolades. The innovations and discoveries described below have already achieved notable success and some media attention, but have yet to reach their full potential. The scientific trajectories and technical paths of these innovations suggest that they are achievable if they stay on track. If successful, these solutions will transform the nature of medicine and greatly improve human healthcare – we need to keep them on our collective radar screens and do what we can to fulfill their promise.

10.1
Smart Pills and Wearable Sensors = Digital Medicine

Nanoelectronics and nanofluidics are raising the I.Q. of medicine and putting more information in the hands of doctors and nurses as well as patients and their families. Patients are beginning to benefit from digital medical feedback systems that diagnose and release pharmaceutical drugs, proteins, enzymes, hormones, and variety of therapeutic chemicals. These systems monitor doses, track patient responses, and wirelessly communicate the results.

A great example is the "smart pill" depicted on the cover of this book. The picture shows a sensor made entirely of ingredients found in the human diet, embedded in a drug tablet. Think about swallowing a pill that contains a computer chip that can communicate wirelessly with a diagnostic laboratory or care provider. It may seem like science fiction to embed an ingestible sensor in a drug tablet and connect it wirelessly to the Internet, but this FDA-approved innovation exists today. This remarkable innovation came from the convergence of therapeutic drugs, semiconductors, and wireless communication.

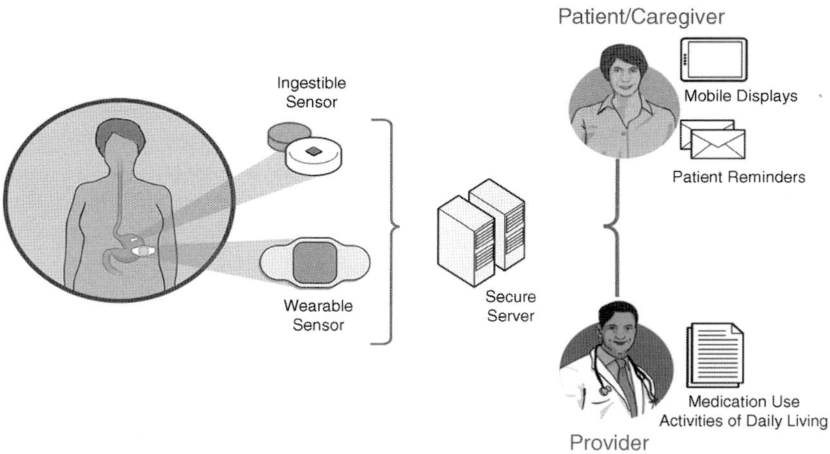

Figure 10.1 The Proteus Digital Health Feedback system. A patient swallows the ingestible sensor (which can be embedded in a pill, a capsule, or other carrier). The ingestible sensor sends data wirelessly to a wearable sensor, which transmits the data via phone/Internet to a secure server. The sensor readings can be accessed by or sent to the patient, family members, caregivers, medical practitioners, researchers and others involved in monitoring the patient's care (image courtesy of Proteus Digital Health).

This unique digital health feedback system was developed by Proteus Digital Health (*previously known as Proteus Biomedical*). This personal monitoring system includes an ingestible sensor that can be embedded in a pill and swallowed without ill effects or rejection by the body. The sensor detects when the pill is swallowed, and wirelessly communicates this to a bandage-like patch, which in turn sends a notification to a caregiver or diagnostic laboratory (see Figure 10.1). The sensor was engineered from materials found in the human diet to prevent rejection by the body's immune system, and is activated by stomach fluid. In large quantities, the sensors cost only a few cents each. This technology exists today for use inside placebo tablets, and will soon be approved for use in active pharmaceuticals.

Innovating the Regulatory Process

Proteus' ingestible sensor represents a huge leap forward in the field of telehealth monitoring and diagnostics that do not neatly fall into an existing medical category. The FDA's *de novo* approval process is reserved for devices that are low risk, but are so novel that there is no existing precedent in the market. The concept of a "digital pill" was an entirely new innovation, which is not unusual in the field of nanoinnovation, where most developments involve technologies and applications that did not previously exist.

Proteus' management team began working with the FDA in 2008 to obtain regulatory approval, which required the designation of a new patient care and medical device category. The first version of Proteus' technology – embedded in a placebo tablet – was approved by the FDA in mid-2012.

According to Andrew Thompson, cofounder and CEO of Proteus Digital Health, the company's goal was to help change our current "sick care" system into a much more responsive "healthcare" system. Thompson recalls how Proteus conducted early clinical trials in Europe and the United States, working with the company's major partners Medtronic, Kaiser Permanente, Novartis, and Otsuka Pharmaceuticals. The involvement of Swiss-based Novartis and Otsuka, based in Japan, gave Proteus market access in Europe and Asia. In May 2013, Oracle made a strategic minority investment in Proteus, which enables them to work together in clinical trials exclusively to provide clinical investigators with the ability to measure information about medication ingestion, dose timing, and associated physiological responses for patients enrolled in clinical trials. The two companies expect to integrate Proteus' ingestible sensor with Oracle's clinical trial products.

What is Unique about Drugs that Communicate?

The Proteus system involves several remarkable innovations that are broader in scope than the product itself. The concept of an ingestible computer, for example, removes concerns about how to make a sensor that can be used in the body without triggering a human immune response. Adding wireless communication allows the device to send information to the Internet or via a smartphone. The timing for this innovation is excellent, given how smartphones themselves are morphing into hybrid forms such as phablets that combine the best features of phones and tablet computers. Another question was how to power the sensor. Proteus' ingestible sensor is activated by stomach juices, which means the human body is the "battery."

The degree of personalization in this digital health feedback system is unprecedented. Knowing when a medicine is being taken can confirm that a patient enrolled in a clinical trial has taken his or her medicine at the proper time. It can also tell a caregiver that someone's elderly grandmother hasn't forgotten to take her medicine. The system can also transmit data on the drug's effect on the patient and a wealth of other information.

Applications for this technology include cardiac care, organ transplants, mental health, and more. For example, doctors will be able to make smarter decisions based on dosages a patient is actually taking. They might see that a cardiac patient isn't sleeping well, which could cause congestive heart failure. The system can also help monitor patients with emotional disorders who need to take their drugs on a prescribed schedule.

The Proteus Story

Proteus Digital Health was created in 2001 (*then named Proteus Biomedical*). The founders were Andrew Thompson (Chief Executive Officer), George Savage, M.D. (Chief Medical Officer), and Mark Zdeblick, a Ph.D. physicist from Stanford (Chief Technology Officer). Proteus has received several awards for the company's groundbreaking innovations, including the Technology Pioneer Award from the World Economic Forum.

Mr. Thompson is glib, thoughtful, and is extremely well credentialed. He holds three master's degrees – an engineering degree from Cambridge where he was a United Kingdom National Engineering Scholar, a master's in education, and an MBA from Stanford University. He has an easy-going manner and speaks with an eloquent British accent. Thompson's enthusiasm is reflected in his voice as he describes how he and his two cofounders came up with the concept for their company.

"We've basically established a universal architecture for our digital health feedback system," Andrew explains. "We can embed sensors in almost any therapeutic product such as implanted devices or drugs. Everything you might use for therapy is potentially a networked device and a carrier of electronics. We can do it without changing the appearance of the drug or device, or how the consumer or medical practitioner interacts."

"The genesis of this company came from the focus in the late 1990s on applications that made things very small, such as circuits, line widths in semiconductors, electromechanical systems, fiber optic switching, etc. We had an interest in making use of these technologies in healthcare. In 2001, we had just experienced an enormous technological cycle around the creation of the global Internet, which essentially created a new utility like water or gas, but with the distinction that it is more accessible, reaching about 6 of the 7 billion people on the planet."

"We began to look for analogies in other industries and innovations that improved value, delivery and economics. We had seen some industries totally transformed by making use of embedded computers and sensors to personalize products. We thought about the evolution of the automobile and the embedding of computers and sensors in cars, which made cars safer and more personal. We thought about how we might embed computers and sensors in therapeutic products that would also make use of the Internet, perhaps by linking patients to caregivers and diagnostic services. When someone has a medical problem, it would be great to have architecture for what we call digital health feedback. If we could embed computers and sensors in almost any therapeutic product – implanted devices, drugs, etc. – we might reduce the need to send out the human equivalent of a tow truck, which is an ambulance."

"Our first challenge was to create a package at the millimeter scale that could interface and communicate and survive inside the body without corroding or causing an immune response. The second challenge was to design an architecture that would create digital signals inside the body, using components found in the food chain. As a result of this process, we created a new discipline which I call food-based computing. This combines advanced technologies with materials found in a banana, a beer or a loaf of bread. It is a very significant technical achievement and a new paradigm that dramatically changes the way we think about computer and sensor-based functionality. We can embed an ingestible sensor in almost any type of device or drug that can monitor and report results. We can network this sensor to the Internet, or to a wireless communication system that reports the results back to a caregiver or diagnostics lab – which is what we might call a networked pill."

Thompson said the company has conducted numerous studies to identify the most effective applications for these technologies. The best applications are those that involve directly observed therapy. These include conditions where patients come to a clinic every day for treatment, need to take their medications frequently, or need to be monitored to have their medications adjusted or changed.

"One of the problems in medicine today is patients who stop taking their medications," he explains. "Another trend in patient care is the need for directly observed therapy. Some patients need to come to a clinic every day. All patients need to keep taking their medication. Using our technology, we can implement electronically observed therapy. If you take a networked pill, this can be observed over the Internet. Nurses like it because instead of watching 10 people a day they can monitor a thousand, or even ten thousand. Patients like it because instead of going to a clinic they can be monitored electronically."

One promising application involves tuberculosis. As many as two billion people may be infected with TB, which is about one-third of the world population. This is a curable disease today, but treatment requires observation and adjusting of medications. Other examples include HIV/AIDS, cardiovascular conditions, infectious diseases, and mental health. Digital medicines can provide information on whether patients are taking their drugs as prescribed and getting the proper response, and can even determine if they are getting enough sleep. For some patients, being awake at night can be a symptom of mania, while sleeping too much during the day can be an early symptom of depression.

"By knowing what medicines people are taking, how they are responding, and being able to intervene, we can build a community health care system that is much more empowering and cost effective than any system we use today," Thompson says. "What we're really doing is opening up an entirely new domain for the delivery of medical products, and a new area for medical therapeutic innovation."

"Twenty years from now," he predicts, "every pharmaceutical in the supply chain will include a sensor and instead of talking about sensors inside pills, we'll be talking about putting pills inside sensors. The sensor is going to be the thing we swallow, not the drug."

10.2
Organs-on-a-Chip

Today, nanomedical researchers can use a patient's own cells to grow a portion of an organ on a scaffold, in a solid or liquid medium, or on a biochip. This innovation is extremely important for many reasons. Imagine being able to test a drug outside the body (*ex vivo*) on a cluster of heart or lung cells instead of testing drugs on living patients in expensive clinical trials. The implications are profound for drug development, screening, and testing.

The use of organs-on-a-chip to test drugs and other therapies could reveal which drugs treat or cure the disease most effectively *before* clinical trials are conducted. Diseased organ tissue could be grown to test different types of drugs and dose levels, which could be custom-tailored for each individual patient. This is a potentially huge advance in the field of personalized medicine.

Today, we know that individual patients respond differently to various types of chemotherapy and different dose levels. Cancer patients may have a weaker response to a particular chemotherapy due to a physical, chemical, or genetic factor. HIV/AIDS patients require new chemical cocktails at different stages of their disease, as the virus mutates or develops resistance. Different types of antibiotics work better for some patients than others. Some groups of patients are allergic to a particular medicine. Growing miniature versions of organs- or organ-tissues-on-a-chip offers a noninvasive way to test drugs and dose levels. With this approach, every patient can have his or her own "high-throughput" screening test, exposing those miniorgans or tissues to a battery of drug and dose combinations. It is the ultimate personalized medical test.

Lung-on-a-Chip

One of the best publicized innovations in this field is the "lung-on-a-chip" (Figure 10.2) developed by cell biologist Donald Ingber, M.D., Ph.D., Founding Director of the Wyss Institute for Biologically Inspired Engineering at Harvard University; and his colleague Dongeun Huh, Ph.D., Technology Development Fellow at the Institute. Their lung-on-a-chip incorporates living human lung and blood vessel cells that mimic breathing and lung functions. A variety of

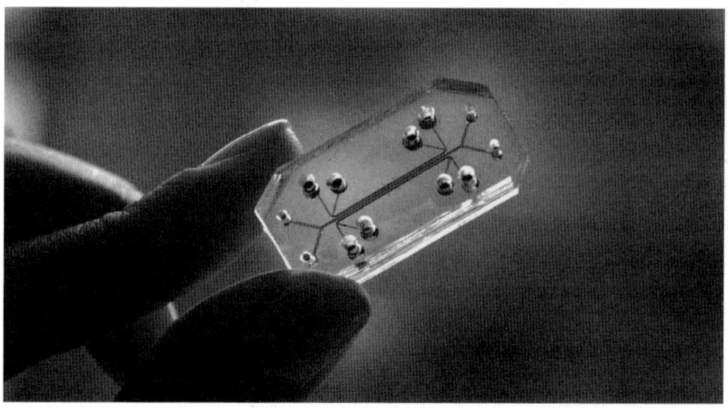

Figure 10.2 Lung-on-a-chip. The pioneering chip designed by Don Ingber and Dongeon Huh is made from a crystal clear, flexible polymer about the size of a memory stick and borrows parts of its design from computer microchips. Two of the hollow channels are separated by a thin, flexible, porous membrane lined on one side with human lung cells from the air sac exposed to air. Human capillary blood cells are placed on the other side with a medium flowing over their surface. A vacuum applied to side channels recreates the way human lung tissues physically expand and retract when breathing (image courtesy of the Wyss Institute, Harvard University).

important tests have been performed using the lung-on-a-chip, including studies related to the effects of microbial pathogens, environmental toxins, aerosolized therapeutics, the safety and efficacy of new drugs, and how airborne nanoparticles are absorbed.

"Many pharmaceutical companies spend a lot of time and a huge amount of money on cell cultures and animal testing to develop new drugs, but these methods often fail to predict the effects of these agents when they reach humans," Ingber observes. Just 2 years after the initial design was unveiled, the team demonstrated its potential to model therapies for a complex human disease. The lung-on-a-chip was used in conjunction with animal studies for a new class of drug under development by GlaxoSmithKline (TRPV4), designed to treat pulmonary edema. Pulmonary edema is a deadly condition in which the lungs fill with fluid and blood clots. Dr. Ingber believes this and other early examples offer a glimpse of what drug discovery and development might look like in the future [1].

"Our innovation is a human breathing lung on a chip," he explained (Figure 10.3). "The structure includes the lung sac (alveoli) which breathes and stretches as it operates. The whole thing is mechanically active and emulates the functions of human lung sacs with the same degree of function and flexibility. Physically it is transparent. If you hold it in your hand it looks like a tiny transparent eraser."

"In our laboratory, we've been able to use our lung-on-a-chip to capture a real time video of the human inflammatory response to bacteria in the lung. We can watch in real time as white cells engulf bacteria, which 'stick' to the lung cells. We've also been able to achieve predicted results involving nanoparticles in the air that are known to be toxic, and we discovered that breathing increases absorption of nanoparticles, which we described in a *Science* article" [2].

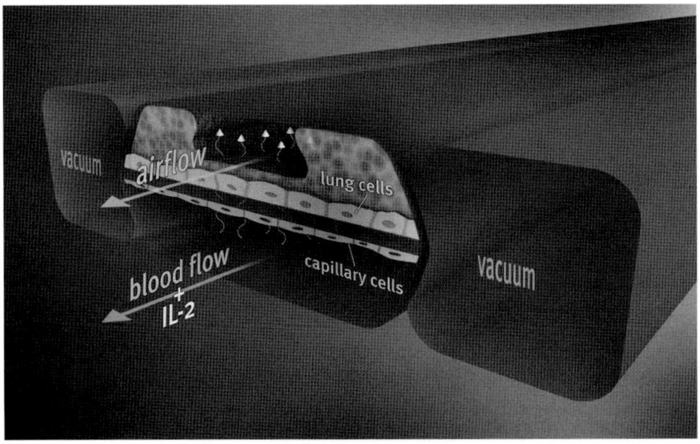

Figure 10.3 The lung-on-a-chip provides the essential functions of a human lung, using a patient's lung cells. This illustration shows how the chip arranges the cells to provide lung functions (image courtesy of the Wyss Institute, Harvard University).

Dr. Ingber's lung-on-a-chip research is funded by the National Institutes of Health, the Food and Drug Administration, and the Defense Advanced Research Projects Agency. His team is expanding the research to develop more than nine other organ chips that together constitute a "human-body-on-a-chip." The team hopes to engineer a system that mimics body physiology in a system that connects a lung-on-a-chip, heart-on-a-chip, and gut-on-a-chip.

In many ways, the story of the lung-on-a-chip sounds analogous to how computer microchips were developed in the 1970s and 1980s. At first, computer functions such as processing functions, memory, input/output, and so on were contained in separate chips, but eventually these functions were combined in a single microprocessor chip. If organs-on-a-chip follow the same development path, it is likely that organs-on-a-chip will evolve into "human-systems-on-a-chip." This combination of organs will allow researchers to study the effects of toxins, diseases, and pathogens on multiple organs simultaneously.

Collaboratories and Other Themes that Drive Innovation

Many pioneers in the field of nanoinnovation have developed strategic and thematic frameworks to guide their work, and Don Ingber is a terrific example of this. Under Dr. Ingber's leadership, the Wyss Institute has cultivated a highly effective innovation culture that he calls *collaboratories*. This concept provides a framework for collaborative innovation across a spectrum of technologies and applications.

"One of the goals at the Wyss Institute is to develop *biomimetic microsystems*," Dr. Ingber explains. "This involves engineering microchips containing living cells that reconstitute human organ functions for drug screening, diagnostic and therapeutic applications. This technology can accelerate drug development and replace animal testing." Over time, he believes, the technology has the potential to replace and automate human tests for toxicity, pathogens, and drugs.

"We are working at the convergence of life sciences, engineering and physical sciences," Dr. Ingber explains. "Our mission is to transform healthcare, industry and the environment by emulating the way nature builds. One of the first things we did was to establish what we call 'Collaboratories' that encourage industry professionals as well as faculty and students to collaborate around well defined projects. Instead of simply giving out grant money, we decided to fund enabling technology platforms that we call 'cores that create other cores'. This focuses us on research that develops technological capabilities that lead to other innovations."

An important goal is to develop "anticipatory medical devices" – portable wireless devices that sense life-threatening medical events and transmit signals that detect and treat disease. These types of devices are key to the future of autonomous home healthcare. "For example," Dr. Ingber says, "an iPhone can sense multiple dynamic physiological rhythms. Preliminary data can be collected that can predict seizures before they happen, using remote sensing."

Another Wyss Institute platform is called "Programmable Nanomaterials." Here, Dr. Ingber explains, the goal is to create smart nanotechnologies for regenerative medicine and drug delivery applications. The focus is on designing injectable medical devices rather than implantables. Projects include research on injectable cancer vaccines and differentiation therapies, and stem cell reprogramming for tissue regeneration.

Dr. Ingber described how the group's ideation process influenced key innovations: "When we started this research, we asked ourselves, 'Why do we take stem cells out of the body and grow them, then implant them back into the body at the site where we want them to do something? Could we find a way to inject them and have them go to the intended site and do something therapeutic at the injury site?' This might involve building a scaffold at the injury site that 'recruits' the stem cells to go there and do something."

"We're also studying the structural design of cells, which has led us in some new directions," he said. "The old view is that cells are structured like balloons. Actually, when cells are anchored they are not round, but flat. That realization led to our idea that cells are structured more like tents. This allows us to use what we call 'Tensegrity Architecture' where we can use tent poles to keep the tent – in this case, the cell – rigid in a shape such as something that resembles a geodesic dome. Using tensegrity, we have used DNA origami to infuse nanometer sized hairs in a sea of flexible polymer. When the structure tenses, it makes them move differently which can be used to sense or filter light, change colors and harness energy."

The combination of conceptual platforms and research platforms described by Dr. Ingber provides a distinctive way of visualizing and developing innovations. These kinds of conceptual mindsets are a vital factor in innovation, along with technical competency, a problem-solving orientation, and shared vision. This is the kind of thinking that creates innovation cultures, innovation teams, and innovation breakthroughs such as organs-on-a-chip.

Next Steps

Organs-on-a-chip are currently being used to supplement animal studies. At some point, organs and tissues grown on chips will be approved for use in testing certain reactions, safety factors, and even efficacy of experimental drugs. If successful, applications will migrate from supplementing animal studies to full testing in animals, and from there to testing in human clinical trials.

Currently, organs grown on a chip are not large enough to replace full-sized organs. That's not their intended function. However, one possibility – which might be considered an "upgrade" – would be to design and engineer full-featured replacement organs using our own cells. Growing our own replacement organs is not as futuristic as it sounds. Actually, the ability to grow replacement organs already exists.

10.3 Growing Your Own Replacement Organs

The failure or malfunctioning of a major organ such as the heart, liver, kidney, lungs, bladder, and so on is a global medical problem. Every day, in every region of the world, patients experience organ failure due to disease, injuries, genetic defects, and old age.

In the past decade, medical pioneers have been learning how to "grow your own" replacement organs to replace or repair a failing bladder, damaged blood vessel or damaged skin tissue. Virtually any human cell can now be cultured in a laboratory. Many human organs and body parts can be regenerated from stem cells harvested directly from patients. Today, researchers have made impressive progress engineering replacement organs in four categories: *flat structures* such as skin and organ "patches," *tubular tissue* that includes blood vessels and urethras, *hollow organs* including most notably the human bladder, and *solid organs* such as the liver and kidney.

These innovations are being enabled by nanoinnovations that allow bioengineers to develop scaffolds for growing organs using nanomaterials as well as conventional materials. Thanks to nanotechnology, bionano imaging systems allow researchers to study the structure and behavior of cells and tissues grown in the laboratory, and observe interactions needed to regrow human organs.

The Organ Replacement Crisis

Millions of people need organ transplants or major organ repairs to save their lives or maintain their health. While the science of transplanting organs from donors to recipients is well developed, *there are not enough organ donors available to meet the need.* Suitable organ donors come from a limited pool of live donors (kidney and bone marrow) or from brain-dead donors (heart, lung, and liver). After a transplant operation, organ recipients need to undergo an expensive and risky lifetime regimen to prevent rejection by the body's immune system. The need to suppress the body's immune system opens the patient to additional risks.

Successful transplants have been performed to replace the heart, kidney, bladder, liver, lung, cornea, trachea, and other vital organs. Organ transplants have been performed for more than 50 years and the number of operations has been increasing steadily. In December 2012, the University of Pennsylvania commemorated its 1000th heart transplant, with 3 year survival rates >80%. In 2013, more than 14 000 kidney transplants were performed in the United States, according to the National Kidney Foundation.

There are numerous examples of people who are living full lives after receiving a major organ transplant. Many famous celebrities – actors, politicians, even famous innovators – have received organ transplants. Among these are American actor Larry Hagman, star of the "Dallas" TV series, who received a liver transplant in 1995 and died at age 81 in November 2012. Entrepreneur and car designer Carroll Shelby received a heart transplant in 1990 and a kidney from his son in 1996. He lived 22 years after his heart transplant and died in May 2012 at the age of 89. Other organ recipients include US Vice President Dick Cheney and Apple's late founder/CEO Steve Jobs. The results of these operations vary widely. Some patients survive years or decades, while others succumb a few days after the operation.

The number of patients who need organ transplants far exceeds the availability. In the United States, ~120 000 people are on the waiting list, but only 30;000 transplants are performed each year. Many people who would benefit from transplants are not registered. In some countries, organ donation is fairly common, while in other countries it is rare. Spain has the highest organ donation rate in the world, with 35.1 donors per million people, compared with estimates of 26 per million in the United States, 24.8 per million in Austria, 22.2 per million in France, and 10.8 per million in Australia; and single digit percentages in most countries – yet all countries have a shortage of organ donors.

In the United States, candidates are registered and placed on a waiting list. In some countries, organs are provided to those who can afford them, which discriminates against poorer patients. An organ transplant and follow-up care can cost $1 million or more. Most patients need to be placed on a lifelong regimen of drugs to suppress their immune systems. This regimen needs to be monitored and adjusted over time, levying an additional cost on individuals, families, and government agencies. In some areas of the world, there is an illegal black market for organs, raising ethical and legal concerns. Clearly, alternative solutions are needed.

The optimum solution to the replacement organ crisis is to *use each patient's own cells* to grow a new organ. This is already being done. There are patients alive today who are living with organs grown from their own cells. Regeneration scientists have used stem cells biopsied from patients to grow new skin, bone, cartilage, corneas, windpipes, arteries, and urethras, as well as human bladders. Today, medical researchers are currently experimenting with more complex organs such as the liver, heart, and kidney.

The First Replacement Organs Grown in the Laboratory

The first complex human organ to be fully regenerated using a patient's own cells was the human bladder. Human bladders were grown and successfully implanted in several patients more than a decade ago by Dr. Anthony Atala, who pioneered the field of regenerative medicine.

Anthony Atala (Figure 10.4) was born in Peru and was raised in Florida. He is an excellent example of the new generation of physician innovators who are developing technical solutions to problems they see in their medical practice. Dr. Atala studied medicine at Harvard and in 1990 he joined Children's Hospital Boston in 1990 as a pediatric urologist, where he became Director of the Laboratory for Tissue Engineering and Cellular Therapeutics. In 2004, Dr. Atala joined Wake Forest Baptist Medical Center, where he founded the Wake Forest Institute for Regenerative Medicine.

"Every 30 seconds, a patient dies from diseases that could be treated with tissue replacement," Dr. Atala explains, speaking earnestly and with a calm sense of determination. "Wouldn't it be great if our bodies could regenerate? Wouldn't it be great if we could actually harness the power of our bodies to heal ourselves? Salamanders have an amazing ability to regenerate their limbs. So salamanders can do it. Why can't we?"

Figure 10.4 Dr. Anthony Atala.

"Actually, we *can* regenerate. It happens every day. Our bones regenerate every 10 years. Our skin regenerates every 2 weeks. When there is an injury, the body's first reaction is to seal itself off, whether it's an organ inside your body or skin or an arm or leg, scar tissue forms to seal the wound."

He goes on to explain that smart biomaterials can be used to repair tissues that are damaged, to form a bridge that allows cells to regenerate. The body can regenerate itself, but cells can only regenerate and bridge gaps at short distances, typically about 1 cm. The solution he and his team pioneered involves building biodegradable scaffolds and repopulating those with stem cells, or decellularizing donor organs and using the remaining protein/collagen matrix as the scaffold.

In 1999, Dr. Atala performed the first regenerated bladder implant. At the time, the conventional solution to repairing a damaged or malfunctioning bladder was to graft a section of intestine to the bladder, which posed a cancer risk to the patient. Frustrated with this approach, Dr. Atala pioneered a technique for growing organs and tissues from the patient's own cells, which are applied to a scaffold shaped like the organ.

One of his first patients was 10-year-old Lucas Massella, who was born with spina bifida and had already had a dozen corrective surgeries. In 2001, the boy was experiencing a life-threatening kidney failure due to a deteriorating bladder that was backing up fluids into his kidney and causing incontinence. To treat the problem, Dr. Atala used Lucas' own stem cells to grow a new bladder, which was implanted to replace his failing bladder. To create the bladder, Lucas' own cells were sprayed onto a baseball-sized, biodegradable mold shaped like a bladder. After the mold was implanted to replace his bladder, the mold dissolved, leaving in place a fully functioning bladder. Today, 23-year-old Lucas is completing his studies at the University of Connecticut where he is majoring in communications. Six other spina bifida patients who suffer from similar complications also volunteered to have regenerated bladders implanted to replace or repair their damaged organs. After the first 7 years, the patients all reported that their regenerated bladders were still functioning and providing improved urinary control.

This groundbreaking research was conducted by Dr. Atala and his colleagues Stuart B. Bauer, Shay Soker, James J. Yoo, and Alan B. Retik. The team waited 7 years to confirm the long-term results of the first laboratory-grown organs, which was described in a historic article in the April 2006 issue of *The Lancet* [3].

Today, the Wake Forest Institute for Regenerative Medicine is the world leader in replacement organ research. Wake Forest is also involved in extensive collaboration with other research centers and medical schools, working with a network of 250 scientists from 13 different fields who are regenerating bones, heart valves, muscles, and two dozen other body parts. This network forms an important and valuable "innovation ecosystem" that supports global research on tissue engineering and organ regeneration. This network also provides a knowledge-sharing community that is constantly building on the achievements of individuals and teams to keep expanding the frontiers of regenerative medicine.

A Recipe for Growing Your Own Bladder

Dr. Atala's group at Wake Forest is currently working to engineer more than 30 different replacement tissues and organs, and to develop healing cell therapies. As part of this extensive research, Dr. Atala is also compiling a "menu" of regeneration techniques, which is important since each type of cell exists in a particular environment. The goal, Dr. Atala explains, is to develop a menu of regeneration approaches to identify the best possible technique for each type of organ and patient situation. More than one type of cell may be needed for each organ.

Various types of nutrient cocktails have been formulated to coax each type of cell to regenerate, using cell culturing techniques Dr. Atala developed while he was on the medical school faculty at Harvard, and perfected at Wake Forest. The nutrient mix contains growth factors, enzymes, proteins, and other cell nutrients. Getting the nutrient mix just right is critical. The cells need to be induced to grow, and ideally to grow and multiply faster than normal. When the cells reach the right concentration in the growth medium (which is billions of cells), they are checked with an atomic force microscope to ensure the individual cells are normal, healthy, and at the right maturity level.

According to Dr. Atala, it takes ~6 weeks to grow a human organ. Each organ is unique. In addition to the nutrient growth cocktails, Dr. Atala has created scaffolds in small, medium, large, and extra large sizes to fit patients of different ages and weight profiles. For example, a trachea might need a more rigid scaffold

Figure 10.5 This baseball sized, biodegradable scaffold provides a template for growing a human bladder in the laboratory at the Wake Forest Institute for Regenerative Medicine (image courtesy of Wake Forest Institute).

structure, while an esophagus requires a softer more flexible scaffold. Most of the organs being regenerated today require layers of different types of cells on the inside and outside surfaces of the organ.

To grow a human bladder, a baseball-sized scaffold is created, which is shaped like a bladder (Figure 10.5). The scaffold is made from biodegradable collagen that dissolves after the organ is implanted. The cells grown in the laboratory are sprayed or "printed" on the scaffold, using an inkjet printer or 3D printer. The specially modified bioprinters apply the cells one layer at a time. Urothelial cells are applied to the inside of the bladder, and muscle cells are applied to the outside. Finally, the bladder is surgically implanted in the patient, in place of the failing bladder, and connected.

There are several variations being developed from this core recipe. For example, researchers at other hospitals are experimenting with different types of scaffold materials. Dr. Christopher Chen at the University of Pennsylvania has used scaffolds made of sugar to regenerate blood vessels. Dr. Seifalian at Royal Free Hospital uses a nanocomposite material modeled on the honeycomb structure of a butterfly's wing – the nanocomposite is used because it is resistant to infectious bacteria and its pores are the right size to hold cells.

The "recipe" varies for each organ, explains Shay Soker, Professor of Regenerative Medicine at Wake Forest and a senior member of the Wake Forest research team. Dr. Soker has been conducting cutting-edge research on regenerating blood vessels. To create a human artery, he explains, you need epithelial cells that line the artery and muscle cells that cause the artery to pulse. The cells are applied to a tube-shaped scaffold made from flexible collagen, but it isn't enough to simply construct the tube. As the cells grow, the vessel is "exercised" by a machine that stretches and strains the tube to condition it. Regenerating muscle tissue involves a similar process, including exercising the muscle to condition the regenerated cells.

During the growth and conditioning process, individual cells are monitored with an atomic force microscope to help determine their progress, functionality, and maturity. Nanoparticles may be injected into the artery to enhance the imaging for monitoring and quality control. In 2001, Wake Forest was the first research group to engineer functional blood vessels that were implanted preclinically and survived long term.

Some of the other intriguing research projects underway at Wake Forest include the growth of a regenerated liver that incorporates a vascular network so that blood can reach the deepest parts of the organ – a major challenge in regenerating dense, complex organs.

Another long-term research project involves regeneration of human fingers and limbs to treat amputees. This initiative is exploring what is required to induce the regrowth of amputated organs, as well as regeneration of human digits and limbs that could be reattached to amputees. A federally funded project to study how regeneration technologies can be applied to treat battlefield injuries is being pursued by the Armed Forces Institute of Regenerative Medicine, which is co-led by the Wake Forest Institute.

Printing New Skin Directly on Burn Victims

Of the more than 30 projects underway at the Wake Forest Institute, one of the most interesting involves an innovative technique for bioprinting regenerated skin directly onto human patients (Figure 10.6). For this application, the inkjet printer is loaded with immature human skin cells and applies one layer of skin at a time, at the burn site. This revolutionary innovation could change how burn victims are treated. Researchers are currently working on a portable device to treat battlefield burns that uses their inkjet printer technology. This remarkable device is modeled on technology developed by a research team at Clemson University.

Different formulations are being developed that offer the ability to combine skin and fat cells, or change the thickness or pigmentation. The conditioning process may involve stretching the skin before it is grafted to the burn site. Bioprinting skin on a burn victim at a field hospital in a combat zone, or at the scene of a fire, can seal and heal the wound faster and less painfully than current burn treatments. Burn treatments that currently take months or years could shrink to a few weeks [4].

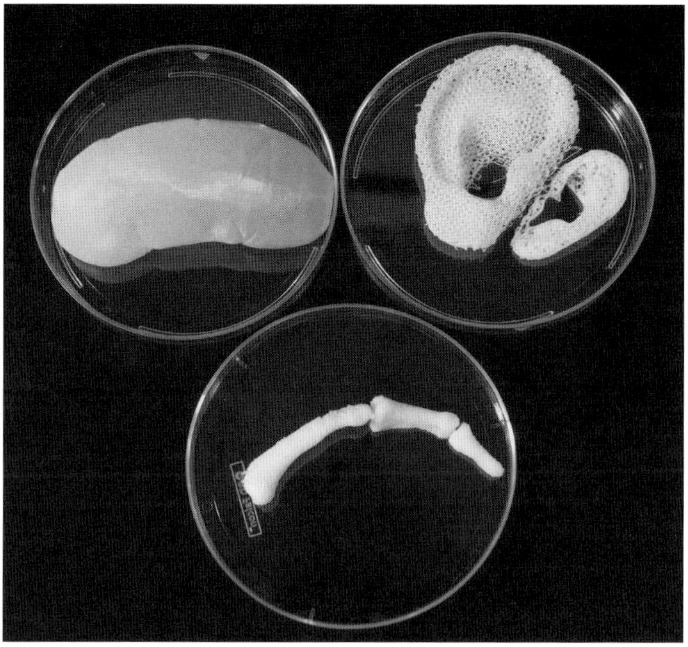

Figure 10.6 These are a few examples of integrated scaffolds being used by Wake Forest researchers to bioprint body parts using modified inkjet and 3D printers (image courtesy of the Wake Forest Institute for Regenerative Medicine).

Global Research in Regenerative Medicine

Advances are being achieved at hospitals and universities around the world, as surgeons and specialists experiment with regeneration as a solution to some of their patients' most frustrating and dire medical conditions. Many of these achievements are experimental, developed to save the life of a patient or provide a solution when there was no other alternative.

Alex Seifalian at Royal Free Hospital in London has transplanted laboratory-grown tear ducts and an artery into patients. In 2013, he bioengineered an artificial nose for a 53-year-old man who lost his nose to skin cancer. When the bioengineered nose was grown in the laboratory, it did not have a layer of skin cells, so Dr. Seifalian implanted the nose under the skin of the patient's arm, to get the skin to coat the surface of the nose. This is still experimental, but reflects the kind of innovative thinking that is going into this type of research [5].

Doris Taylor has grown a beating heart of a rat in the laboratory at the University of Minnesota and worked on growing human hearts on donor scaffolds using decellularized hearts. Dr. Taylor, who now works at the Texas Heart Institute and also collaborates with Dr. Seifalian, has led pioneering research on development of a bioartificial heart grown on a decellularized donor heart scaffold. Her seminal research was reported in *Nature Medicine* in 2008 [6]. Her seminal work on decellularizing cadaver organs has stimulated a surge of research on this.

Laura Niklason's research team at Yale and, independently, Harald Ott at Massachusetts General Hospital (now at the University of Innsbruck), have used bioscaffolds to build lungs that functioned in rats for 2–6 h. The creation of fully functional lung tissue is an early achievement that paves the way for development of engineered lung tissue for humans. This is significant, Dr. Niklason has observed, because the only current way to replace damaged adult lung tissue is to perform lung transplantation, which is highly susceptible to organ rejection and infection. Lung recipients have a survival rate of 10–20% after 10 years.

"For this technology to be applicable to human patients," she says, "it is likely that years of research with adult stem cells will be needed to repopulate lung matrices and produce fully functional lungs" [7]. Dr. Niklason is Vice-Chair for Research and a Professor in the Department of Anesthesiology and Department of Biomedical Engineering at Yale.

Stem cell biologists Hideki Taniguchi, Takanori Takebe, and their colleagues at Yokohama City University in Japan have reported that they had engineered and implanted laboratory -grown human liver cells into mice. They used an innovative process to coax stem cells to self-organize into what they call "liver buds" that exhibited some liver functionality including blood vessels [8].

On April 9, 2013, a 2-year-old Korean–Canadian girl became the youngest person ever to receive a bioengineered organ when she received a bioengineered trachea (windpipe). The surgery was conducted at the Children's Hospital of Illinois and was the sixth surgery of this type performed in the United States It was performed as an experimental life-saving procedure approved by the Food and Drug Administration. The 9 h surgery was performed by a team led by Dr. Paolo Macchiarini, a regenerative medicine specialist from the Karolinska Institute in Stockholm, Sweden, who developed the trachea. It was the sixth surgery of this type performed by Dr. Macchiarini. The bioengineered tracheas are created using a 0.5 in. diameter tube made from plastic fibers, bathed in a solution containing stem cells from the patient's bone marrow and incubated in a bioreactor that is about the size of a shoebox.

Martin Birchall, a professor at University College in London, is creating tracheas using donor organs from cadavers, stripped of their original cells, and repopulated with stem cells. Dr. Birchall is currently leading a project called RegenVox, which includes the first clinical trial of a tissue-engineered larynx grown from stem cells on a collagen framework. This area of research is still highly experimental and risky, and the next step will be to conduct extensive clinical trials to refine the technology.

These are only a few examples that illustrate the wide range of progress in the field of organ regeneration. In many of these projects, stem cell research is playing an extremely important role, which makes it especially important to find ways to fund and authorize stem cell research (Figure 10.7).

Recycle Donor Organs or Grow Complex Organs?

As the regenerative medicine field evolves, scientists are seeking ways to grow more complex organs such as the heart, liver, kidney, and lung. These complex

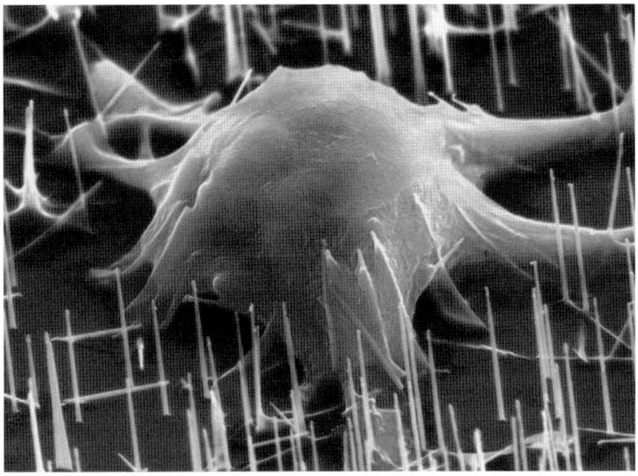

Figure 10.7 While growing human tissue from stem cells sounds fairly straightforward, there are numerous challenges involved in this process. One challenge involves inducing the stem cells to keep reproducing as long as possible, which is needed to "grow" an organ or a tissue. One solution is to use a stencil to cluster the cells, since groups of stem cells divide and "grow" for a longer time than individual stem cells. Positioning the stem cells to anchor them is another approach (image courtesy of Dr. Peidong Yang).

organs are difficult to grow, for many reasons. Their shape is more complex, so the scaffolds need to be dense and three dimensional. Their cells have complex functions. In addition to surface (epithelial) and muscle cells, they may require fat cells and other types. The thickness of these organs makes it difficult to deliver nutrients to individual cells. For example, a thick, dense organ such as a liver will have a complex vascular network of blood vessels and capillaries to provide blood to the deepest parts of the organ. Scientists need to consider how to maintain, or grow, a vascular system to deliver blood to the deepest and densest parts of the organ.

One of the most promising approaches involves recycling organs and tissues from deceased donors, by removing all of their cells, leaving only the protein matrix that serves as a scaffold. Special chemicals are used to cleanse the entire organ of cells so that the protein structure retains the shape of the organ, the compartments, as well as the veins and arteries. When stem cells harvested from a patient are applied to this decellularized organ, the stem cells differentiate and repopulate the organ. This is an elegant and promising method for recycling organs from deceased donors. Currently, donor organs only remain viable for a few hours – for example, a donor heart only remains viable for about 4 h and needs to be transplanted quickly. Recycled organs with their cells removed could potentially be preserved and stored for longer periods in organ banks.

In June 2012, a 10-year-old girl received a vein transplant to treat a potentially fatal vein obstruction. A 3.5 in. long vein from a deceased donor was decellularized, leaving only a protein framework. The framework was covered with stem cells extracted from the patient's blood, which formed a new vein in 2 weeks. The new vein was implanted and the patient's condition improved significantly. A year later, an obstruction developed in another part of the vein requiring a duplication of the procedure, using another vein grown in the laboratory. Again, the patient improved.

The significance of this example is that it allowed the patient to receive a replacement body part without the need to suppress the immune system. Normally, when a vein or other body part is transplanted, the body rejects the implant that requires drugs to suppress the immune system – which makes the patient susceptible to other diseases. By using a body part harvested from a deceased donor, and using the patient's own stem cells to replace the cells in the matrix, there is no need to suppress the immune system because the transplant is made from the patient's own cells.

Research teams at Wake Forest and other institutions have already used this technique to grow a rudimentary animal heart, portions of a liver, kidney, and other complex organs (Figure 10.8). Scientists have found that the structure of decellularized blood vessels remains intact after the cells are cleansed from the organ, and are trying to recellularize those vascular networks with stem cells. In some studies, stem cells are finding their way to various locations in the organ and developing into functional cells. Repopulating cells on the outside and inside of a network of veins and arteries in a decellularized heart or liver remains a challenge.

Scientists have also been studying the feasibility of using decellularized animal organs such as a bladder or heart from a pig, which can be seeded with human stem cells to produce transplant organs. This approach would provide a constant, reliable source of organs for regeneration. In theory, pigs bred for the food industry could provide organs for regeneration, which would provide a reliable and economical source of supply.

It remains to be seen whether complex human organs for transplant will be provided from "recycled" organs taken from deceased donors, from organs grown using a patient's own stem cells, or from animals bred to produce organs that can be seeded with human cells. Which approach prevails as the best long-term solution depends on what researchers are able to achieve in the coming decade. Public sentiment and acceptance will also play a critical role, since using embryonic stem cells or animal organs raises a variety of issues.

Scientists may have a solution that addresses sensitivities about using embryonic stem cells. In 2011, Dr. Atala and colleagues at Harvard University announced in the journal *Nature Biotechnology* that pluripotent amniotic stem cells can be harvested from the amniotic fluid of pregnant women – and can be used to regenerate human organs [9]. The pluripotent stem cells can be manipulated to become almost any type of cell such as epithelial, fat, muscle, bone, nerve, or other cells. They can thrive for years in a laboratory. They do not cause benign tumors. They can be harvested from the amniotic fluid without harming

Figure 10.8 This regenerated liver tissue grown at the Wake Forest Institute was engineered to grow a vascular network of blood vessels, which can be seen in this image. A blood vessel network is needed for dense organs such as liver, heart, and kidneys, to provide nutrients to cells located throughout the organ (image courtesy of the Wake Forest Institute for Regenerative Medicine).

the mother or fetus. These stem cells can be harvested from amniotic fluid discarded from amniocentesis tests used to check fetuses for birth defects.

Next Steps

Medical researchers have demonstrated that it is possible to regenerate functioning human organs from stem cells and to implant them in patients to replace or repair failing organs. Patients in several countries have survived fatal conditions and have been living healthy lives for more than a decade, thanks to regenerated organs. The next step is to translate these experimental procedures into clinical trials. This will enable them to be tested, refined, and certified for use by surgeons and specialists worldwide. However, important steps remain before this new field of medicine can be commercialized and made available to all patients who need it. Expensive clinical trials need to be funded and conducted. Regulatory approvals are needed. More tests are needed, and different types of patient populations may need to be evaluated.

To fund clinical trials for regenerated bladders, urethras, and related tissues, regeneration medicine technologies developed at Wake Forest, Harvard and other institutions have been licensed to Tengion, a biotech venture based in Winston-Salem, North Carolina. Tengion is currently conducting clinical trials, including a phase I clinical trial for a neourinary conduit, and in 2013 received approval for a phase I clinical trial in Sweden to evaluate the safety and delivery of a new-kidney augment in patients with chronic kidney disease. The initial trial is designed to evaluate the safety and delivery of Tengion's Neo-Kidney Augment^{TM} product in up to five patients with advanced chronic kidney disease (CKD). Tengion's long-term goal is to commercialize the production of regenerated organs/tissues, although the company has struggled financially and flirted with bankruptcy in 2013 before receiving a strategic investment of $15 million from Celgene and an additional $18.6 million in investment capital. These investments enabled the company to move forward with its two lead clinical programs. As previously noted in this book, funding long-term research leading to commercialization has been a major challenge for many nanotechnology ventures and presents a special challenge in bionanotechnology.

10.4
Tumor-Seeking Nanoparticles

Nanotechnology is becoming one of the most potent weapons in the cancer-fighter's toolkit. Nanotechnology discoveries have moved the fight from finding and removing or destroying tumors to detecting the very earliest precancerous cells and fighting them at the individual cell level. Large-molecule cancer drugs and nutraceuticals are being nanosized to make them more soluble and small enough to penetrate the walls of cells and blood vessels. Nanoscale biomarkers and binding agents are being engineered to find and target cancer cells. Gold and iron nanoparticles are being used to improve cancer imaging. These particles can be heated to kill heat-sensitive tumors. Some nanoparticles have been engineered to act like nanobots, delivering a drug directly to cancer cells. Another innovation acts like a biological nanofactory to produce cancer-fighting proteins at the disease site. These are just a few examples.

The Cancer Challenge

Cancer is a disease that has touched all of our lives, wherever we work and live. According to the World Health Organization, cancer is the leading cause of death worldwide. Cancer kills more people globally than AIDS, tuberculosis, and malaria combined. It is also one of the most heartbreaking medical conditions. One in two men and one in three women will contract cancer during their lifetime – which puts all of us at risk. Virtually everyone knows someone who has died of cancer. This makes cancer fighting a top priority for nanomedicine, and science and medicine in general.

The American Cancer Society has indicated that there are a total of 1.6 million new cancer cases annually in the United States with more than half a million US deaths each year from this disease. In the European Union, −1.2 million people die from cancer every year. Worldwide, more than 14 million new cancer cases are diagnosed each year, according to the World Health Organization (WHO); this number is expected to rise to 19 million cases a year by 2025, 22 million by 2030, and 24 million by 2035. The WHO has called this a looming "tidal wave" of cancer.

Most cancer deaths result from lung, stomach, liver, breast, and colon cancer. The frequency of these cancer types varies between men and women, among age groups, and from country to country. For example, of the 7 million cancer deaths in 2002, half were in Asia and nearly one-quarter were in China. More than 70% of all cancer deaths are in low and middle income countries. Most of the cancer cases in the coming decade are expected to occur in developing nations.

Cancer can be caused by genetic factors, bacterial or viral infections, environmental toxins, particulates, chemicals, and lifestyle choices such as smoking, drinking alcohol, chewing tobacco, poor diet, or overeating. This greatly complicates the ability of oncologists (cancer specialists) to diagnose and treat this ubiquitous disease.

Although we use the term "cancer" that sounds like one disease, there are actually more than 200 forms of cancer. Cancer can infect 60 different organs in the human body. Cancer cells are constantly forming in our bodies. In most cases, the body's immune system destroys cancer cells and prevents them from forming tumors. At the cell level, cancer occurs when a cell's DNA is damaged. The cancer cell typically repairs the damage, or the cell dies – but if the cancer cell replicates itself and makes more cells with the same damaged DNA, it can surge out of control. The uncontrolled proliferation of cells and unregulated cell growth create life-threatening tumors. A single tumor can be caused by multiple mutations that can occur in different locations in the tumor.

Nanotechnology plays a special role in diagnosing cancer early, by improving imaging systems to detect early formation of tumors, tumors that metastasize, and very small amounts of free-roaming cancer cells that may exist in the blood. To increase the sensitivity of imaging systems used to detect cancer and other diseases, magnetic and fluorescent nanoparticles are being used to enhance the imaging systems. Nanobiomarkers are also being engineered to identify and target individual cancer cells that may be free-roaming in the bloodstream. At the cutting edge of cancer therapy, nanoparticles are being created that seek out and bind to cancer cells – conceptually, this can be considered a form of biological cancer-seeking nanobots.

Fighting cancer cell by cell represents a shift in how we treat this dread disease. In the past, oncologists flooded an organ or the body in general with cancer-fighting drugs. Unfortunately, this approach tends to destroy healthy cells as well as cancer cells; it weakens the immune system and lead to life-threatening complications.

Today, medical researchers are attempting to move from this shotgun approach to a sniper approach by engineering tumor-seeking nanoparticles. These magic bullets use several methods to evade the body's immune system while actively seeking the cancer cells. Scientists use nanoparticles like biological Federal Express couriers to deliver cancer-fighting drugs directly to cancer cells and tumors. Nanoparticles are small enough to move through the pores of blood vessels that feed cancer tumors, while in most cases they cannot move through the much smaller pores of vessels that feed healthy tissues. They can be engineered to include a shell that can slide past the body's immune system without triggering a hostile response. Nanoparticles can be designed to carry proven anticancer drugs such as Docetaxel. They can be equipped with binding agents that seek out and bind to receptors on the surface of a certain type of cell, including a cancer cell. Typically, the drug causes the cancer cell to self-destruct, which is called apoptosis (*cell death*). Attached to the outside of the nanoparticles are molecules called ligands that bind to receptors on the surface of cancer cells.

The nanoparticles are often encapsulated or disguised to trick the body's immune system into letting the nanoparticles reach and enter the target cells. In 2011, a group of researchers at the University of California, San Diego, developed a method to disguise nanoparticles as red blood cells. They wrapped the particles in the membranes from red blood cells, like an invisibility cloak from Harry Potter. These nanoparticles can deliver a cocktail of small-molecule drugs without triggering an immune response [10].

In 2013, scientists at Methodist Hospital Research Institute in Houston coated drug-carrying nanoparticles with the membranes of white blood cells (leukocytes), which they call "leuko-like vectors" (LLVs) (Figures 10.9 and 10.10). The research was led by Ennio Tasciotti, Ph.D., the study's principal investigator and Interim Co-Chair of the Department of Nanomedicine. Dr. Tasciotti's team tested the LLVs with healthy and inflamed blood vessels and found that the particles selectively targeted inflamed tumor blood vessels. In addition to delivering nanoparticles to targeted sites, the researchers demonstrated that the particles provided controllable release of a drug without triggering an immune response.

"Our cloaking strategy prevents the binding of opsonins – signaling proteins that activate the immune system," explained Dr. Tasciotti. "I expect this technology to become a new player in the crowded world of drug delivery systems thanks to the opportunities it offers for the personalization of drug therapies."

He added that this cloaking method can be accomplished using human cell membranes harvested directly from a patient to create personalized LLVs, or could use synthetic/artificially created membranes that they plan to explore in future research. This approach has the potential to use membrane-cloaked nanoparticles to deliver cancer-killing drug molecules directly to tumor cells without damaging the surrounding healthy cells – which is needed to fight hard-to-treat cancers such as prostate and head and neck cancers.

Dr. Tasciotti's innovations include "Fracture Putty," a material for stabilizing bone fractures and to promote bone tissue regeneration. His early research laid the groundwork for two major nanotechnology platforms – protein nanochips for early detection of disease, and mesoporous silicon particles for targeted delivery of therapeutics and contrast agents. The latter was selected as one of the "five big ideas for nanotechnology" by *Nature Medicine* in 2008.

Dr. Tasciotti grew up in Italy and earned a B.S. in Biological Sciences, M.S. in Molecular Biology, and Ph.D. in Molecular Medicine from Scuola Normale Superiore in Pisa, Italy, where his research focused on adeno-associated virus (AAV)-based gene therapy, stem cell therapy, molecular imaging for the assessment of targeted drug delivery, and biodistribution of injected agents such as viral vectors. He moved to the United States in 2006 and expanded his research to include material science and the use of nanotechnology for early diagnostics, drug delivery, and regenerative medicine.

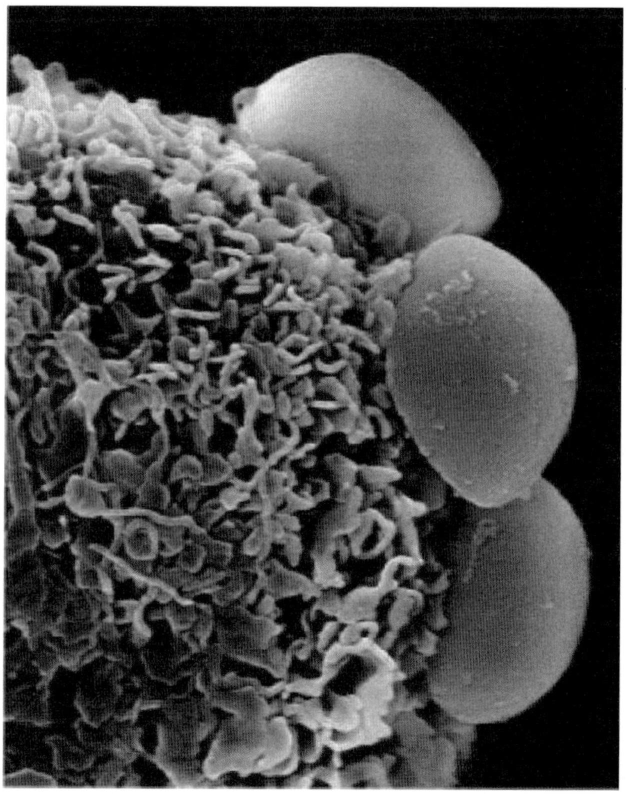

Figures 10.9 and 10.10 These two nanoscale images show three camouflaged nanoparticles cloaked in leukocyte membranes resting on the surface of an immune system cell (phagocyte) without being recognized and ingested/destroyed. These images help to prove that the cloaking strategy developed by Methodist Hospital Houston researchers does not trigger an immune response to the nanoparticles they are using to deliver cancer-fighting drugs (images by Ennio Tasciotti, Houston Methodist Research Institute).

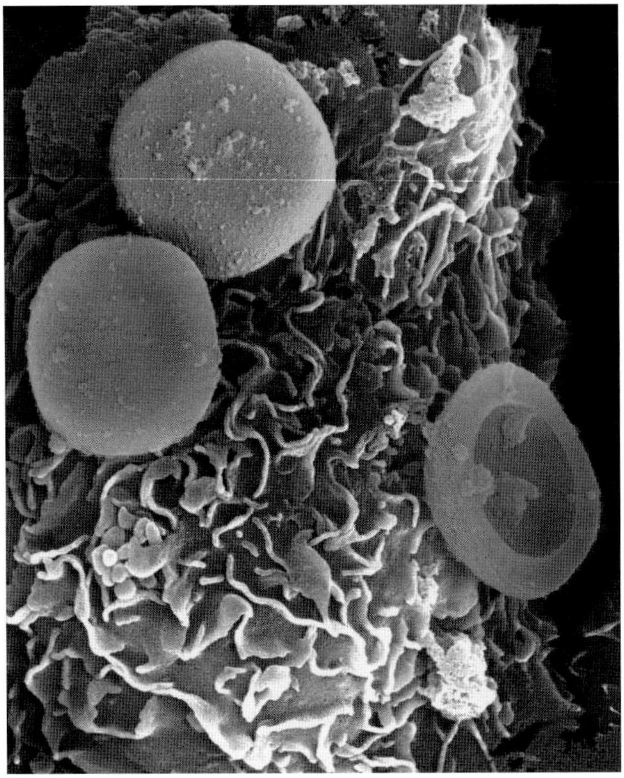

Figures 10.9 and 10.10 (*Continued*)

Fighting Cancer with Targeted Nanoparticles

Some of the most important and successful research on cancer-fighting nanoparticles have been done by research groups led by Robert Langer at MIT and his colleague Omid Farokhzad.

Robert Langer is one of the most widely recognized pioneers in this field. He has written more than 1200 articles and is the most cited engineer in history. He has ~815 patents issued and pending, of which 250 have been licensed or sublicensed to pharmaceutical, chemical, biotech, and medical device companies. He is the codeveloper (with Dr. Farokhzad) of the first targeted and programmed nanomedicine (As) to enter human clinical trials. For more than 20 years he has been pursuing his vision for remote-controlled drug delivery using microchips, polymers, nanoparticles, and other mechanisms. In February 2013, Dr. Langer was awarded the National Medal of Technology and Innovation by President Barack Obama.

When asked what excited him most about nanomedicine, Dr. Langer said, "I'm particularly excited about combining nanotechnology and drug delivery to fundamentally change how we treat cancer." He described this as a combination of

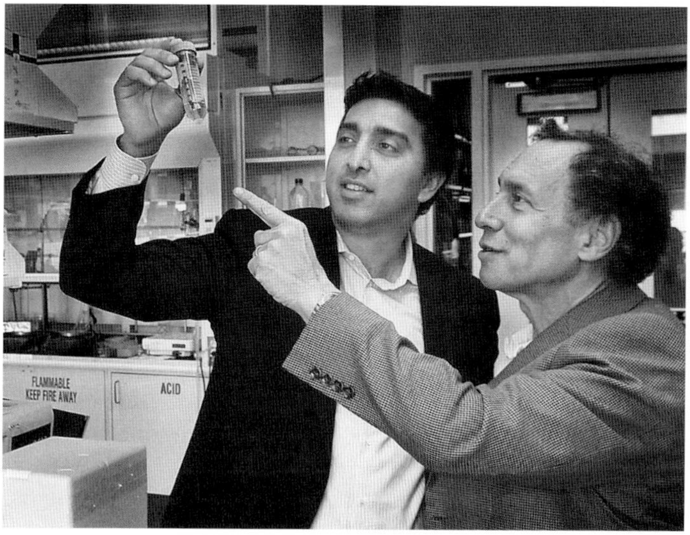

Figure 10.11 Drs. Omid Farokhzad and Robert Langer examine a vial of BIND-014, the first nanoparticle drug delivery system to enter human clinical trials (photo by Ted Fitzgerald/ Boston Herald Photo Department).

magic bullet and homing device, noting that "the very small size of nanoparticles enables you to get into cells, which is why it's so powerful for drug delivery."

Omid Farokhzad is Associate Professor at Harvard Medical School and a physician-scientist in the Department of Anesthesiology at Brigham and Women's Hospital where he directs the Laboratory of Nanomedicine and Biomaterials. The Iranian-born scientist has authored more than 95 papers and holds more than 125 issues/pending US and international patents. In 2013 he was one of 12 people to be recognized among the top innovators in Massachusetts by the Boston Globe [11].

Drs. Langer and Farokhzad began collaborating to develop tumor-targeting particles in the early 2000s and in 2006 they demonstrated that targeted nanoparticles can shrink tumors in mice (Figures 10.11). This led to the development of a targeted molecule called BIND-014, which began clinical trials in January 2011. In this study, the researchers coated the nanoparticles with targeting molecules that recognize a protein called PSMA (prostate-specific membrane antigen), found on the surface of most prostate tumor cells and on the tumor vasculature of other solid tumors [12].

The challenge, Langer says, is to engineer the particles so that they evade the body's immune response and still reach their intended target. The BIND-014 nanoparticles have three components: one that carries the drug, one that targets PSMA, and another that helps evade the macrophages and other immune system cells. This technology is based on innovations developed by Langer and Farokhzad that enables them to manipulate these properties very precisely, to create large collections of diverse particles that can be tested to determine the ideal composition. To develop BIND-014, the researchers screened more than 100 different nanoparticles of various sizes, surfaces, and drug-release capabilities.

In 2006, Drs. Langer and Farokhzad cofounded BIND Biosciences Inc. in Cambridge, Massachusetts, to fund clinical trials and commercialize their innovations, and to organize a network of research partners. BIND is a privately held, venture-backed biopharmaceutical company focused on developing therapeutic targeted nanoparticles for the treatment of cancer and infectious and cardiovascular diseases.

In January 2011, BIND-014 became the first targeted and programmed nanomedicine to enter human clinical studies. The clinical study was conducted at the Virginia G. Piper Cancer Center at Scottsdale Healthcare in Scottsdale, Arizona, in collaboration with Phoenix-based TGen, the Scottsdale Healthcare Research Institute, Karmanos Cancer Institute in Detroit, Marin Specialty Care in Greenbrae, California, and the Samuel Oschin Comprehensive Cancer Institute at Cedars-Sinai Medical Center.

The Principal Investigator was Dr. Daniel Von Hoff, Physician-in-Chief at the Translational Genomics Research Institute and Chief Scientific Officer at Scottsdale Healthcare's Clinical Research Institute. Drs. Langer and Farokhzad were co-senior authors of the study, which was published in the April 4 issue of *Science Translational Medicine.*

The first phase of the study involved 28 patients with advanced or metastatic tumors who had previously gone through traditional chemotherapy, and were treated with BIND-014 to deliver the cancer-fighting drug Docetaxel. The study was designed to determine safety, tolerance for various dosages, and preliminary efficacy.

Docetaxel is a clinically well-established chemotherapy drug that has been shown to kill cancer cells, including breast, colorectal, lung, ovarian, gastric, renal, and prostate cancer cells, according to the American Cancer Society. The drug is an extract from the bark of a rare Pacific yew tree. Docetaxel is a cytotoxic drug that works by interfering with microtubules that are part of the internal structure cells needed when dividing, which leads to cell death. The drug kills cancer cells because cancer cells divide faster than normal cells and are more likely to be affected by this drug.

In April 2013, the researchers reported that their phase I studies confirmed the safety of BIND-014 and also indicated tumor shrinkage in several patients at significantly lower doses of Docetaxel than normally given.

Commenting on the results, Dr. Von Hoff said, "This Phase 1 trial has successfully established the safety and tolerability profile and maximum tolerated dose of BIND-014 in patients with advanced or metastatic solid tumor cancers," adding that "There is a critical need for targeted treatment options for patients with difficult-to-treat solid tumors, and we look forward to further evaluating the potential of BIND-014 in patients with specific solid tumor types in the near future."

In April 2012, Dr. Farokhzad told the Boston Herald: "What this technology does, for the first time, is allow an administered dose of a drug to concentrate in cancers. Previously, it was not possible for drugs to accumulate in cancers in a preprogrammed and deliberate way" [13].

"In addition to confirming the safety, tolerability and maximum tolerated dose of BIND-014, these data also provide encouraging signs of anti-tumor activity in a variety of solid tumors," said Dr. Gregory Berk, Chief Medical Officer at BIND. He indicated that BIND is "moving expeditiously" to advance BIND-014 through multiple phase 2 clinical trials that will include treatment of non-small cell lung cancer, prostate cancer, and bladder cancer.

In April 2013, based on the promising results of their phase I clinical trials, BIND Biosciences changed its name to BIND Therapeutics – reflecting the expectation of transitioning from a "clinical-state biopharmaceutical company" to a biopharmaceutical therapy provider. The company currently has in place a strong network of researchers and collaborators at several organizations who are conducting research and clinical trials. In 2013, the company formed three major collaborations, with Amgen, Pfizer, and AstraZeneca, with a total deal value of up to US$1 billion. This is a strong indication that the biopharmaceutical sector may have reached an inflection point and is now embracing nanotech.

In 2008, Drs. Langer and Farokhzad and their colleague, Dr. Ulrich Von Andrian, an immunologist and Professor at Harvard Medical School, formed Selecta Biosciences to develop nanoparticles for directed antigen-specific immune suppression and antigen-specific immune induction. The company has a pipeline of product candidates for treatment of autoimmune diseases such as juvenile diabetes and food allergies and cancers. In 2012, Selecta formed a partnership with Sanofi-Aventis valued at up to US$900 million. In 2011, Drs. Langer and Farokhzad teamed with Dr. Stephen Lippard, a chemist and professor at MIT who was among the recipients of the 2004 National Medal of Science, to form Blend Therapeutics; the venture has developed an integrated nanotechnology and new drug molecule design platform to provide personalized medicine for treatment of cancers.

Related Innovations

Dr. Langer's research group at MIT and his global network of academic research partners are continuing their groundbreaking research on a broad spectrum of nanotechnology-enabled systems. Current research projects include systems to enhance the long-term delivery of insulin, anticancer drugs, growth factors, gene therapy agents, and vaccines. They are developing controlled release systems that can be magnetically, ultrasonically, or enzymatically triggered to increase release rates as well as biodegradable polymeric delivery systems, the delivery of proteins and genes across complex barriers in the body, new polymer systems for tissue engineering, and new biomaterials with shape memory or surface switching properties.[1)]

1) Robert Langer, Langer Lab/Research, Department of Chemical Engineering, Massachusetts Institute of Technology, 2012.

Cancer-Fighting Nanofactories

Another exciting development involves using *biological nanofactories* to produce cancer-fighting proteins. In 2012, a team of nine researchers reported how they engineered injectable nanoparticles that can move to a specific location in the body – such as a cancer tumor – and autonomously synthesize proteins in response to a signal or stimulus. This innovation represents an important advance toward the development of protein nanofactories that will be able to produce disease-fighting drugs on demand in the human body.

The research was led by Daniel Anderson, Associate Professor of Chemical Engineering at MIT's Koch Institute for Integrative Cancer Research, and was reported in the March 20, 2012 issue of *Nano Letters*. Robert Langer is a coauthor of the research study [14].

Protein-based drugs have shown promise in treating cancer; however, the body tends to break down proteins before they can be delivered. To overcome this challenge, the team engineered a nanoparticle that contains the key elements needed to produce cancer-fighting proteins. Their solution includes a combination of DNA, ribosomes, and amino acids wrapped in a chemical cage made from a compound called DMNPE. Although amino acids can also be conjugated to DMNPE, in this study the DNA was wrapped by the DMNPE, making it inactive until exposed to UV light. The particles are encased in a lipid shell to keep them from degrading and to avoid an immune response. The particles are programmed to start protein synthesis – to manufacture the cancer-fighting proteins at disease sites – when exposed to UV light. This allows nanoparticles to be delivered and controllably triggered to synthesize proteins "on demand." The types of proteins synthesized are determined by DNA sequences, which act as blueprints for the "nanofactory."

In their paper, Dr. Anderson and his coauthors describe how they tested the nanofactory concept *in vitro* and *in vivo* (in mice) to successfully deliver and activate a green fluorescent protein using UV light. To test a range of sizes, the researchers engineered particles at 100, 170, and 400 nm and found that the two larger sizes produced proteins, while the smallest (100 nm) size did not.

Dr. Avi Schroeder, lead author of the article, said: "This is the first proof of concept that you can actually synthesize new compounds from inert starting materials inside the body" [15]. Schroeder was a postdoc and researcher at MIT and is currently Assistant Professor at Technion-Israel Institute of Technology.

This ability to switch on a protein-making process opens the door to further developments for remotely activated drug delivery, and also raises many questions for researchers. For example, what types of cancer-fighting proteins and other drugs can be synthesized using this process? What other stimuli besides UV light can be used to activate these nanofactories? Could specific biochemical conditions such as acidity levels be used to activate the nanoparticles?

Using the body's own cellular machinery in controllable nanofactories to produce cancer-fighting compounds is one example of the many disease-fighting bionanoinnovations being developed in cutting-edge laboratories around the world.

10.5 Nanosizing Drugs

As previously mentioned, one of the early successful uses of bionanotechnology involved the development and delivery of nanosized versions of drugs to tumors. Nanosizing a drug can make it more soluble (and more bioavailable), and allow it to remain intact until it reaches the desired organ. The drug can also be engineered to provide a "time release" feature that keeps it working longer. Some drugs need to be nanosized to reach an organ protected by a membrane, or to penetrate the walls of human cells. Nanosizing drugs and other biological materials expands the available surface area, which is what gives nanocatalysts their potency. Nanosizing a reagent that is expensive or difficult to produce reduces the amount needed and lowers the cost.

Many therapeutic drugs cannot be taken orally because they do not penetrate membranes and tissues well. Nanosizing drugs to make them soluble is one way to convert injectable drugs to a form that can be taken orally.

Medical scientists have been nanosizing drugs since the 1990s to make them more soluble and able to pass through membranes and barriers in the human body. Early examples of nanosized drugs include two cancer-fighting drugs, Doxil and Abraxane. Doxil is a first-generation drug approved in 1995 that is coated with a liposome. Abraxane is a nanosized formulation of the cancer drug Paclitaxel that uses nanoparticles made from the human protein albumin to treat metastatic breast cancer. In 2009, Abraxane was approved for the treatment of pancreatic cancer. Abraxane minimizes toxic side effects and delivers a 50% higher dose in 30 min than standard Paclitaxel, which originally needed to be administered over several hours. Another example mentioned in the discussion of cancer-fighting therapies (earlier in this chapter) is Docetaxel, which induces cancer cells to commit suicide.

Some cancer-fighting compounds exist in common foods and seasonings and need to be downsized to make them more bioavailable. A good example is curcumin – a version of the seasoning cumin used in cooking. Some cumin formulations are available today in the form of serums and diet supplements, but this technology has not yet been refined and the true unlocking of nanosized curcumin is still underway.

Another promising area for nanosizing includes the development of nanosized neuroprotectants and diagnostic compounds to fight ischemic stroke and regenerate neural cells. Virtually all neuroprotectants have failed in clinical trials because of safety issues or lack of efficacy, mostly due to the inability to effectively penetrate the "blood–brain barrier." Nanosizing is expected to help secure methods for delivering neuroprotective compounds as well as methods for neural regeneration and treatment [16].

While there are only a few examples of "nanodrugs" on the market today, dozens of nanodrugs and nano-enabled therapies are currently in clinical trials in the United States and overseas.

Nanosizing drugs is not only a good therapeutic technique but it also makes good business sense. Patent Attorney Jeffrey Rosedale makes a strong case for nanosizing pharmaceutical drugs to extend their patent life, which can be critically important for companies and ventures that are trying to recoup billion-dollar R&D costs. Dr. Rosedale is a US Patent Attorney and Partner at the law firm of Baker and Hostetler LLP in Philadelphia, and is a Ph.D. chemical engineer. He believes that more biopharma firms should consider nanosizing drugs to extend expiring patents. In addition to providing a more bioavailable version, nanosized drugs that convert injectable drugs to a form that can be taken orally may also extend a drug patent.

10.6 Gene Therapy: The First Breakthroughs (At Last)

Few areas of medical innovation capture the imagination more than gene therapy. Being able to control the expression of genes, the blueprints of life itself, sounds like a miracle. More than 30 years of intense research and well over a billion dollars have been invested in this complex effort.

As a manager, you should know that gene therapy is poised to play an increasingly important role in medicine. Gene therapy is finally working. In 2012, the European Commission certified the commercial use of Glybera, which became the first gene therapy approved in the western world. With more than 1900 clinical trials now underway, it is likely that more gene therapies will be approved in the next few years. This is especially encouraging for the many patients and medical practitioners who have been waiting for genetic cures.

Gene therapy involves the treatment of a medical disorder or disease by replacing a missing or defective gene, correcting a genetic sequence, or using genes to produce therapeutic proteins to prevent or treat a disease. Gene therapy falls squarely in the domain of nanomedicine, since most of the structures and processes involved are nanoscale in size to have nanoscale features. These include DNA, genes, and the viruses used as vectors to deliver therapeutic genes to cells.

If we think of gene therapy in terms of innovation – which is the *implementation* of inventions and discoveries – we can ask: What are the innovation goals for gene therapy, and how far have we come toward fulfilling those goals?

Most people would agree that gene therapy would succeed if it allows us to (i) read and interpret the specific gene(s) that need to be replaced to prevent or cure disease, (ii) provide gene therapies that treat and cure diseases better than existing methods, (iii) make gene therapies commercially available – which requires successful completion of clinical trials and regulatory approval, and (iv) apply gene therapy not only to "orphan diseases" but also to common diseases (e.g., cancer, heart disease, and influenza) that affect larger populations.

The good news is that progress is being made in each of these areas.

During the past few years, a growing number of people with inherited genetic disorders have been successfully treated and cured, and the list of successes is growing. Despite some setbacks including the death of an 18-year-old patient named Jesse Gelsinger in 1999 and patients who contracted leukemia and other complications, the field of gene therapy has marched inexorably forward, creating maps of genomes, exomes, and proteomes associated with disease, identifying delivery vectors, overcoming immune responses, and dealing with all sorts of unexpected complications. Most important, a growing number of children and adults are being cured with gene therapy.

The Urgent Need for Gene Therapy Solutions

Gene therapy was made possible by the mapping and sequencing of the human genome, and was developed to address an urgent medical need. Millions of people die each year from fatal and debilitating diseases that are linked to genetic causes. In the United States, 1 out of every 200 children is born with a single-gene disorder. Some of these mutations are harmless, while others are life threatening or debilitating. Examples include cystic fibrosis, sickle cell anemia, Marfan syndrome, Huntington's disease, and Hemophilia A and B. In addition, hundreds of medical disorders are caused or influenced by multiple genes. These include heart disease, high blood pressure, Alzheimer's, arthritis, diabetes, cancer, and obesity. Gene-related diseases can also come from viral and bacterial infections, chemicals and pollutants, radiation, and other factors.

Gene therapy has benefited a few hundred people – mostly victims of hereditary single-gene disorders for which there was no treatment or cure. Many gene therapy patients are alive and healthy today as a result of experimental treatments administered years ago in preclinical trials. Broader applications now under development include treatments for cardiovascular disease and a "spray on" gene therapy that could help prevent the spread of pandemic flu strains. These applications could benefit billions of people.

So why hasn't gene therapy been mainstreamed into general medical practice?

There are several reasons. While some patients responded well to their therapy and were effectively cured, others were partially cured or received no benefit at all. In some human organs, the therapy works for a period of time, then diminishes. In many patients, antibodies attack the viruses delivering the genes.

Investigating why gene therapy works for some patients and not for others has engaged the medical community in the past decade. Slowly but steadily, the mysteries are being unlocked. Some of the stories read like a Sherlock Holmes novel. Thanks to a great deal of detective work by molecular scientists, the most daunting and frustrating genetic mysteries are being solved.

How Gene Therapy Works

Genes are the building blocks of DNA, the programmable blueprint that regulates all biological functions. The genes that make up the human genome provide the set of instructions that determine human characteristics such as hair and eye color and blood type and also control functions that are not visible or obvious – such as regulating more than 100 000 human proteins that regulate virtually every aspect of human biology.

Without going into the detailed science, here is a very basic description of genes and how they work: A gene is a set of chemical units (nucleic acids) designated A, C, G, and T, which are arranged in different combinations and lengths. A (adenine) is complementary to T (thymine) and C (cytosine) is complementary to G (guanine). Structurally, these base pairs form the rungs on the DNA molecule, which is shaped like a twisted ladder called a double helix. The four chemical nucleotides form more than 20 000 genes that occupy ~6 billion possible combinations/locations on a DNA molecule.

Every cell contains the full DNA code including all the genes, although gene expression can vary according to the organ or tissue. For example, a liver cell may express or repress different genes than a lung or muscle cell. DNA molecules form chromosomes that determine our sex and other characteristics and serve as our genetic blueprint. To transcribe and translate this blueprint to make functional proteins, four types of RNA are involved in expressing gene information and using this to produce proteins.

As you can see, our genetic system is like a very complex chemical computer. Genes are analogous to bytes that contain the instructions that tell a computer what to do. Genes form the code that tells our body what to do. This complex coding system has evolved over millions of years, creating an elegant method for preserving the identity of humans, animals, plants, and microbes. Just as bugs can appear in a software program, defects and mutations occur in the genetic code. Many of these defects are associated with disease.

In most cases, genes express their function by producing *proteins*. Proteins are large molecules made of amino acids that are shaped like tangled spaghetti. There are more than 100 000 different proteins in the human body. Proteins are present in every cell and play a role in virtually all biological structures and processes in the human body, from forming tissues and organs to transmitting signals between cells. Which proteins are produced and when are determined by genes? Proteins can play a role in causing, enabling, or inhibiting diseases and disorders. For example, protein deficiency can cause hemophilia and other diseases. Misfolded *proteins* can cause mad cow disease and Parkinson's disease [17].

Human genes are not lined up neatly like a line of computer code. They can be scattered along the DNA molecule. One gene or group of genes may govern the expression of a different gene at another location. Some genes are programmable and can be turned on or off. Some genes are dormant and become active when needed. Some provide spacing on the molecule. Other genes and sequences are duplicated many times, and these repetitions may also have a function or not. Many genes that were previously thought to be evolutionary dead-ends with no function – called "junk genes" – have been found to play a role in several genetic disorders. Some of these spacer genes don't seem to have a function, but recently, scientists have found that these gene segments help prevent viral DNA strands from infiltrating or hijacking the DNA molecule.

So, how does gene therapy work?

To begin with, there are several types of gene therapy. Gene therapy is sometimes called gene replacement and is often done in conjunction with gene splicing. A replacement gene is called a *transgene*. To replace a defective or missing gene, a scientist can remove a gene from a patient's DNA, using a special enzyme (a form of protein) that moves along the DNA strand until it encounters a certain chemical sequence. The enzyme then cuts the gene out of the molecule like a chemical scissors, leaving two "sticky" ends. Another enzyme can then be used to deliver and insert the therapeutic gene into another DNA molecule (using genetic *engineering*). It's sort of like editing an old-fashioned piece of movie film, where frames are cut out of the filmstrip and replaced with other frames, and then reinserted into the projector. The transgene may be cut and pasted into a DNA molecule, or it can be added without removing the defective gene (assuming the defective gene doesn't cause problems if it is allowed to remain). In some approaches, a DNA strand carrying a functional gene is added to the cells as a separate entity without replacing or editing the existing genome, typically to produce a functional benefit; this DNA may reside in the cytoplasm of the cell instead of in the nucleus.

Gene replacement can be done outside the body (*ex vivo*) or inside the body (*in vivo*). If this is done *ex vivo*, the engineered DNA with the healthy gene(s) can be inserted in a patient's own cells and cloned to create enough cells to provide a therapeutic effect when they are injected back into the body.

In vivo gene therapy typically involves using a virus that has been rendered harmless and genetically altered to deliver human DNA to target cells where they are needed. This method can deliver genes to cells that do not divide or where the cells divide slowly. The genes express their function as long as the cell is alive, and the function is related to the life of the cells, which can vary widely. For example, skin and epithelial cells that form layers in the body are replaced often and typically do not last long, while muscle cells tend to last as long as the body is alive.

Viruses are a preferred delivery vector because they have evolved many creative methods for delivering their genes to human cells. Typically, a virus has a binding agent or structure on the outside of the virus that enables the virus to seek out and attach to a particular type of cell, which is then "infected" with the viral genes. Most viruses transfer their DNA to the cell where it reproduces more viruses, while other viruses (retroviruses) enter the cell first and then produce their DNA.

Using viruses as delivery vectors can trigger the body's natural immune system, which has evolved to create antibodies to find and destroy harmful viruses. Selecting the right virus or version (serotype) of the virus is critical. It's important to match the gene-carrying virus (adenovirus, adeno-associated virus, retrovirus, lentivirus, etc.) to the cell type (liver, lung, nasal, muscle, etc.). Also, one variant or *serotype* of a virus may be more effective than other serotypes of the same virus. Some virus families such as AAV have more than 100 known serotypes.

If the virus comes from a family that may have already infected a patient sometime during the patient's life, the patient would develop an immunity to the virus being used to transmit the gene and reject the virus. For example, many people develop weak infections that never take hold but form antibodies against a particular virus – this could trigger an immune response against engineered harmless versions of that virus that are used to deliver a gene.

Researchers have tried to solve this problem by using the outside "shell" of the virus (called a *capsid*), with only the bare minimum "ingredients" needed to transfer the gene. This has been an effective approach, although in some cases the binding structures on the surface of the virus may still trigger an immune response, requiring additional engineering to make the viral package more "bio-friendly."

This is a very basic explanation, but generally how it works.

Measuring the Success of Gene Therapy

There have been thousands of gene therapy preclinical and clinical trials, although only one gene therapy has received regulatory approval in the western world. Despite many notable successes, including cures for medical conditions, gene therapy has struggled to move from laboratories to commercial use. One reason is that gene therapy has been producing varying results. In preclinical and clinical trials, some patients were totally cured – that's the good news. However, others were partially cured, or not at all, while some patients developed complications or relapsed. Scientists are striving to understand exactly why some patients benefit, while others do not. Results can vary from one individual to another based on age, health, what stage the disease is in, whether an individual was exposed to a virus family earlier in life, the influence of multiple genes, and other factors. In the meantime, the list of gene therapy patients who have been cured and whose conditions have improved continues to grow.

Gene therapy requires a high degree of personalization. For example, we need to know if a patient has already developed antibodies that will attack a virus being used to deliver a transgene. We need to know if a disorder results from one gene or multiple genes. Is a condition inherited or is it showing up for the first time? Which molecular structures – proteins/enzymes, chromosomes, mitochondria, and so on – play a role?

As personalized medicine gains more traction, tests for genes, antibodies, proteins, and other factors will come down in cost and become more routine, although it may be a decade or longer before we enter an era when genetic profiling becomes routine for everyone and gene therapy becomes routine.

If you have an interest in gene therapy in general, or in a specific disease, more detailed updates are available on the Web sites of the NIH, FDA, and European Commission and advocacy groups such as the National Cancer Society. Detailed information on worldwide clinical trials can also be obtained from an excellent online tracking library maintained by the *Journal of Gene Medicine*, published by

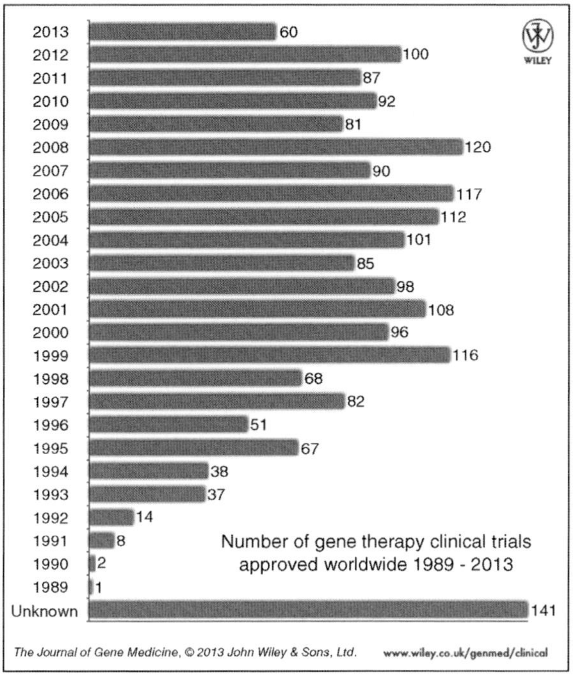

(a)

Figure 10.12 These gene therapy statistics are updated periodically by the *Journal of Gene Medicine*, published by John Wiley & Sons, Inc. Charts shown here include (i) number of gene therapy clinical trials approved worldwide from 1989 to 2012, (ii) geographical distribution by country, (iii) delivery vectors used, and (iv) disease targets (indications) (used with permission from the *Journal of Gene Medicine*, Copyright 2013, John Wiley & Sons, Ltd.).

Wiley.[2)] In 2013, the Wiley database documented more than 1900 gene therapy clinical trials. The accompanying charts provide some current insights into the distribution of gene therapy clinical trials internationally, disease targets, and vectors used to deliver the genes covering the period from 1989 to 2012 (Figure 10.12).

Pioneers and Perseverance

There are many pioneers in gene therapy who have been working for decades to advance the field, despite the daunting array of obstacles. A notable example is Dr. James Wilson, a gene therapy pioneer who has experienced every aspect of gene therapy, from the frustration of seeking viral vectors that are safe and effective to the death of a patient in an experimental trial, to the development of

2) *Journal of Gene Medicine*, Gene Therapy Clinical Trials Worldwide, http://www.abedia.com/wiley/

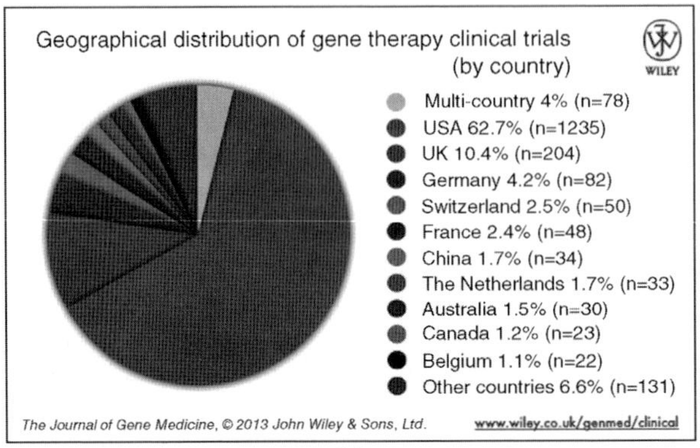

(b)

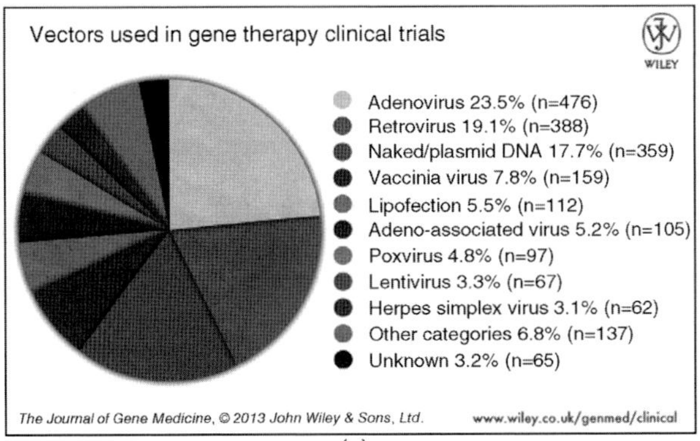

(c)

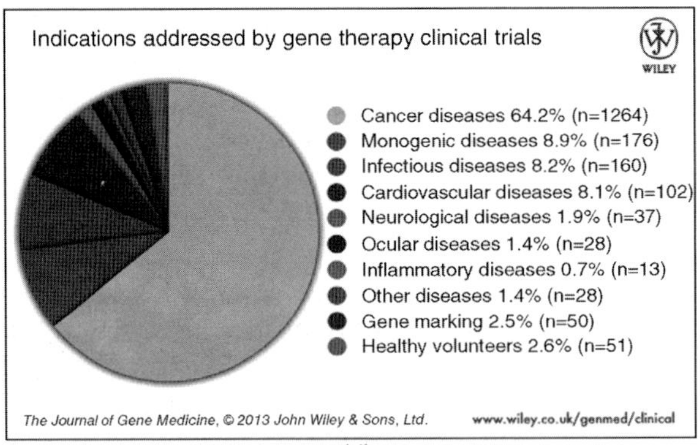

(d)

Figure 10.12 (*Continued*)

cutting-edge tools and vectors that are now being used to make gene therapy – finally – a commercial reality.

I've known Jim Wilson since 1996 when I helped organize a Wharton workshop on gene therapy being led by Dr. Wilson and Professor George Day (Figure 10.13). For that event, I coauthored (with Aris Persidis) a white paper entitled "Obstacles to Commercialization of Gene Therapy," which was summarized in a *Nature Biotechnology* article. Aris and I identified and predicted some of the hurdles that gene therapy would likely encounter before achieving commercialization. We cited several possible pitfalls that could slow the pace of commercialization, including the possibility that a patient might die in a clinical trial, the slow participation of biopharma firms, and the challenge of creating a business model for a therapy that could cure a disease with a single injection. To the disappointment of everyone involved in the field, virtually all of the obstacles we identified came to pass.

Figure 10.13 Dr. Jim Wilson has persevered tirelessly to fulfill the promise of gene therapy (photo by Michael Tomczyk).

Despite these obstacles and many industry failures along the way, Jim Wilson has persevered to learn from each setback and has kept driving forward. He has worked tirelessly with research teams at the University of Pennsylvania and other institutions and companies to find many of the formulas that are being used today to administer successful treatments. He leads the gene therapy research at the University of Pennsylvania and serves as Editor-in-Chief of the journal *Human Gene Therapy*. He has authored or coauthored more than 500 papers, contributing important insights to all aspects of gene therapy and related areas of science, including the use of viral vectors to treat diseases of the eye, lung, heart, liver, skeletal muscles, brain, and other organs/tissues.

Some of Dr. Wilson's most notable research involves the development of novel recombinant adeno-associated viruses for use in gene therapy (see Figure 10.14). AAVs are widely represented in the human and primate populations, although no pathology has been associated with infection from these viruses. AAVs are a type of *parvovirus*. It should be noted that an *adeno-associated* virus is different from an "adenovirus," which is a separate type of virus that has been used with some success in gene therapy experiments. AAV serotypes 1, 7, 8, 9, and rh10 were originally isolated at the University of Pennsylvania in Dr. Wilson's laboratory, which has identified ~100 AAV serotypes.

AAVs are being used in a growing number of gene therapy clinical trials, due to the unique characteristics of this type of virus – including low pathogenicity and a typically weak immune response. AAV is one of the smallest viruses, ~22 nm. AAVs infect humans and primates, are not known to cause disease,

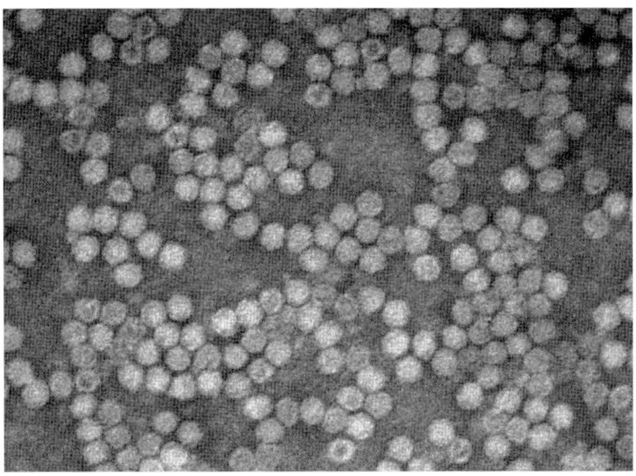

Figure 10.14 Gene therapy trials using adeno-associated viruses (AAVs) as vectors (shown here) are producing significant advances in the fight to cure disease. These viruses, which measure 22 nm in diameter, are favored because there are more than 100 different serotypes, they generally are not pathogenic, and they generate a relatively low immune response (image courtesy of the James Wilson Laboratory, Perelman School of Medicine, University of Pennsylvania; image by Peter Bell).

and cause a minimal immune response. The first AAV was discovered in 1965. Since then, >100 versions of AAVs have been engineered.

More than 70% of the population is infected with a naturally occurring (wild-type) AAV, which means that they have developed antibodies that bind to the AAV and prevent gene transfer and expression. This means that strategies are often needed to address this potential immune response. For patients who have developed antibodies against the wild-type strain, gene therapy strategies include using novel AAV serotypes that won't trigger the antibodies, or administering immunosuppressant drugs with the gene-carrying viruses to allow more vectors to reach the target cells. AAVs work best in infants and children who have not developed antibodies to AAV and who are more likely to tolerate AAVs as a gene delivery vector. By matching the serotype to the disease and patient, AAVs can provide a delivery vector for 90% of the population, according to Dr. Wilson.

Another strategy that is being investigated involves using "empty" AAV capsids that trick the antibodies into attacking nonfunctional viruses that are not carrying genes. This may allow more of the gene-carrying capsids to slip past the immune system [18]. As scientists try to understand why some gene therapies have had more success than others, one theory suggests that therapies may work better if they have more empty capsids as opposed to those that have almost all gene-carrying viruses. Ironically, it was originally thought that virus samples with empty capsids were impure and less desirable; however, it may be that the empty capsids are needed to absorb and sidetrack antibodies to allow the gene carriers to reach the cells. Scientists are now studying the empty capsids of AAVs and other vectors to see if this applies more to some serotypes and not others, and also to ensure "empty" capsids don't mutate or move to locations on the DNA/chromosome that can cause other problems. A related strategy involves directing gene-carrying AAVs to a specific location on the DNA/chromosome where they can lie dormant without apparent effects, until triggered by a disease that the gene can help combat [19].

The gene therapy vectors developed by Dr. Wilson and his colleagues are made available for license to both for-profit and nonprofit research organizations. Nonprofit organizations and government agencies involved in preclinical studies and basic research can obtain access to these vectors through the Penn Vector Core at the University of Pennsylvania. Biotech/biopharma organizations can license technology from ReGenX BioSciences, which was cofounded by Dr. Wilson. ReGenX BioSciences has more than 100 patents and patent applications, including fundamental patents on novel AAV serotypes.

Translating Therapy into Practice

As gene therapy begins to mature and longer-term successes with patients are verified, the next question becomes: How will gene therapy affect traditional medical practices and what industry sectors will be disrupted or transformed? Dr. Wilson was one of the earliest experts to recognize that gene therapy is a potentially disruptive technology that could eventually replace many existing medical treatments. A disruptive technology is a science-based innovation that has the potential to replace an existing incumbent technology, which means that it could "destroy value" as well as "create value."

The most commonly cited example is the potential for a single injection of genes to provide such a strong and lasting therapeutic effect that it totally does away with drugs, infusions of proteins or enzymes, and radiation or surgery – which could represent billions of dollars in revenues. This begs the question: *What is the business model for gene therapy if the treatment involves only one injection, and it does away with other treatments that would normally cost hundreds of thousands or millions dollars over the life of the patient?*

For example – as Dr. Wilson observes – in clinical trials involving hemophilia B, a single gene therapy injection has eliminated or minimized the need for weekly protein injections to control bleeding episodes. These protein infusions can cost hundreds of thousands of dollars a year for each patient. If a gene therapy shown to work with hemophilia B is commercialized and applied to hemophilia A and other related diseases that have a much larger population of patients, it could disrupt the current $6.5 billion market for protein replacement therapies. There are many areas where gene therapy could replace expensive and painful injections with a single shot. Gene therapy could replace drugs and even some types of surgery. The implications for some sectors of the medical industry are profound.

Dr. Wilson summarized these critical issues in a thoughtful January 2012 essay published in Ref. [20]. In his commentary, he wrote: "Insurance companies currently reimburse for each service/infusion using existing protein replacement products, which in the aggregate cost up to $300,000/patient/year for prophylactic dosing. An obvious question concerns how to charge for gene therapy in which one injection of vector confers stable expression for an extended period of time, which, for purposes of argument, I will say is 10 years. I have heard this kind of therapy referred to as 'one and done' with the very legitimate question asked: How do you price a cure? It is unlikely that reimbursement for the one-time injection of vector will be equivalent to the costs of treating a hemophiliac with 10 years of protein (i.e. $10 \times \$300{,}000 = \$3$ million). Unfortunately, there are no data available to answer these pricing questions with any certainty. The critical path to the development of gene therapy for diseases like hemophilia may reside in the creation of new business models that reimburse for the long-term efficacy afforded by a single gene therapy injection/treatment. One approach is to charge the insurance company an annual fee for gene therapy as long as it continues to work." He goes on to observe that this approach may not work in a country where private insurance is the dominant form of reimbursement, especially when people change jobs and insurance providers. A new provider may not agree to honor the long-term payments provided by (or negotiated with) the previous insurer.

"Commercial development of gene therapy has been proceeding slowly but is likely to quickly gain momentum," Dr. Wilson predicts. "We have the keys to the genetic locks. Many of the mysteries are solved. We're so close to commercial application in several areas that we now need to consider how gene therapy will disrupt and transform existing methods of treatment. We need to define the new business models that will be needed."

A Growing Portfolio of Gene Therapy Successes

The important thing we all need to know is – Where is gene therapy succeeding and why? Where are the demonstrable results? Which diseases have responded to the therapy? The following examples show where some of the most notable early successes are occurring.

Europe's First Commercial Gene Therapy

The first commercial gene therapy in the western world was approved by the European Commission in October 2012. The treatment is called Glybera, which is the trade name for Alipogene tiparvovec, a gene therapy treatment that compensates for lipoprotein lipase deficiency (LPL). LPL is an enzyme protein deficiency that restricts the transport and breakdown of fat (lipids) in the body, which can cause acute pancreatitis and other complications. This family of protein disorders affects approximately one out of every million people, with a much higher incidence in some geographical and ethnic populations. As the first gene therapy approved in the western world, the approval of Glybera was a groundbreaking milestone in the history of gene therapy.

The corrective gene is delivered using a modified form of adeno-associated virus AAV-LPL S447X. Treatment involves placing the patient under spinal anesthesia and administering several intramuscular injections. During the clinical trials, fat concentrations in the blood were reduced within 3–12 weeks after the injections in nearly all patients, with minimal immune response or other complications.

Glybera was developed and commercialized by a Netherlands-based company called Amsterdam Molecular Therapeutics – which had to be rescued by investors to keep it from going bankrupt during the complicated, 2 year approval process. During the approval process, the application was delayed partly because there were only 27 patients in the clinical trials; however, this ultraorphan disease only affects one in a million people that limited the number of available patients. No alternative treatment was available.

During the regulatory process, AMT had to slash its staff and was reorganized as uniQure B.V., which saved the enterprise and enabled it to survive. The company began commercialization and marketing of Glybera in Summer 2013. uniQure's largest shareholders are Forbion Capital Partners and Gilde Healthcare, two of the Netherlands' leading life sciences venture capital firms. In February 2013, uniQure was named to the "50 Disruptive Companies List" published by the MIT *Technology Review*. This is a story of perseverance, near-failure, financial hardship, regulatory challenge, and hard-won success as the first approved gene therapy in the western world.

On their Web site, uniQure describes Glybera as "the first stable and scalable AAV manufacturing platform, capable of delivering gene therapies to treat many other diseases." uniQure and its partners in Europe and the United States are conducting or planning clinical trials that target other single-gene disorders, including acute intermittent porphyria, Parkinson's disease, SanfilippoB, and hemophilia B. All of these studies use adeno-associated viral-derived vectors to deliver therapeutic genes.

Regulatory approval does not mean that the therapy is certified as totally safe and foolproof. In the type of gene therapy represented by Glybera, the viral genome carrying the therapeutic gene enters the cell and persists in the cytoplasm as a separate structure, and rarely integrates into the existing cell genome. There have been cases where genes carried by viruses have integrated at sites in the genome that can activate cancer-promoting genes; however, studies to date of this viral vector have shown this to be rare [21].

China's First Gene Therapy

Glybera was not the world's first approved gene therapy. Actually, the first gene therapy, called Gendicine, was approved in 2003 by the China State Food & Drug Administration (SDFA) – making China the first country where gene therapy became commercially available. The treatment uses a recombinant (serotype 5) adenoviral vector to administer a naturally occurring tumor-suppressing gene known as p53 used in conjunction with radiotherapy and chemotherapy to treat head-and-neck squamous cell carcinoma. This type of cancer represents 10% of the 2.5 million annual new cancer cases in China. According to company press releases, other potential applications include liver, abdominal, and pancreatic cancers. By 2008, more than 6000 patients had been treated with Gendicine, including 400 non-Chinese patients [22]. The primary benefit, where successful, is tumor shrinkage. The main side effect is a fever that subsides over time. No claims are made concerning extending longevity of cancer patients.

Gendicine was developed by Shenzhen SiBiono GeneTech Co. Ltd., which was founded by Dr. Peng Zhaohui. A majority stake in SiBiono was acquired by Benda Pharmaceuticals in 2007. Dr. Peng resigned as chairman of SiBiono in 2008. Dr. Peng and major investors have subsequently been embroiled in a patent dispute and legal conflicts with Benda Pharmaceuticals.

X-Linked SCID ("Bubble Boy" Disease)

One of the first successful gene therapy trials was conducted to treat X-Linked SCID, an immune deficiency that requires patients to be isolated, sometimes in physical "bubbles" to avoid contracting even common viruses and other infections that can be lethal. SCID is a genetic disease that occurs in babies who are born lacking an enzyme called *adenosine deaminase*. These infants, if untreated, can die within a year from severe infections for which they have no immunity. About 1 in 50 000 babies are born with this disease. As many as 1 in every 2500 children in some American Indian populations inherits a version of this disease. Conventional treatments are extremely expensive, involving twice-weekly enzyme injections or bone marrow transplants. A related disease, called adenosine deaminase deficiency (ADA), is a similar inherited disorder that causes severe immunodeficiency. Over 120 different gene mutations have been identified that can impair the immune system [23].

From 1999 to 2009, 20 patients with SCID-X1 disease and 27 with ADA deficiency received *ex vivo* gene therapy using γ-retroviral vectors. The trials were performed at hospitals in Paris, London, Milan, Los Angeles, and at the National Institutes of Health. Of the 20 SCID patients, 5 developed a T-cell leukemia 2–5 years after receiving the gene therapy. The cancer was fatal for one patient and cured in four others. This was widely publicized and – at the time – caused some sources to question the viability of gene therapy. Of the 19 ADA patients who were successfully treated, none showed any adverse effects. This suggested that the leukemia may have resulted from disease-specific factors [24].

Based on what was learned from the SCID experience, viral vectors have been made safer by engineering the integration of the transgene (*ex vivo*) to a "neutral" region of the genome. The next step is to refine the therapy so that it can be administered *in vivo*.

In 2010, ~11 years after the first treatments, 18 out of 20 SCID-X1 patients and all 27 patients treated for ADA were alive, with the immune deficiency corrected in 17 of the SCID patients and in 19 of the ADA patients.

Hemophilia B

One of the most significant gene therapy achievements involves the treatment of hemophilia B. The "first unequivocal success in treating hemophilia B using gene therapy" was reported in Ref. [25]. Researchers used AAV serotype 8 to deliver genetic material to the liver of four patients treated at the Royal Free Hospital in London who were under the care of Professor Edward Tuddenham. This therapy eliminated the need for expensive injections of blood-clotting protein factors in the trial subjects for up to 18 months. Hemophilia B is a disorder caused by a gene mutation where a protein needed to form blood clots is missing or reduced, causing prolonged bleeding that can be fatal. It affects ~3000 men in the United States Typically, a protein-clotting factor is injected three times a week and can cost as much as $200 000 a year for each patient. If this therapy proves safe and effective in larger trials, it means that essentially one injection of therapeutic genes can save millions of dollars in protein injections over the life of each patient. This is an excellent example of how nanomedicine – and gene therapy in particular – has the potential to disintermediate traditional medical practices.

Fighting HIV/AIDS

In May 2012, Researchers Bruce Levine and Carl June at the University of Pennsylvania's Perelman School of Medicine reported that T cells genetically modified to fight HIV persisted up to 11 years in 41 patients, without serious side effects. All of the patients were also taking antiretroviral drugs. Drs. Levine and June and their teams have tested several ways to treat HIV using gene therapy. Achieving "persistence" of therapeutic genes in modified cells is an important achievement since most therapies that involve inserting DNA into T cells triggered an immune response that neutralized them. Their approach involved harvesting the patients' T cells, inserting a gene that boosts their ability to recognize and kill HIV-infected cells, and multiplying the super T cells and infusing them back into the patient.

Treating Leukemia

Leukemia is a cancer of the blood cells that starts in the soft tissue of bone marrow where blood cells are produced. Abnormal white blood cells are produced that do not perform the disease-fighting function of normal white blood cells. The cells keep growing and crowd out normal blood cells, which leads to anemia, bleeding, and infection. Causes can be genetic or environmental (such as exposure to radiation, chemicals, or an adverse reaction to chemotherapy). In the United States, about 40 000 patients are diagnosed with leukemia and about 22 000 leukemia victims die annually.

In December 2012, the University of Pennsylvania reported the results of a clinical trial led by Dr. June in which 9 out of 12 leukemia patients treated with gene therapy remained in remission more than 2 years. The patients included 10 adults and 2 children. This breakthrough was cited as the "first successful and sustained demonstration of the use of gene transfer therapy to turn the body's own immune cells into weapons aimed at cancerous tumors" [26]. In the press announcement, Dr. June stated, "It is possible that in the future, this approach may reduce or replace the need for bone marrow transplantation."

In 2012, Dr. Carl June was recognized by the American Society of Hematology as corecipient of the Ernest Beutler Lecture and Prize; in May 2013, Dr. June received the Philadelphia Award from the Perelman School of Medicine for "extraordinary advancements in gene therapy aimed at treating HIV and cancer, specifically chronic lymphocytic leukemia and acute lymphoblastic leukemia." Dr. June is the Richard W. Vague Professor in Immunotherapy in the Department of Pathology and Laboratory Medicine at the University of Pennsylvania's Perelman School of Medicine and Director of Translational Research at Penn's Abramson Cancer Center.

Inherited Blindness (LCA)

Some gene therapies have reported initial success, but over time the condition worsens or is only cured for a temporary period. An example is Leber congenital amaurosis (LCA), which is an inherited form of blindness that affects 3000 people in the United States, with an incidence of approximately 1 person in 80 000. LCA is caused by a mutation of a gene (RPE65) that encodes an enzyme in the retina of the eye, needed to maintain retinal health. In LCA, a mutated enzyme fails to convert vitamin A into the form needed for photosensitive cells in the retina to detect light. Children with LCA typically lose their vision in the first 6 months of life. Since this disease is progressive and most photoreceptors are destroyed by the time a patient reaches adulthood, gene therapy has worked best and most effectively in children.

In 2009, 12 people received an experimental gene therapy as part of an experimental trial led by the husband and wife research team of Jean Bennett, MD, Ph. D., a physician and molecular geneticist at the University of Pennsylvania and her husband Albert Maguire, a retinal surgeon [27]. An AAV was used to deliver the RPE65 gene along with a string of DNA that switched on the gene. A drop of solution the size of a pea containing the genes was injected by syringe through the white portion of the eye, directly to the retina. This injection contained ~15 billion copies of the AAV virus carrying the RPE65 gene. The treatment was first used to restore the sight of dogs, and there was some apprehension whether the treatment would be as effective in humans since results in animal trials often do not translate to human trials. Also, there was the fear of an immune reaction to the therapy; however, the immune response in the human eye is typically weak, so this was not a factor.

When the trial began, all the patients were legally blind, and only 2 weeks after their injections, half of the participants improved enough to lose their legally blind status (although their vision was not yet totally normal) and most of the patients improved their ability to negotiate an obstacle course under low light conditions and improved vision was retained for 3 years. Subsequent treatment and tests verified that the gene therapy not only prevented the death of photoreceptor cells but also activated vision-processing areas of the brain.

One of the insights gained is the importance of administering this therapy as early as possible, since it is too late to restore photoreceptor cells that are already destroyed. Much of the retinal tissue is lost by the time an individual becomes a teenager and total vision loss typically occurs by age 40, so gene therapy for LCA should ideally begin in childhood. Unfortunately, recent evidence has shown that visual function in some patients may continue to deteriorate despite the early visual improvement provided by this gene therapy, which may require additional treatments or therapies to slow retinal degeneration into adulthood [28].

Gene Therapy for Heart Disease

While the first wave of gene therapy successes have helped treat and cure rare inherited disorders among fairly small populations, there is evidence that gene therapy has the potential to treat and prevent large population diseases such as heart disease or the common flu.

In 2013, more than 200 people with heart problems enrolled in a clinical trial that targets a common form of heart failure. The trial, called CUPID2, will use a genetically modified virus to insert extra genes into heart muscle cells to increase a protein called SERCA2a. This therapy is based on research at Imperial College London that discovered that SERCA2a protein levels are lower in heart-failure patients. The research is being led by Dr. Alexander Lyon, British Heart Foundation Senior Lecturer at Imperial College London and Consultant Cardiologist at the Royal Brompton Hospital, and will be conducted at these and other hospitals in the United Kingdom

"Our goal is to fight back against heart failure by targeting and reversing some of the critical molecular changes arising in the heart when it fails," Dr. Lyon said in a BBC interview [29].

Heart failure affects more than 750 000 people in the United Kingdom and infects more than 120 000 victims each year, killing one-third within 12 months. A heart attack or progressive heart disease damages heart muscles and affects the heart's ability to sustain a normal heartbeat. As heart cells deteriorate, the damaged cells are unable to recycle calcium, which is essential to producing heartbeats at a steady rhythm.

The UK gene therapy trial will attempt to restore the heart's ability to recycle calcium and normalize the heart beat. The gene will be delivered to heart muscle cells by a harmless virus that will be injected through a tube inserted in the patient's leg. If successful, this gene therapy will reset the calcium control, and restore the production of calcium needed to sustain healthy heartbeats.

A second trial called SERCA-LVAD started recruiting patients in 2013 to test the gene therapy in 24 UK heart failure patients who are fitted with mechanical heart pumps (ventricular assist devices). The CUPID2 trial is funded by US biotech company Celladon and the second trial is funded by the British Heart Foundation (BHF). The trials are part of a BHF campaign called "Fight for Every Heartbeat" that promotes heart disease research.

The implications are profound, not only for the millions of people with cardiovascular disease who could benefit from this innovation but also for millions of victims of other common diseases who are waiting and hoping for a day when gene therapy will move beyond the treatment of rare, exotic diseases to include mainstream diseases that affect enormous numbers of people.

A Gene Therapy Strategy for Pandemic Flu Control

In 2013, Jim Wilson announced a novel strategy for using gene therapy to produce antibodies that can prevent and neutralize a virus such as influenza. Influenza is the seventh leading cause of death in the United States and causes approximately half a million deaths worldwide every year, according to the Centers for Disease Control.

He predicts that this strategy could be used to prevent the spread of pandemic viruses that are one of the looming threats that keeps the global medical community on constant alert. Dr. Wilson recalls that he was inspired to pursue this research by Microsoft founder Bill Gates, who posed a question during a conversation they were having about the prospects for gene therapy. Gates asked if the AAV-mediated gene therapy Dr. Wilson was studying could be used to prevent or cure the flu, especially a flu pandemic.

Translating Gates' question into a research approach, Dr. Wilson started thinking about how gene therapy could be used to deliver genes to the nose and throat to prevent flu infections, especially the lethal strains of influenza most likely to cause a pandemic. Dr. Wilson assembled a team that included Anna Tretiakova, Ph.D., Director of Translational Research, and Maria P. Limberis, Ph.D., Director of the Animal Models Core at the University of Pennsylvania. They were joined by colleagues from the Public Health Agency of Canada, Winnipeg; the University of Manitoba, Winnipeg; and the University of Pittsburgh.

Typically, flu vaccines are developed to protect against seasonal flu viruses that emerge each year. Vaccines generally work by injecting the body with an incapacitated version of the virus that causes the body to produce antibodies that will attack full-strength versions of the disease virus. Once blueprinted, these antibodies will remain effective and can be produced by the body's immune system for decades.

Creating a vaccine against new flu strains involves growing the flu virus in eggs and cleaving off the encapsulating proteins from the exterior of the virus to create a harmless vaccine that will stimulate the immune system to produce antibodies against that virus. When the vaccine is administered, the people receiving the vaccine need time to build up antibodies – as much as 2 weeks – so the effect of the vaccine is not immediate.

Most vaccines do not guard against flu viruses that originate in animals and unexpectedly jump to humans. This includes animal-originated strains that are expected to mutate into forms that can be transmitted from human to human – strains that have the potential to cause a global pandemic. The risk is that a contagious flu pandemic can spread internationally before a vaccine is developed. During past pandemics, the only way to control the spread was to destroy enormous numbers of livestock and to quarantine anyone showing symptoms.

Dr. Wilson had an idea for a prophylactic approach that would take advantage of the gene's ability to instruct cells to make proteins. An antibody is a protein. If an antibody could be found that attacks a broad range of viruses – including pandemic strains – genes could be delivered to the throat and nose that would instruct the cells to produce antibodies that could fight off the flu. This could be developed faster than a vaccine, which is really important if a major pandemic suddenly appears.

Since most viruses enter the body through the nose and respiratory passages and are replicated in the mucous linings, Dr. Wilson and his team reasoned that delivering antibody-producing genes to these locations would prevent viruses from attacking the epithelial cells.

The first challenge was to identify an antibody that could be used against influenza viruses that have the highest potential of causing a human pandemic. A discovery by an immunologist working in Switzerland gave them a key piece of the puzzle. In 2011, a rare antibody named F16 was isolated by immunologist Antonio Lanzavecchia at the Institute for Research in Biomedicine in Bellinzona, Switzerland. This is the only antibody discovered to date that binds to all 16 serotypes of the influenza A family of viruses. This family includes some of the most potent viruses associated with pandemics such as the H1N1 "swine flu" and the H5N1 and H7N7 "bird flu" as well as emerging strains. In 2013, the H7N9 influenza virus killed more than one-quarter of people infected, and in January 2014, a new virus called H10N8 was identified in Jiangxi Province, China. These influenza strains exist primarily in poultry and other animals, but have the ability to leap to humans. The fear is that these strains will mutate into a form that can be transmitted from human to human.

These high mortality virus families are considered to be the most threatening candidates for a future global pandemic. The H5N1 virus that can be caught by humans from infected poultry has only infected 650 people detected in 15 countries during the past decade. However, this strain kills 60% of those it infects. The first person to die from this strain in North America died on January 3,

2014 after returning from a trip to China [30]. The findings by Dr. Lanzavecchia and his colleagues were reported in Refs [31,32].

To deliver the F16 gene, the researchers in Jim Wilson's group used a modified version of the AAV serotype 9 vector that was discovered in the Wilson laboratory and previously studied by Dr. Limberis as a possible vector for treating cystic fibrosis. They engineered the virus to deliver a gene that would instruct cells to produce the F16 antibody. The first step was to conduct animal studies to confirm the safety and efficacy of the approach. The gene therapy was administered to mice and ferrets that were then exposed to lethal quantities of H1N1 viruses.

The flu viruses tested were previously associated with some of the world's most deadly human pandemics. H1N1 (see Figure 10.15) is implicated as the primary virus responsible for the 1918 Spanish Flu epidemic that killed as much as 3–5% of the entire world's population. A new version of H1N1 resurfaced in 2009 that is estimated to have killed as many as half a million people worldwide (*including estimates of unreported deaths*).

In the animal trials, pretreatment of mice and ferrets virtually shut down virus replication and provided complete protection against all strains of flu exposed to the treated animals.

Dr. Limberis observed, "The novelty of this approach is that we're using AAV to deliver a prophylactic genetic vaccine to the nose and throat in a non-invasive

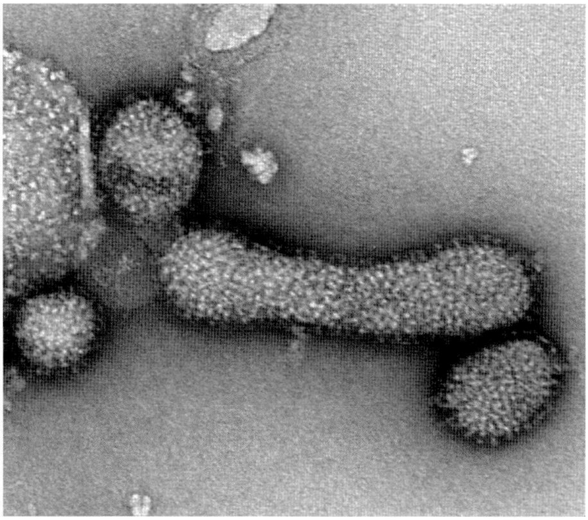

Figure 10.15 The 2009 H1N1 influenza outbreak was the first pandemic of the twenty-first century. Approximately 95% of those afflicted were younger than 51 years of age, suggesting that previous strains of H1N1 that circulated during 1930–2000 may have conferred immunity to older people to the 2009 virus strain while leaving younger people unprotected. This nanoscale image shows H1N1 Influenza A/PR8 (image courtesy of SAIC-Frederick, Inc. Copyright SAIC-Frederick).

manner, not a shot like conventional vaccines that passively transfer antibodies to the body's general circulation."

Because the genes were delivered to epithelial cells that are constantly being replaced, the protection only lasted 3 months or less, which is not long enough to provide seasonal protection, according to Dr. Wilson. Epithelial cells line the surfaces of the body – in this case, the nose and throat linings – and are constantly being sloughed off and replaced by new cells. The new cells do not contain the antibody-producing genes. The next step is to explore strategies for extending the life of the therapy. Dr. Wilson hopes that further research may lead to a stopgap defense against a fast-moving pandemic, which could help control the outbreak until traditional vaccines can be developed and administered.

Initial results of this ongoing research were reported in Ref. [33], and described in University Press releases [34]. The ongoing research is supported in part by ReGenX, the Public Health Agency of Canada, Canadian Institutes of Health Research, and the National Institutes of Health. If successful, this approach could lead to the development of a "spray on" gene therapy to treat flu pandemics.

10.7 DRACO: Designing a "Kill Switch" for Viral Diseases

Being able to use genetic engineering to prevent 15 types of influenza viruses would be remarkable, but imagine a medical treatment that could prevent *15 or more entirely different diseases* caused by viruses? That's what Dr. Todd Rider has been working on. Dr. Rider is Senior Staff Scientist at the Chemical, Biological and Nanoscale Technologies Group at MIT's Lincoln Laboratory.

In July 2011, after more than a decade of research, Dr. Rider reported the development of a broad-spectrum antiviral treatment called DRACO. This therapeutic works by identifying a type of RNA that occurs only in cells infected by viruses and activates a natural "suicide switch" inside those cells, causing the infected cells to self-destruct without harming uninfected cells.

DRACO has been hailed as the world's first broad-spectrum antiviral drug – potentially capable of treating or preventing almost all disease viruses. Animal studies in mice have shown that DRACO destroys cells that have been infected by a virus, effectively stopping the replication of the virus. In experimental tests on human and animal cells cultured in the laboratory, DRACO was effective against 15 different viruses. Mice infected with the H1N1 influenza virus were treated with DRACO and completely cured of the infection [35]. More recently, DRACO has also been effective against three hemorrhagic fever viruses in preliminary mouse trials. As intended, DRACO has been shown to be nontoxic to healthy, uninfected cells and nontoxic in mice.

Based on these early indications, Dr. Rider believes that DRACO has the potential to treat nearly any type of virus. While the treatment is far from commercialization – perhaps a decade away – this innovation has created a new strategy for treating viral diseases, and opened the door to an entirely new class of antiviral drugs.

When Todd Rider submitted his findings for publication, the first scientific journals rejected his article. They didn't seem to realize the magnitude of the discovery. This was extremely frustrating to Dr. Rider and his research team, which included Scott Wick, Christina Zook, Tara Boettcher, Jennifer Pancoast, and Benjamin Zusman. The researchers knew they had developed a major new strategy for fighting viral diseases, although they had to fight to get recognition by the scientific community and to report their findings.

In July 2011, their research was reported in the journal *PLoS One* [36]. The article created an immediate sensation. DRACO has since been featured in such publications as *Business Week, Nature Biotechnology, Science Illustrated, New Scientist, Popular Science*, the *Wall Street Journal, Wired*, and dozens of other publications. The discovery has been featured on television (BBC Horizon and Discovery Channel) and radio (BBC and NPR). DRACO was named one of the top inventions of the year by *TIME Magazine* and cited by the White House in the National Bioeconomy Blueprint (April 2012) as an example of work that should be fully supported. Today, DRACO is listed in Wikipedia and a Google search for "DRACO antiviral" yields more than 25 000 results.

How a Rocket Scientist Became a Nanomedicine Pioneer

Like many nanoinnovators, Todd Rider did not start his career in nanotechnology. His early interest focused on other areas such as rockets and nuclear science. In elementary school, he was inspired by the Tom Swift Jr. novels, in which a young scientist masters many different fields of science and creates inventions in all of them. In middle school, Todd designed and built radios and a robot dog from scratch. In high school, he built, test-fired, and patented a novel rocket staging system that could enable launch vehicles to carry up to 50% more payload into orbit with no extra fuel.

With a penchant for acronyms, he named his rocket staging system CIGAR (continual improvement for greater advantage in rocketry) "because it gets shorter the longer it burns," he explains. That project won him the grand prize at the International Science and Engineering Fair, a free trip to Stockholm, Sweden where he met that year's Nobel Prize winners.

Todd Rider started college at the age of 17 at MIT where he earned four degrees, including a Ph.D., in electrical engineering and computer science (Figure 10.16). In his Ph.D. thesis, he proved that any working hot fusion reactor would either consume more energy than it generates, or else produce almost as much radiation as existing fission power plants. As part of his Ph.D. program, he minored in biomedicine (including coursework at Harvard Medical School), relativistic quantum field theory, and solid-state and optical physics. He also studied nuclear, mechanical, and aerospace engineering, chemistry, advanced math, and several years of languages, including French, German, Chinese, and Japanese.

Although he had studied numerous fields, Dr. Rider was most interested in applying an engineering approach to biology. "In the early 1990s when I was studying both fields, everyone in each field at M.I.T. and Harvard thought I was crazy," he recalls with a smile, "but now they have entire departments dedicated

Figure 10.16 Dr. Todd Rider, shown here at MIT's Lincoln Laboratory, is surrounded by an array of sophisticated technologies used to develop the DRACO broad-spectrum antiviral system (photo reprinted with permission of the MIT Lincoln Laboratory, Lexington, Massachusetts).

to combining biology and engineering." He adds, "The biologists thought I was a physicist, the physicists thought I was an electrical engineer, the electrical engineers thought I was a nuclear engineer, and the nuclear engineers thought I was the devil incarnate."

After graduating in 1995, Dr. Rider worked for a short-lived biotech start-up, then in 1997 he joined MIT's Lincoln Laboratory, which is dedicated to using advanced technology for national security. In order to combine his interest in bioengineering with Lincoln Lab's general mission, he started thinking about new ways to detect viruses and other pathogens. Less than 3 months after joining Lincoln Labs, he invented a biological sensor that glows when pathogens are present, providing rapid detection and identification of harmful pathogens at a time when no rapid sensors were available.

He named his biosensor technology CANARY after the caged canaries that once warned people of environmental dangers. Rather than trying to invent a pathogen assay from scratch as others had tried, Dr. Rider borrowed from Nature. "From immunology in grad school, I knew that B lymphocytes or white blood cells were very fast, very sensitive, and very specific at detecting pathogens," he says. He found a way to genetically engineer the cells to make a jellyfish protein called aequorin. He explains: "When antibodies on the surface of the B cells detect the specific pathogen they have been designed to recognize, they send a signal inside the cells, which activates the aequorin and produces light, all within seconds." Dr. Rider engineered the first CANARY cells and demonstrated that they worked as well as he had predicted. The results, published in *Science* in 2003, showed that CANARY cells could recognize bacteria and viruses in less than 1 min, could detect as little as 50 particles of a pathogen, could distinguish among different pathogens or even different strains of a pathogen, and could be used with real-world samples such as air samples or nasal swabs.

The latest version of CANARY technology is capable of detecting more than 24 airborne pathogens such as anthrax and smallpox in less than 3 min even at dilute concentrations, with much higher sensitivity than competing technologies that are based on polymerase chain reaction (PCR) or immunoassays. CANARY was licensed to Innovative Biosensors, Inc., in Rockville, Maryland to commercialize the technology. Today, the government is using this system to sample the air for the presence of biological threats. CANARY assays using nasal swabs, blood samples, water samples, and agricultural specimens have also been developed that could quickly detect and control outbreaks of infectious diseases.

The Need for an Antivirus Solution

The need for a simple diagnostic test to identify a pandemic form of influenza is illustrated by a classical – and tragic – example. In 2002, the deadly and highly infectious SARS virus originated in Foshan City, located in the Guangdong Province of China, and spread to 37 countries in just a few weeks. The virus gained worldwide attention after a Chinese-American businessman contracted the disease after staying at a hotel in Hong Kong, and then traveled to Vietnam where he died of the disease. The SARS virus was recognized and reported as an "atypical pneumonia" by Dr. Carlo Urbani, a brave and perceptive WHO physician based in Vietnam, who subsequently died of the virus.

The SARS virus killed more than 50% of people over the age of 65 who contracted the virus. This compares with a fatality rate of 0.03% for a typical influenza virus in this age group. The SARS virus was especially dangerous because the symptoms were difficult to distinguish from a common flu virus or pneumonia and often the diagnosis came too late to save at-risk victims. As soon as this deadly viral strain appeared, the world medical community rushed into action. Researchers in Hong Kong, Singapore, China, Germany, France, the Netherlands, Canada, and the United States examining the virus with atomic force microscopes compared the viral genetic material with genetic libraries, which helped them determine that this was a type of coronavirus. A combination of quarantines, air travel checks, chest X-rays, temperature monitoring, and several diagnostic tests helped stop the outbreak.

The availability of a nasal swab of the type being developed by Dr. Rider and his colleagues potentially could speed detection of victims and carriers of pandemic diseases to provide faster treatment and quarantine of patients. The world medical community is currently working to develop better diagnostics and vaccines for SARS as well as other types of coronaviruses. In 2013, the effort to develop a robust SARS vaccine continues.

While working on his CANARY system to detect pathogens, Dr. Rider began thinking about how to treat viral infections (Figure 10.17). He noticed that there are broad-spectrum antibiotics such as penicillin that can be used to treat any

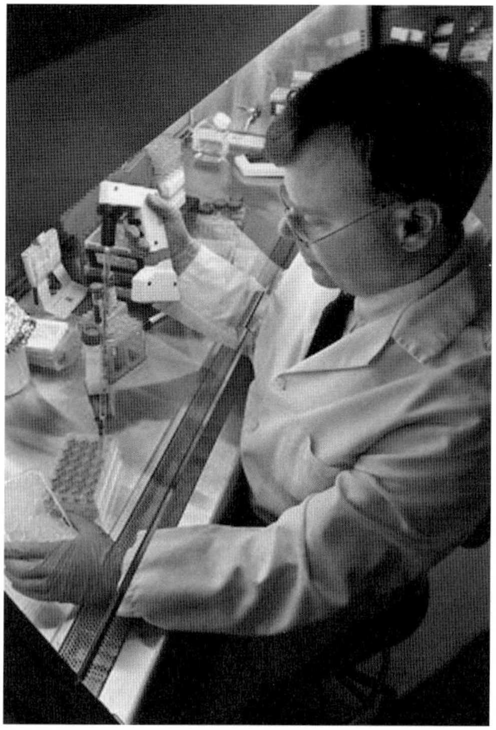

Figure 10.17 Todd Rider wondered why there was no broad-spectrum drug to treat many different viruses, the way that antibiotics can treat many different bacteria. He identified a type of RNA produced only in cells that have been infected by viruses and used this to develop drug that could selectively target infected cells and induce them to self-destruct (photo reprinted with permission of the MIT Lincoln Laboratory, Lexington, Massachusetts).

bacterial infections that CANARY might detect, but there weren't any broad-spectrum treatments for viruses. Existing vaccines are specific for a particular virus or strain, are usually effective only if given before infection, and thus far have proven difficult to develop for major viruses such as HIV and dengue. There was no broad-spectrum solution that could be used to directly attack all viruses.

Dr. Rider dubbed his approach to broad-spectrum therapeutics the PANACEA project. He reasoned that if a mechanism common to all viral infections could be identified and targeted, it might be possible to send a signal to infected cells that would cause the cells to self-destruct. Destroying infected cells would stop the replication of viruses and halt the progression of a viral disease. In theory, this approach could be used to treat virtually all viruses. As he was contemplating the mechanisms that viruses use to replicate themselves, he realized that viruses use many different strategies. What would be common to all viruses? What changes would signal that a cell is being infected? How could those changes be detected? And what strategy could stop the virus from using a cell to make more viruses?

DRACO: The First Broad-Spectrum Antiviral

Viruses cannot reproduce on their own. They need to infect a cell where they hijack the cell's ability to reproduce structures – turning the cell into a virus factory. As part of this process, a double-stranded ribonucleic acid molecule (dsRNA) is created. This double-stranded RNA is produced by virtually all known viruses.

Dr. Rider reasoned that if a method could be found to identify and hone in on this feature, it would be possible to identify cells that were infected with a virus and produce new viruses – which information could then be used to target and destroy those cells, preventing further infection.

To borrow from and reengineer *Nature* as he had previously done with CANARY, Dr. Rider designed a biological therapeutic that combines two natural proteins – one that would bind to the viral RNA and another that would cause a cell to "commit suicide." The first protein detects that the virus has entered the cell and the second protein acts as a "kill switch" that causes the cell to self-destruct. When a cell self-destructs, it is called *apoptosis*. Destroying cells as they are infected prevents the continued reproduction and spread of the virus. This original approach delivered a one–two knockout punch to virally infected cells. This particular PANACEA therapeutic is dubbed DRACO. The DRACO molecule is also designed with a built-in "delivery tag" derived from naturally occurring proteins, which allows it to cross cell membranes and enter any human or animal cell; however, DRACO does not harm a cell if no viral dsRNA is present [37].

Dr. Rider and his team tested the treatment in animal and human cells in the laboratory, and then tested it in mice. In their initial laboratory tests, DRACO penetrated the cells within minutes. DRACO was also shown to be effective when administered at least 11 days before the virus was introduced. Cell populations treated with DRACO were also shown to remain "cured" when DRACO was withdrawn. The tests demonstrated that DRACO is nontoxic in 11 different types of cells.

Most important – in the first round of laboratory tests, DRACO was shown to significantly reduce or destroy 15 viral strains from 7 virus families. These viruses cause a wide range of debilitating or fatal diseases, including dengue hemorrhagic fever, polio, flu, stomach problems, and the common cold. The DRACO research team hopes to work their way through other viruses and virus families to test the efficacy against HIV/AIDS, Ebola, and other disease viruses.

There was no toxicity or interference with efficacy by immune responses in the animal laboratory tests, although this does not guarantee that there won't be an immune response or complications in humans. Dr. Rider notes that considerable research will be needed to determine safety and efficacy, dosage, treatment duration, and techniques for optimizing and monitoring the treatment.

As Dr. Rider has noted, "M.I.T. is not a pharmaceutical company," so he's hoping to collaborate with pharmaceutical companies to fund and conduct large-scale animal trials that typically involve testing in several different species from mice to monkeys, before human trials can begin. That process could take about a decade, Dr. Rider estimates.

If DRACO works as well in animal preclinical trials and in humans as it worked in mice in the laboratory, DRACO could provide the first broad-spectrum antiviral treatment for a host of deadly diseases. Even if the treatment works for only one disease, it would be significant.

Dr. Rider is currently working on other PANACEA broad-spectrum antivirals, although DRACO is at the most advanced stage of testing. He is also developing PANACEA treatments to treat bacteria that are resistant to existing antibiotics.

"We could do much more with additional funding," he says, noting that the economic climate and reduced funding for scientific research require him to do a lot of grant writing to support his team's research – a constant struggle for research scientists.

More Information about Viruses and Vaccines

Most people don't understand how a virus works, so here is a quick overview that you may find enlightening. Virus means "poison" in Latin – and for good reason. Diseases caused by viruses include seasonal maladies such as the common cold and influenza; pandemic strains of influenza that have killed hundreds of thousands of people; childhood ailments such as chicken pox, measles and mumps; mononucleosis; rabies; chronic conditions such as hepatitis and herpes; deadly epidemics such as HIV/AIDS and exotic hemorrhagic diseases such as Ebola and Marburg. Viral diseases transmitted by insects include West Nile Fever and Yellow Fever. Viruses can also cause certain kinds of cancer such as leukemia. The human papillomavirus can cause cervical cancer. Many viruses such as the rhinovirus (which causes the common cold) and the human papillomavirus have 100 or more different variations or *serotypes*.

The human and economic cost of these diseases is enormous, including the cost of preventive vaccines, medical care, hospitalization, sick days taken by workers, and fatalities.

Most viruses spread from person to person, but they can also be transmitted through contaminated blood, by sexual activity, in saliva, through contaminated food or water, and by mosquitoes and other insects. All viruses turn healthy host cells into "virus factories" that cause them to manufacture new viruses. A virus cannot reproduce outside a cell, so it has to seek out and attach to the cell – to "infect" the cell – and hijack the cell's reproductive machinery. They can instruct the cell to make a few viruses at a time, or to make thousands of new viruses that literally burst open the cell to release a swarm of viruses. The fastest viruses can infect a cell, force the cell to make new viruses, and burst the cell to release the new viruses in as little as 20 min.

A typical virus consists of nucleic acids encapsulated in a protective protein shell called a *capsid* that has a physical or chemical structure on the surface that allows it to dock with a matching receptor on the surface of a human cell. Viruses develop "key and lock" surface structures that allow them to attach to and enter target cells. They can lie dormant until they encounter a host cell (a dormant virus is called a *virion*). Viruses can use single or double strands of

RNA or DNA to transmit the instructions that cause the cell to produce more viruses. Some viruses penetrate the nucleus and others exist in the cytoplasm outside the nucleus.

Viruses are insidious pathogens. They come in many different shapes and sizes. Most viruses are between 20 and 300 nm. Some filament-shaped forms such as the Ebola and Marburg virus are less than 100 nm across, but can be more than 1000 nm long.

The best way to understand the complex nature of viruses is to look at the four different viruses shown in the accompanying images (Figure 10.18a–d). Viruses can be shaped like spheres, rods, bullets, bricks, bottles, coils, spindles, helixes, filaments, polyhedrons, and teardrops. Some viruses such as the T4 bacteriophage (Figure 10.18d) that infects bacteria – shown here in a nanoscale reconstruction – have shapes that can only be described as “alien.”

Viruses use many different strategies to reproduce, and have the ability to mutate to change their strategies, which is how viruses develop resistance to vaccines. Some viruses such as the “swine flu” and “avian flu” originate in animals and can be transmitted to humans who come into contact with infected animals such as poultry or livestock. A critical concern in the world health community is how to prevent viruses that jump from animals to humans from mutating into a form that can be transmitted from humans to humans, since humans haven’t developed a resistance to many of these strains.

The complex strategies used by viruses to infect cells require equally complex strategies to defeat them. For example, to defeat a viral infection we can (i) stimulate the body’s immune system to create antibodies that can *identify and destroy* (*or resist*) a specific type of virus, (ii) create a particle that harmlessly binds to a receptor on the cell that *blocks the virus* from anchoring to receptors, (iii) create a viral analog (an empty or “fake” virus) with a receptor that *tricks the virus* into binding to the nonfunctional structure, or (4) *interfere with* one or more steps in the disease cycle of the virus.

Today, the process of diagnosing new viruses and developing vaccines is a relatively fast and efficient process. For example, in April 2013, the Centers for Disease Control and Prevention in Atlanta, Georgia, received their first samples of a new bird flu virus from China called H7N9. H7 viruses normally do not infect humans and those cases reported occurred among people who had direct contact with infected birds. Approximately one-quarter of the first 130 human cases resulted in death. Fortunately, the strain was not a full-fledged human virus that can be transmitted from humans to humans. The CDC used the samples from China to prepare seed strains of the virus, and distributed them to vaccine manufacturers so that they could develop experimental flu vaccines. In addition, preventive measures were taken at the source by Chinese officials who destroyed more than 90 000 chickens in one province to keep the virus from spreading [38].

Science is still unlocking the secrets that make viruses such an effective and nefarious disease agent. Given the propensity of viruses to keep mutating, the fight against these nanosized pathogens will continue indefinitely.

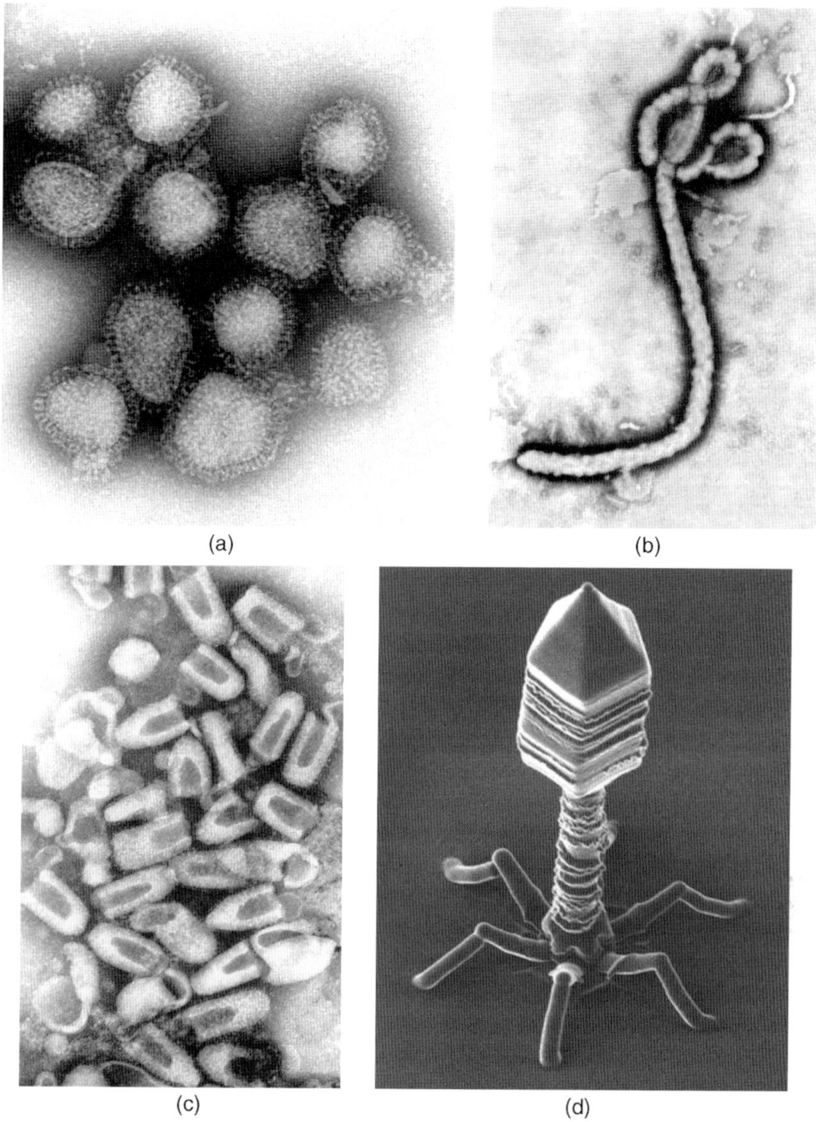

Figure 10.18 Viruses come in many different shapes and sizes. (a) Influenza A virus (which caused the 1968 Hong Kong flu outbreak). (b) Ebola Virus (filament shaped). (c) Rabies virus (bullet shaped). (d) Nanofabrication of a T4 bacteriophage (a virus that infects bacteria) (Figures 10.18a–c by Frederick A. Murphy, Department of Pathology, University of Texas Medical Branch; Figure 10.18d by Dr. Shinji Matsui and Dr. Reo Kometani, Laboratory of Advanced Science and Technology for Industry, University of Hyogo).

10.8
Nanoinnovation in the Decade of Diagnostics

In 2005, I wrote in a research report that we are entering a "Decade of Diagnostics" – a golden decade when our ability to diagnose any disease is being realized [39]. According to the National Institutes of Health and other government sources, biomedical research has revealed the molecular basis of more than 4000 individual diseases. However, only about 250 of these diseases have available treatments. It is a sad irony that we can diagnose virtually any disease, but still can't cure many of them.

During the past decade, 17–34 new molecular entities became available each year to treat disease. Given the amount of research that goes into finding these solutions, this is a modest result. For every 10 000 new compounds discovered, only one becomes a new drug – about 95% of drugs fail in clinical trials. More than 80% fail in phase 2. Even promising drugs that work in the laboratory and in animal tests fail after expensive human trials are begun.

While we can identify the existence of virtually all diseases, science is still struggling to understand the molecular basis for disease as well as aging, obesity, addiction, and other conditions. Better nanodiagnostics are needed to identify diseases that have many variations such as the papillomavirus that has more than 100 different variants, of which about half a dozen are most prevalent and likely to cause cervical cancer. More than 65% of people infected with HIV/AIDS live in sub-Sahara Africa, where expensive and sophisticated antiviral cocktails are not available.

Every year, medical researchers fight to identify new forms of viruses, so flu vaccines can be developed. For two decades, HIV/AIDS researchers have been developing an ever-changing mix of "cocktails" designed to suppress the effects of this pandemic disease, which now infects more than 35 million people worldwide.

Making Gene Tests Affordable and Available

Gene testing is critical not only to genetic diseases and gene therapy but is also becoming a critical component of cancer and other common diseases. This makes genetic screening increasingly important. Unfortunately, the quality of genetic tests by test providers remains highly variable and inconsistent. Some tests only check for a few conditions. Most are prohibitively expensive, although the cost has been falling to the $1000 price point. The $1000 level is important because this is the cost of an MRI test, which most insurance companies currently reimburse for, so presumably, genetic tests in that price range will also be covered.

More than 1000 genetic tests are currently available. The number of reports in one test has grown from about a dozen in the mid-2000s to anywhere from 100 to 250 from some testing services. The cost of individual genetic tests has dropped steadily, from $100 000 to $5000 to $1000 and now some specific tests are available for about $100. Most tests accept blood or saliva samples for evaluation. According to a March 2012 report by United Health Group, the market for genetic tests is expected to increase from $5 billion in 2010 to $25 billion in the next decade.

One of the best known gene testing services is 23andMe, a California-based company that markets genetic tests over the Web. Customers send saliva samples to be evaluated. When 23andMe started offering its test in 2007, they provided 14 reports for about $1000. In 2013, the company was offering more than 240 reports for $99. The company's personal genome test kit was named "invention of the year" by *Time Magazine* in 2008. Company investors include Google (Anne Wojcicki, cofounder of 23andMe, is the wife of Google cofounder Sergey Brin). Products include the 23andMe Saliva Collection Kit and Personal Genome Service (PGS). The company has provided DNA tests to more than half a million customers. In December 2013, the FDA banned 23andMe from selling its personal genetic test kits due to concerns about "inaccurate results." The ban generated extensive media coverage and sparked a grassroots petition from consumers. The company indicated that it is continuing to provide genome tests with "ancestry information" and raw genetic data, but without interpretation or analysis of health implications. Customers have been working around the ban by analyzing the raw genomic data using software provided by other companies. 23andMe has indicated that they intend to work with the FDA to meet regulatory requirements and validate their tests.

Another gene testing pioneer based in California is Navigenics, which developed the Personal Genome Machine and Ion Proton DNA sequencer. Navigenics was acquired in 2012 by Life Technologies, which in turn was acquired by Thermo Fisher Scientific in February 2014. These acquisitions reflect a consolidation and restructuring that has been occurring in the biosciences industry as industry leaders explore new frameworks to streamline medical research, incorporate genetics into medical diagnostics, and develop novel healthcare solutions.

In January 2014, Illumina, a biotech company based in San Diego, California, announced a new sequencing machine called the HiSeq X, which can process multiple genomes simultaneously and can deliver five human genomes per day (up to 20 000 genomes a year) for a cost of $1000 each. The first models of the new supercomputer cost $1 million each. Illumina also produces smaller desktop machines.

The availability of affordable genome processing brings us closer to a day when all patients can have their genomes sequenced to screen and diagnose diseases, and to pinpoint genetic causes of cancer and other diseases. In the United Kingdom, anyone with cancer will eventually have access to genetic testing, under the Mainstreaming Cancer Genetics Program. The program is supported with £2.7 million ($4 million) in funding provided by the Wellcome Trust. This pioneering initiative is being developed by a consortium that includes the Institute of Cancer Research, the Institute of Cancer Research and Cancer Genetics Clinical Unit at the Royal Marsden, and the Wellcome Trust Centre for Human Genetics and Illumina, Inc. The initiative is anchored by a new genetic test called the TruSight Cancer panel, created in collaboration with Illumina. The new test claims to be able to evaluate 97% of known genes that predispose people to develop cancer. First availability in clinics was planned for 2014 [40].

It is likely that genetic testing will eventually become universal and routine (for consenting patients). The advent of universal gene profiling poses a major challenge to the computer industry. Storing the genome of everyone in the world, or even everyone in a large country, will strain the capacity of even the largest supercomputers. One solution is to store only those portions of the genome that are directly associated with disease, instead of storing the entire genome. Another approach is to analyze the genome to detect anomalies, but not store the entire genome.

A Landmark Court Decision May Lower the Cost of Genetic Tests

When companies started patenting genes, they discovered, for therapeutic or research applications, there was a fear that companies were going to patent and theoretically "own" genes that are naturally occurring biological components. In June 2013, the US Supreme Court unanimously ruled that naturally occurring human genes cannot be patented. The ruling impacted patents held by Myriad Genetics that had enabled them to be a sole source for the BRCA gene evaluation used to determine susceptibility to breast and ovarian cancer. This test was used to determine that Angelina Jolie – the popular actress/humanitarian and UN special envoy/goodwill ambassador – carried the hereditary BRCA1 gene that made her susceptible to breast and ovarian cancer. Ms. Jolie used this information to make her courageous decision to have a mastectomy and breast reconstruction, which helped publicize the cause of breast cancer, the role of genes as a causal agent, and the importance of genetic testing.

Using Nano-Biochips to Diagnose Disease

Increasingly, medical diagnosis is being automated by nano-biochips that analyze biological (or biologically active) structures that are typically 100 nm or less, such as DNA molecules or viruses. These chips can be programmed to react to various stimuli and may involve the use of nanochannels, nanopores, nanowire arrays, and other nanostructures [41]. Dr. John McDevitt at Rice University is a leader in biochip research. In 2010, Dr. McDevitt's team developed a portable nano-biochip that tests saliva to detect a heart attack. The chip can be reprogrammed to test for different types of biomarkers to diagnose nearly 40 different diseases. Dr. McDevitt has also developed a biochip that tests oral lesions for early-stage cancer [42]. Oral cancer afflicts more than 300 000 people each year, globally. The 5 year survival rate is only 60%, but if detected early the survival rate can be increased to 90%. The problem is that 95% of oral lesions do not become cancerous, but doing a biopsy and waiting for a laboratory analysis can take several days. A biochip can perform the same function in 15 min. Nano-biochips can also be used to monitor patients with a high probability of contracting oral cancer, and multiple tests can be conducted over time to determine if and when a cancer develops.

Using a Nanofluidic Biochip to Analyze DNA

Nanofluidic chips are also being used to expedite and lower the cost of DNA analysis. One of the most interesting stories about nano-biochips involves Dr. Han Cao who developed and patented a method for coaxing an array of individual DNA molecules to unwind, so that they can be moved through channels in a unique nanofluidic chip that analyzes the length and structure of the molecules. The nanofluidic chip was developed by Dr. Cao when he was a postdoc at Princeton University in the early 2000s, while studying ways to apply silicon chip fabrication technologies to biological research. Dr. Cao believed that it was possible to stream thousands of DNA molecules through a nanochannel array in perfectly straight lines, like a Eurostar train in the Channel Tunnel (the "Chunnel") that connects England and France [43].

"This approach allowed us to distinguish densely packed biomarkers from each other along the DNA molecules. This was difficult to do at the time because DNA molecules in solution tend to coil randomly," he recalls. "The key was designing massively parallel 'chunnels' where many DNA strands could be analyzed using high throughput analysis. Furthermore, if we could place a detector at a point along the nanochannel array, we could isolate and examine specific defects which can be identified by fluorescent dyes attached to the genes."

Dr. Cao's patented innovation included a method for coaxing the DNA molecules to unwind and find a channel on the biochip. This was the first time that very long strands of individual DNA molecules were isolated and imaged rapidly on a nanoscale fluidic biochip, and the first time this was done using massively parallel nanochannel arrays. Pressure and/or electrostatic forces are applied to move the molecule through the chip. This is needed because nanoscale fluids and molecules don't flow like most macroscale liquids. Although you can't normally see DNA molecules moving, the molecules in Dr. Cao's chip are actually visible as they move through the channels because they are tagged with a fluorescent marker that causes them to glow.

Jeffrey Rosedale, who was Dr. Cao's patent attorney at the time, recalls that they almost missed the patent filing deadline. When the young Ph.D. came to his office half a decade ago, the patent filing deadline was only a few days away. Normally this would have been too close to the filing date to complete all the paperwork and the patent would have been forfeited. However, the patent lawyer spent 4 days with the inventor, working 15 h a day to rush the draft of the patent application for Princeton University. This allowed them to meet the deadline and secure the patent, which was critical to licensing and funding the venture.

Dr. Cao notes that the innovation involved a multidisciplinary collaboration involving Princeton faculty from the School of Engineering and Applied Science, the Department of Physics, and the Department of Molecular Biology. The research was supported by a grant from the US Defense Advanced Research Projects Agency (DARPA). Dr. Cao's fellow inventors included Steven Chou and Joseph C. Elgin, Professor of Engineering; Robert Austin, Professor of Physics; and Jonas Tegenfeldt, who is currently an associate professor at Lund University (Figure 10.19).

Figure 10.19 Dr. Han Cao, Founder and Chief Scientific Officer, and Erik Holmlin, President and CEO at BioNanoGenomics, with the single-molecule imaging system (photo courtesy of BioNanoGenomics).

To commercialize the technology, Dr. Cao founded BioNanomatrix in Philadelphia in 2003, which licensed the technology from Princeton. The company was renamed BioNano Genomics and subsequently moved to San Diego, California. Dr. Cao serves as Chief Scientific Officer. The company's first product was an instrument called the nano*Analyzer*®.

In 2008, Dr. Cao received the Ben Franklin Technology Partners Best Management Team Award. Dr. Cao's innovation was featured in Ref. [44]. In September 2010, he was awarded the Chinese Rising Star Entrepreneur Award at the China International Economic and Technical Cooperation Forum in Beijing.

In November 2012, BioNano Genomics announced the commercial rollout of a novel genome mapping system called Irys™ that uses a nanofluidic chip called the IrysChip to capture, image, and visualize the entire genome contained in DNA molecules. The system can analyze several tens of gigabases per hour, which can speed up the genome mapping process. The technique identifies structural variations that underlie phenotype variations and provides high-throughput scanning analysis of DNA for genome mapping.

10.9 In Search of the Star Trek Tricorder

Without question, the most fascinating medical device in TV and film is the medical tricorder used on Star Trek. With a wave of the device, Dr. Spock could determine a person's medical condition, the cause, and whether the person was expected to survive or not.

If the Star Trek tricorder existed today in real life, it would streamline the practice of medicine throughout the world. This device would be as easy to operate as a smartphone. Almost anyone could administer the tests. The data could be transmitted wirelessly to a central processing hub for analysis. People in industrialized countries could buy this device and get a quick self-diagnosis. People in rural villages would be diagnosed and treated sooner. Instead of doctors traveling to remote villages, devices would do the traveling. Entire villages could be tested for diseases using mobile tricorders. Patients would know they're sick before they experience any symptoms.

All we need now is for someone to design a Star Trek-style tricorder.

Actually, this is already happening. In fact, we can reliably expect that by mid-2015, the first versions of a medical tricorder will be revealed – thanks to the XPRIZE Foundation.

In January 2012, the XPRIZE Foundation announced a competition called the Qualcomm Tricorder XPRIZE that will reward $10 million for the best medical diagnostic solution(s) providing many of the functions of the Star Trek tricorder.[3] Up to three winners will share a cash prize of $10 million, thanks to a generous grant from Qualcomm. The winner(s) will be determined based on some very specific performance requirements. The winning solution has to be able to diagnose medical conditions in a group of 15–30 people during a 3 day period. According to the competition guidelines, the tests must be made by a "consumer" independent of a healthcare worker or facility. The winning entries will receive $7 million for first place, $2 million for second place, and $1 million for third place.

According to the guidelines, "the winner will be the team that most accurately diagnoses a set of diseases independent of a healthcare professional or facility and that provides the best consumer user experience." The winner will be the best performing solution able to assess as set of 16 conditions and 5 vital signs, in a pool of people, in 3 days. The entire solution can weigh no more than 5 pounds. The teams who are competing for the prize also have the option to collaborate, reorganize, and/or share technical resources to create solutions with the highest chance to win. More than 250 teams registered in the first half of 2013.

The solution must be able to assess 13 conditions from a core set of conditions; 3 elective conditions; and must be able to monitor 5 vital signs – for a total of 21 conditions and parameters. The 13 core conditions include anemia, lower urinary tract infection, type 2 diabetes, atrial fibrillation, stroke, obstructive sleep apnea, tuberculosis, chronic obstructive pulmonary disease (COPD), pneumonia, ear infection, leukocytosis, hepatitis A, and "absence of conditions." Three conditions can be chosen from an elective list that ranges from whooping cough and mononucleosis to skin cancer (melanoma), and HIV and may also include candidates proposed by the competitors. The five vital signs that need to be monitored include blood pressure, heart rate/variability, body temperature, respiratory rate, and oxygen saturation.

This is a daunting list and it is questionable whether the competitors will be able to diagnose all of these conditions. The goal is to demonstrate that it is feasible to design a multifunction portable diagnostic tool that is usable by

3) TRICORDER is a trademark of CBS Studios, Inc.

average consumers. The Qualcomm Tricorder will not only be a diagnostic tool but will also be a *self*-diagnostic tool.

Among the competitors are universities, companies, and small groups of entrepreneurs. San Diego State University has organized a team of more than 60 people to work on the project. The San Diego team is led by Lambert Ninteman, a 37-year-old graduate student (and self-proclaimed Star Trek fan). The team's approach is to modify an Android phone with attachments. While many of the competitors will be adapting smartphones with sensors to perform the required tests, different form factors are also possible. For example, the device could take the form of a glove – or the solution could combine a glove with a modified smartphone.

One of the prime competitors is Scanadu, a company that already makes portable diagnostic devices that perform some of the functions on the XPRIZE list. Scanadu has announced that it is building a suite of products designed to work with smartphones, reflected in their company slogan: "Send your smartphone to med school." To fund their development of a commercial device they call the Scanadu Scout, Scanadu used a crowd-funding site called Indiegogo. By August 2013 they had raised $1.6 million from 8500 funders. Indiegogo contributors were also invited to specify the medical measurements that were most important to them. Scanadu was listed as one of the crowd-funding site's "most funded" projects. To inspire their team, the group bought some original Star Trek props.

The founder and CEO of the XPRIZE Foundation and guiding spirit behind this and many other innovation initiatives is Peter Diamandis. Peter is an engineer, technology pioneer, author, and philanthropist and one of the world's premier innovation champions. He is also cofounder/chairman of Singularity University, Space Adventures Ltd., Rocket Racing League, International Space University, Planetary Resources, Students for the Exploration and Development of Space, and Zero-Gravity Corporation. Peter was born in the Bronx, New York. His father, a Greek immigrant, was a physician. At the age of 8, Peter had already developed an interest in space exploration. When he was 12, he won first place in a rocket design competition. As a sophomore at MIT, he cofounded Students for the Exploration and Development of Space. He holds a master's degree in aeronautics and astronautics from MIT and medical degree from Harvard. He is the coauthor of the 2012 best-selling book *Abundance: The Future Is Better Than You Think.*

The tricorder competition has been as fascinating to watch as any Star Trek episode or movie, and the strategy of using a cash incentive provided by sponsors to create the future is an innovative strategy in its own right.

10.10
Nanobacteria: The Smallest Life-Form?

Nanotechnology is revealing causes and sources of medical problems that were previously unknown, and some of these discoveries are somewhat controversial. Some sound like they come from a science fiction movie.

For example, *nanobacteria* have been called the smallest forms of life. Nanobacteria (*Nanobacterium sanguineum*) are the smallest self-replicating structures, measuring 20–200 nm in diameter. These structures are only one-tenth the size of the smallest known microbes. Researchers have associated these structures with a wide range of diseases that are caused or complicated by calcification such as kidney stones, gallstones, and atherosclerosis. However, despite their name, they may not be true bacteria.

There is a research community in the field of Astrobiology that is evaluating the possibility that nanobacteria exist on Earth and in space. Structures that resemble "fossil bacteria" have been found on meteorites that are thought to have come from Mars. If nanobacteria do exist, they could pose a threat to space travelers, including the first human colonists on Mars. So-called nanobacteria have also been associated with several human diseases (heart disease, Alzheimer's) where calcium buildup is a problem.

Nanobacteria have a fascinating history. In 1981, Francisco Torrella and Richard Morita from the Department of Microbiology and Oceanography at Oregon State University, documented marine bacteria smaller than 300 nm that they called *ultramicrobacteria*. In 1989, geologist Robert Folke was studying hot springs at Viterbo, Italy where he discovered nanostructures he called *nannobacteria* (spelled with two n's).

In 1996, a team led by NASA scientist David McKay analyzed a Martian meteorite discovered in Antarctica and found rod-shaped structures that appeared to be trace fossils of nanobacteria (spelled with one n). His paper in the August 1996 issue of *Science* became one of the most cited papers in planetary science and helped promote the creation of the NASA Astrobiology Institute, which has been studying nanosized microbes and microbe-like structures that may exist in space and on Earth.

In the early 2000s, Neva Ciftcioglu, a Turkish-born postdoc, was working with Dr. E. Olavi Kajander on a project that involved a study of mammalian cells. During their research, they detected particles that they first thought were "contaminants" killing the cells in their cultures. Further study revealed these to be nanoparticles that behaved very much like bacteria.

Until then, the smallest life-form was thought to be 200 nm since smaller structures are not thought to be able to harbor the essential components of life. These "nanobacteria" were even smaller, which challenged the current thinking that defined the "scale" of life-forms. Ciftcioglu and Kajander had in fact discovered *the smallest culturable autonomously replicating agent on Earth*. These particles were called "calcifying nanoparticles." Some people called them "nanobacteria."

In 2006, Dr. Ciftcioglu, now Docent/Professor of Biotechnical Microbiology at the University of Kuopio in Finland, was lead author of a seminal article in *Journal of Investigative Medicine* entitled "Nanobacteria: fact of fiction? Characteristics, detection, and medical importance of novel self-replicating, calcifying nanoparticles" [45]. Her coauthors were David McKay, Olavi Kajander, and Grace Mathew. In their article, the researchers suggested that nanobacteria may be associated with health problems that are caused or complicated by biomineralization/calcification. Associated diseases could include arterial heart disease, Alzheimer's disease, kidney stone formation, gallstones, malignant tumors, and others. Since the publication of their work, which was also featured by NASA on their Web Site, the particles once thought to be nanobacteria have come to be more commonly known as "calcifying nanoparticles." Many scientists hold that they are not true nanobacteria, but rather self-replicating structures similar to crystals, since nanobacteria have not been proven to contain DNA. Whether they are true bacteria or something else, these particles have been associated with a wide variety of diseases and may pose health risks that are more serious than bacteria, given that so little is known about these structures except that they can cause a buildup in calcium deposits that can lead to serious diseases.

Regardless of what we call them, we need to pay attention to unseen, undiscovered entities that have the ability to cause or complicate disease. A particle that is nanoscale in size has the ability to form colonies and can secrete calcium or other minerals may be a relatively harmless artifact that is interesting to study, or a little-understood pathogen that needs to be diagnosed and treated.

References

1 Wyss Institute Press Release (2012) Wyss Institute models a human disease in an organ-on-a-chip, November 7.

2 Huh, D., Matthews, B.D., Mammoto, A., Montoya-Zavala, M., Hsin, H.Y., and Ingber, D.E. (2010) Reconstituting organ-level lung functions on a chip. *Science*, **328**, 1662–1668.

3 Atala, A., Bauer, S.B., Soker, S., Yoo, J.J., and Retik, A.B. (2006) Tissue-engineered autologous bladders for patients needing cystoplasty. *The Lancet*, **367**, 1241–1246.

4 Johnson, M. and Atala, A. (2009) Grinding out new organs one at a Time. U.S. News & World Report, June 30.

5 Naik, G. (2013) Science fiction comes alive as researchers grow organs in lab. *The Wall Street Journal*, March 22.

6 Ott, H.C., Matthiesen, T.S., Goh, S.-K., Black, L.D., Kren, S.M., Netoff, T.I., and Taylor, D.A. (2008) Perfusion-decellularized matrix: using nature's platform to engineer a bioartificial heart. *Nature Medicine*, **14**, 213–221.

7 Niklason, L. (2013) Building a better lung with scaffolding and cells. Yale Innovators, Office of Cooperative Research, Yale University.

8 Cyranoski, D. (2012) Rudimentary liver grown *in vitro*. *Nature News*, June 20.

9 De Coppi, P. *et al.* (2007) Isolation of amniotic stem cell lines with potential for therapy. *Nature Biotechnology*, **25**, 100–106.

10 Hu, C.-M.J., Zhang, L., Aryal, S., Cheung, C., Fang, R.H., and Zhang, L. (2011) Erythrocyte membrane-camouflaged polymeric nanoparticles as a biomimetic delivery platform. *Proceedings of the National Academy of Sciences of the United States of America*, **108**, 10980–10985.

11 Keene, C.A. (2013) Top innovators in Massachusetts. *The Boston Globe*, May 19.

12 Trafton, A. (2012) Targeted nanoparticles show success in clinical trials. *MIT News*, April 4.

13 McConville, C. (2012) Hub minds, money key to creation of cancer drug. *Boston Herald*, April 5.

14 Schroeder, A., Goldberg, M.S., Kastrup, C., Wang, Y., Jiang, S., Joseph, B.J., Levins, C. G., Kannan, S.T., Langer, R., and

Anderson, D.G. (2012) Remotely activated protein-producing nanoparticles. *Nano Letters*, **12**, 2685–2689.

15 Taylor, K. (2012) Nano-factories produce cancer drugs at tumor sites. *TG Daily*, April 10.

16 Nair, S.B., Dileep, A., and Rajanikant, G.K. (2012) Nanotechnology based diagnostic and therapeutic strategies for neuroscience with special emphasis on ischemic stroke. *Current Medicinal Chemistry*, **19**, 744–756.

17 Perelman School of Medicine (2012) Parkinson's disease protein causes disease spread and neuron death in healthy animals. News Release, University of Pennsylvania, November 15.

18 Vanden Driessche, T. and Chuah, M.K. (2013) Vector decoys trick the immune response. *Science Translational Medicine*, **5**, 194fs28.

19 Henckaerts, E. and Linden, R.M. (2010) Adeno-associated virus: a key to the human genome? *Future Virology*, **5**, 555–574.

20 Wilson, J.M. (2012) It's time for gene therapy to get disruptive! *Human Gene Therapy*, **23**, 1–3.

21 Kaeppel, C. *et al.* (2013) A largely random AAV integration profile after LPLD gene therapy. *Nature Medicine*, **19**, 889–891.

22 Frew, S.E. *et al.* (2008) Chinese health biotech and the billion-patent market. *Nature Biotechnology*, **26**, 37–53.

23 Aiuti, A. and Roncarolo, M.G. (2009) Ten years of gene therapy for primary immune deficiencies. *American Society of Hematology*, **2009**, 682–689.

24 Fischer, A., Hacein-Bey-Abina, S., and Cavazzana-Calvo, M. (2010) 20 years of gene therapy for SCID. *Nature Immunology*, **11**, 457–460.

25 Nathwani, A.C. *et al.* (2011) Adenovirus-associated virus vector–mediated gene transfer in hemophilia B. *New England Journal of Medicine*, **365**, 2357–2365.

26 Perelman School of Medicine (2012) Leukemia patients remain in remission more than two years after receiving genetically engineered T cell therapy. News Release, University of Pennsylvania Health System, December 9.

27 Kaiser, J. (2009) Gene therapy in a new light. *Smithsonian Magazine*, January.

28 Cideciyan, A.V. *et al.* (2012) Human retinal gene therapy for Leber congenital amaurosis shows advancing retinal degeneration despite enduring visual improvement. *Proceedings of the National Academy of Sciences of the United States of America*, **110**, E517–E525.

29 Gallagher, J. (2013) Gene therapy: 'heart-healing virus' trial starts. *BBC News*, April 30.

30 Menon, N. and Viera, P. (2014) Avian flu claims Canadian victim. *The Wall Street Journal*, January 9.

31 Corti, D., Lanzavecchia, A. *et al.* (2011) A neutralizing antibody selected from plasma cells that binds to group 1 and group 2 influenza hemagglutinins. *Science*, **333**, 850–856.

32 Corti, D. and Lanzavecchia, A. (2013) Broadly neutralizing antiviral antibodies. *Annual Review of Immunology*, **31**, 705–742.

33 Limberis, M.P., Adam, V.S., Wong, G., Gren, J., Kobasa, D., Ross, T.M., Kobinger, G.P., Tretiakova, A., and Wilson, J.M. (2013) Intranasal antibody gene transfer in mice and ferrets elicits broad protection against pandemic influenza. *Science Translational Medicine*, **5**, 187ra72.

34 Perelman School of Medicine (2013) New gene therapy shows broad protection in animal models to pandemic flu strains, including the deadly 1918 Spanish influenza. Press Release, May 29.

35 Trafton, A. (2011) New drug could cure nearly any viral infection. *MIT News*, August 10.

36 Rider, T.H. *et al.* (2011) Broad spectrum antiviral therapeutics. *PLoS One*, **6**, e22572.

37 Trafton, A. (2011) New drug could cure nearly any viral infection. *MIT News*, August 10.

38 Knox, R. (2013) Officials prepare for another flu pandemic – just in case. *NPR Health News*, May 8.

39 Schoemaker, P.J.H. and Tomczyk, M. (2006) The future of BioSciences: four scenarios for 2020 and their implications for human healthcare. BioSciences

Crossroads Initiative, Mack Center for Technological Innovation and DSI.

40 Glynn, S. (2013) UK aims to make genetic testing available to all cancer patients. *Medical News Today*, May 21.

41 Valiokas, R. (2011) Nanobiochips. *Cellular and Molecular Life Sciences*, **69**, 347–356.

42 Weigum, S.E., Floriano, P.N., Redding, S.W., Yeh, C.-K., Westbrook, S.D., McGuff, H.S., Lin, A., Miller, F.R., Villarreal, F., Rowan, S.D., Vigneswaran, N., Williams, M.D., and McDevitt, J.T. (2011) A new bio-nanochip sensor aids oral cancer detection. *International Society for Optics and Photonics*, 28 March.

43 Mack, J. (2005) Nanotechnology: What's in it for Biotech? *Biotechnology Healthcare*, **2**, 29–36.

44 MIT Technology Review (2009) 10 Breakthrough technologies/TR10: $100 genome, March/April.

45 NASA News (2006) Tiny particles may open new ways to study calcification, December 28.

11
Nanomimicry: Cool Things We Can Do with Nanobiology

> A biological system can be exceedingly small. Many of the cells are very tiny, but they are very active; they manufacture various substances; they walk around; they wiggle; and they do all kinds of marvelous things – all on a very small scale. Also, they store information. Consider the possibility that we too can make a thing very small which does what we want – that we can manufacture an object that maneuvers at that level!
>
> – *Richard Feynman, "There's Plenty of Room at the Bottom"*

In his classic 1959 presentation to the American Physical Society at Caltech, Richard Feynman mentioned both nanocomputing and maneuverable nanorobots. This is a great example of what a visionary pioneer can do with an educated mind, an innovative spirit, and a sense of what's possible. Some of Dr. Feynman's challenges were accepted during his lifetime. Some are just beginning to be realized now.

By turning his vision into a challenge, Dr. Feynman extended his personal reach more than half a century into the future. Today we are accepting his original challenge, crediting him with jump-starting the field of nanotechnology, and fulfilling a dream that started with "thinking small" and continues even farther into the future.

One of the most intriguing concepts originally raised by Dr. Feynman involves engineering and designing functions that are common in biological systems. This is known today as "biomimicry" or "biomimetics."

While many of the innovations mentioned in this classic 1959 talk were not possible to achieve at the time and wouldn't be possible for decades to come, they were very real in Dr. Feynman's imagination. Dr. Feynman offered some intriguing targets for the bionano research community that would actually be achieved – although most people didn't know it at the time. If we could have traveled forward to the year 2015 from December 29, 1959 when Dr. Feynman gave his speech, we would be amazed to learn that most of his predictions have come true, and progress has been made on all fronts.

Of course, when Dr. Feynman presented his challenge, he probably wasn't thinking that one day we would be creating new functions based on engineering

NanoInnovation: What Every Manager Needs to Know, First Edition. Michael Tomczyk.

human DNA, borrowing ideas from the feet of a lizard, or creating structures that replicate the properties of a butterfly wing.

11.1
Turning DNA into Nanocomputers

Two of the innovations that were mentioned in Dr. Feynman's 1959 talk were nanocomputers that "store information" and nanorobots that "maneuver." This prediction relates to biological as well as electronic computing.

This prediction is being brought to life by researchers from the Weizmann Institute of Science in Reghovot, Israel, who in 2002 demonstrated a *programmable molecular computing machine* that used DNA molecules and enzymes. In the April 2004 issue of *Nature*, the team suggested that their DNA computer could theoretically diagnose cancerous activity within a cell, and release an anticancer drug. These functions could be performed autonomously. The team members included Ehud Shapiro, Yaakov Benenson, Binyamin Gil, Uri Ben-Dor, and Rivka Adar.

In 2009, Shapiro and his colleague Tom Ran reported the implementation of simple logic programs that enabled a DNA computer to answer "yes" or "no" and correctly answer a question such as "Is Socrates mortal?" Their research included development of a program that linked a computer programming language to DNA computing code, which further extended the utility of their DNA computing technology [1]. The ultimate applications are in programmable autonomous computing devices that can operate in a biological environment, Dr. Shapiro suggested.

In June 2011, Caltech researchers Lulu Qian and Erik Winfree built the largest DNA computational circuit to date, using an approach that is both simple and scalable. The researchers formed 130 synthetic DNA strands, which they used to create a 74 molecule, 4-bit circuit that was able to compute the square root of any number up to 15, and round the answer to the nearest integer. Their DNA circuit used biochemical logic gates to produce binary (on–off) signals, similar to silicon-based integrated circuits. They had previously tested a DNA computer comprised of 12 DNA molecules.

In the eloquent introduction to their June 2011 article in *Science*, they wrote: "The power and mystery of life is entangled within the information processing at the heart of all cellular machinery. Engineering molecular information processing systems may allow us to tap into that power and elucidate principles that will help us to understand and appreciate the mystery" [2].

In a commentary that accompanied the researchers' *Science* article [3], Duke University science professor John Reif observed that the execution of a single gate could take 30 min–1 h, and computing a 4-bit square root could take as long as 6–10 h, although the researchers indicated that they can increase the speed of their circuit up to 100 times or more using higher concentrations and other strategies. They also recognized that implementation challenges remain such as "increased spurious binding" at larger scales that could slow the reaction rates and affect the processes. A possible solution might be to utilize DNA origami techniques invented by Caltech colleague Paul Rothemund, they suggested.

It is understandable that the first DNA computers can only perform a few calculations, and very slowly – but the fact that they can make these calculations at all is a step in the right direction.

11.2
Turning DNA into "Walking" Nanorobots

While Dr. Feynman did not use the term "nanorobot" or "nanobot," it was clear that he was referring to the capabilities of what we now call nanobots when he talked about nanoscale objects that can "maneuver." A great deal has been written about whether nanobots would be mechanical or biological, with an edge going to biological systems since most self-replicating systems in Nature are biological. One of the amazing innovations to come out of the convergence of biology and nanotechnology has been the use of DNA to create very early versions of nanobots.

Some structures that have been called nanobots are actually nanoscale containers that can hold, transport, and release toxins and other agents to kill fast-growing cancer cells. These take the form of nanoshells made of gold or carbon, boxes made of DNA – even boxes with "locks," and clamshell-shaped containers with "latches" that release when they come into contact with a specific type of cell. These nanoshells are more like packaging than maneuverable nanoscale robots. However, there are several streams of research that are creating maneuverable nanoscale devices, using the principles of DNA nanotechnology and DNA origami.

In the past half decade, DNA has been used to develop *molecular robots* <4 nm in diameter that have the ability to "walk" toward a target. In 2010, Milan Stojanovic at Columbia University led a team that developed a molecular robot made from DNA. The team included Erik Winfree (Caltech), Han Yan (Arizona State University), and Nils Walter (University of Michigan, Ann Arbor).

Their nanoscale walking structures have four "legs" made from biotin-labeled DNA strands that attach and detach to the surface of a pathway or track until the strands reach a patch of DNA that they can bind to but cannot cut, that stops them. The researchers call their DNA robot a "spider" and refer to the stopping surface as "flypaper."

Their first molecular robots could take only three steps. In subsequent tests, the DNA spider took $\sim$50 "steps" to travel 100 nm [4]. The number of legs and their length can be varied, and their motion, gait, and "stickiness" can be adjusted to help them locate targets on a two-dimensional lattice. This concept was elaborated on by Dr. Stojanovic in a paper coauthored with Oleg Semenov and Darko Stefanovic from the University of Mexico [5].

These examples are *very* early-stage indications that illustrate the potential to design biological computers and nanoscale biorobots.

11.3
Nanomimicry: Learning from Nature at the Nanoscale

Biomimetics – also known as biomimicry – is the use of models, systems, and elements found in Nature, to design and engineer materials and machines. While scientists have been learning from Nature for centuries, the term "biomimetics" was coined during the 1950s by American biophysicist Otto Schmitt at the University of Minnesota. The concept received renewed attention in 1997 with the publication of Janine Benyus' book, *Biomimicry: Innovation Inspired by Nature.* Janine is the founder of the Biomimicry Guild and the Biomimicry Institute and the author of *Nature's 100 Best.*

"Biomimicry is innovation inspired by nature," Janine declared in her classic book. "In a society accustomed to dominating or "improving" nature, this respectful imitation is a radically new approach, a revolution really. Unlike the Industrial Revolution, the Biomimicry Revolution introduces an era based not on what we can *extract* from nature, but on what we can *learn* from her . . . ."

In a radio interview with TreeHugger, Janine said: "Nature is nano. It's one of life's principles, really, that life builds from the bottom up. What that means is that nanotechnology basically has to do with scale. It's a term that refers to scale – how small something is. Your body builds out of a complete nano construction" [6].

While scientists and engineers have been drawing lessons from Nature for decades, it was only in the past decade that nanotechnology has enabled scientists to mimic Nature at the scale of atoms and molecules. In this sense, we might extend the term "biomimicry" to create a new term, "nanomimicry" – emulating and replicating structures and processes that occur in Nature *at the nanoscale.* Two of the most intriguing examples of nanomimicry are a dry adhesive based on the footpads of a lizard, and a colorfast waterproof material inspired by a butterfly.

11.4
Mimicking Geckos to Create Glue

Most people are familiar with a small friendly looking lizard called a gecko. In many countries of the world, this cute-looking insect-eating lizard is a common sight in homes and offices. When I was in my 20 s, as a young Army officer serving in Southeast Asia, there always seemed to be one or two geckos on the wall, the ceiling, or the window – how they managed to stick to the ceiling or the smooth surface of a glass mirror was a fascinating mystery.

Geckos are lizards that vary in size from a few grams to half a kilogram, with large eyes that many people find endearing. Geckos are common in tropical regions of the world. Some people keep them as pets. In the United States, the gecko is the symbol of a well-known insurance company. Geckos are among the few lizards that vocalize sounds, making clicks and trills and chirping noises to communicate. Geckos have flattened disk-shaped toe pads that help them climb and stick to surfaces from rocks to windows. They are able to stick to a surface using a form of dry adhesion, as opposed to frogs and other creatures that excrete a sticky glue-like substance and leave sticky trails where they walk.

Geckos have an uncanny ability to walk on vertical window glass, on walls, and ceilings. But how does a gecko stick to a smooth glass surface or keep from falling when it's walking upside down? That's what a team of researchers at the University of Massachusetts, Amherst, asked themselves when they decided to try to replicate how a gecko's footpads allow it to cling to smooth surfaces (Figure 11.1). The result is called Geckskin^{TM}, which replicates the gecko's secret in an entirely new dry adhesive technology that allows heavy objects to be stuck to smooth surfaces. The objects can be removed without leaving any residue.

Geckos can weigh up to 17 ounces, although most are less than an ounce. When a 5 ounce gecko scrambles across a wall or ceiling, the pads on their feet exert an adhesive force of about 9 lb. They can walk upside down on a smooth surface, suspend themselves by 1′, and cling to surfaces in high winds.[1] This is made possible by structures on the pads of their feet called *scansors*, which are lined with rows of *lamellae*, which in turn are covered with tiny elastic hair called *setae* that have branched nanoscale tips called *spatulae*. Controlling the flow of blood to and from the footpads allows the structures to change shape, attach, and detach. These structures are self-cleaning and water repellant. They are one of Nature's most incredible innovations (Figure 11.2).

So what does this have to do with nanoinnovation?

First, it's important to understand that Geckskin was not an overnight innovation. Biologists had studied the ecology and anatomy of geckos for over 50 years, carefully mapping many aspects of anatomy that enable them to climb. However, trying to replicate gecko-like adhesion was not easy.

In 2011, a team of polymer scientists collaborated with a biologist at the University of Massachusetts, Amherst, to mimic the structure of the gecko's footpads to create a dry adhesive they call "Geckskin." The Geckskin team demonstrated that a patch of this material only 100 cm^2 can support a 700 lb object on a smooth surface. The adhesive patch can be peeled off without leaving a residue. The researchers have shown that a small Geckskin patch can hold a 42″ large-screen television to a living room wall (Figure 11.3).

1) **Gecko trivia:** One of the few surfaces that geckos *cannot* stick to is Teflon® – ironically, a human-made material.

Figure 11.1 (a) A Gehyra gecko relies on nanoscale structures in its footpads to walk on glass. (b) A close-up of a Tokay gecko's foot showing a macroview of the toe pads (images courtesy of Duncan Irschick at the University of Massachusetts, Amherst).

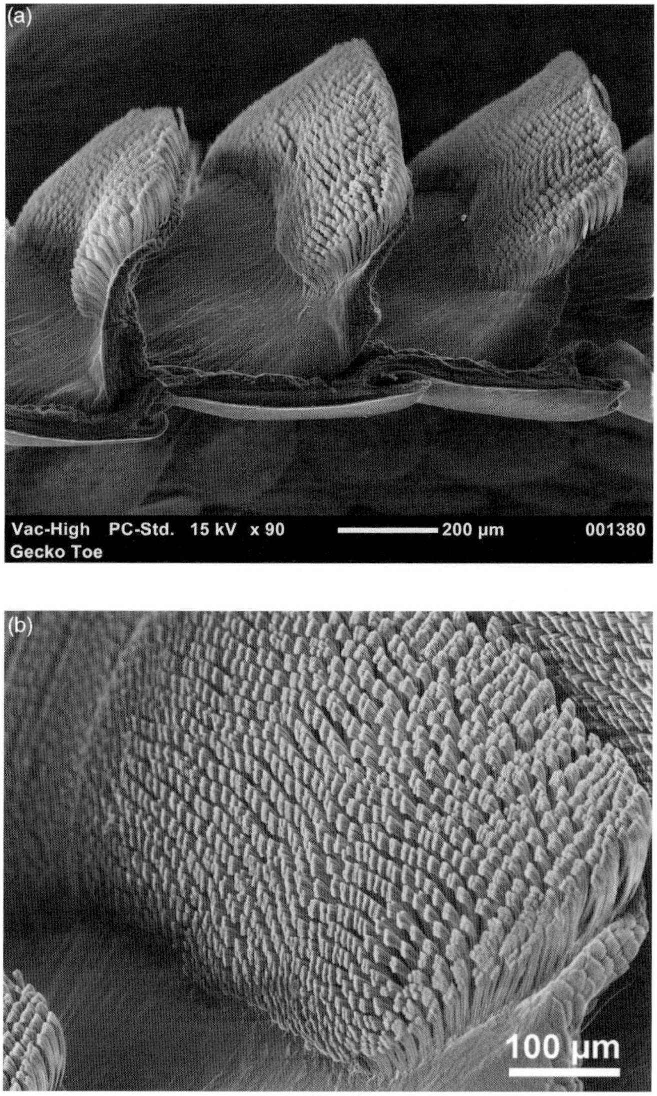

Figures 11.2 These nanoscopic images reveal just how intricate and hard to replicate these gecko foot structures are. (a) A close-up of scansor pads. (b) A field of setae. (c) Bundles of structures called scapulae. (d) A close-up of the scapulae – they do look a bit like pancake spatulas (images courtesy of Michael Bartlett, Alfred Crosby, and Duncan Irschick at the University of Massachusetts, Amherst).

The research team was led by Al Crosby, a polymer scientist, and biologist Duncan Irschick, a functional morphologist who has been studying the gecko's climbing and clinging abilities for over 20 years. Their team united by happenstance: Al Crosby's laboratory had been working for several years on a simple,

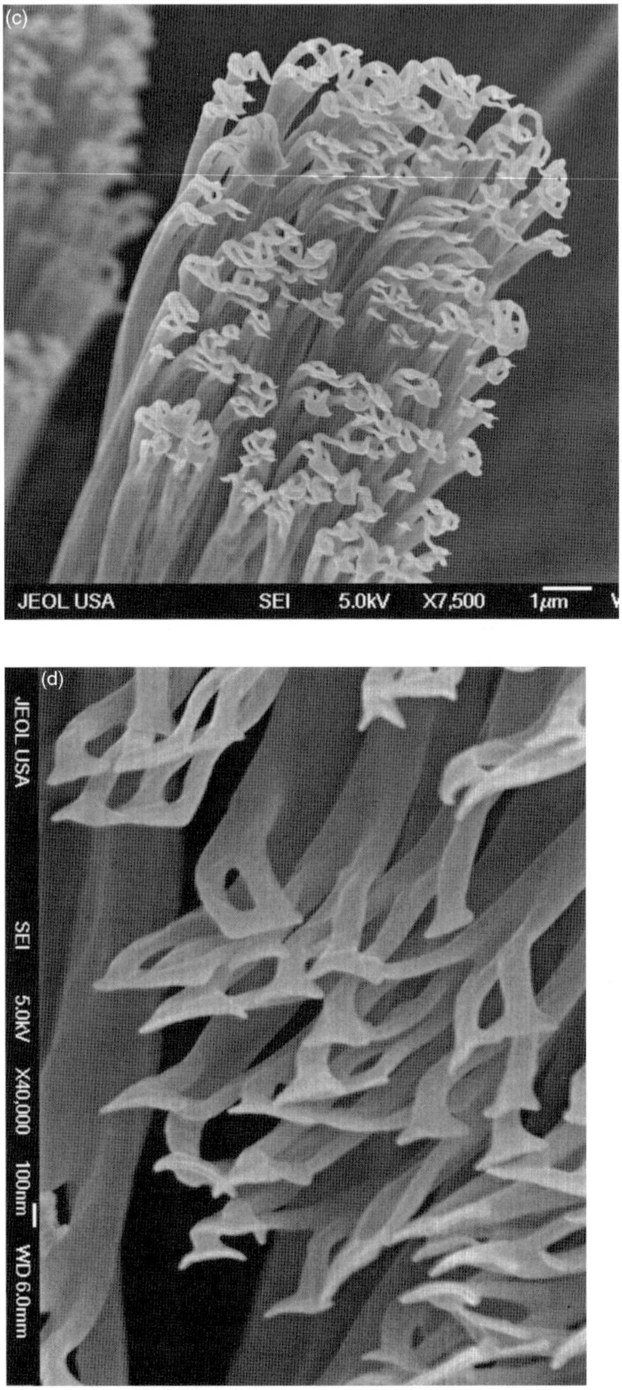

Figures 11.2 (*Continued*)

Figure 11.3 (a) Geckskin is an innovative dry adhesive that mimics the structure of the footpads on the feet of a gecko. It was developed by a team of polymer scientists and a biologist at UMASS, Amherst. (b). A small patch of Geckskin can support an enormous amount of weight as shown – this little patch is holding as much weight as a heavyweight wrestler can lift (images by Michael Bartlett, University of Massachusetts, Amherst).

yet powerful way to mimic gecko feet in a synthetic adhesive. They derived a new theory for how adhesives work, which relied on the principle of draping, much like a tablecloth drapes over a table. They believed that, contrary to most understanding of how adhesives work, making adhesives increasingly stiff would enable powerful adhesion. They also wanted to know if this theory was the secret to the geckos' adhesive success. To do so, they wanted to team up with a gecko biologist. Not knowing of any on campus, they Googled "geckos UMASS" and – *voilà!* – to their surprise, one of the world's experts in gecko adhesion – UMass biologist Duncan J. Irschick – popped up in their search. Team members had also been reading his work. Duncan Irschick first began studying gecko adhesion as an undergraduate, and performed many studies on the lizard's remarkable ability to climb surfaces.

Mike Bartlett and Dan King, two graduate students in the Polymer Sciences Department, joined the team and the researchers delved deeper into the mysteries of "gecko adhesion." The team verified that geckos have stiff tendons that branch into their soft toe pads. These tendons had been previously ignored as being unimportant in geckos, but as it turned out, this feature is vital to the success of both geckos and Geckskin. The stiff tendons enable the soft pad of the gecko foot to achieve high levels of adhesion over a large area, thus enabling even large geckos (half a kilogram) to climb. Similarly, a stiff tendon in Geckskin, made of fabrics such as carbon fiber or Kevlar, enabled the soft pad to adhere across a very wide area and with great effectiveness. This ability to create powerful adhesives at large scales could open up many new opportunities for useful human products, including new approaches for wall mounting objects, and applications in medicine, apparel, and industrial assembly and design.

Their research was published in the February issue of *Advanced Materials* [7]. Their innovation was named one of the top five science breakthroughs of 2012 by CNN. The innovation has received widespread media attention as an inspired example of biomimetic design.

11.5 Biomimicking the Waterproofing Properties of Butterfly Wings

For more than a decade, nanoscientists have studied butterfly scales. The nanosized scales of butterflies are well known for their ability to convey color by size and orientation rather than pigment. Many butterflies that appear to have brightly colored wings achieve their colors by a phenomenon that occurs when their scales refract light at specific wavelengths. Butterfly scales are also known for their waterproofing capabilities, which allow butterflies to survive rainstorms.

In 2012, Dr. Shu Yang at the University of Pennsylvania announced the development of a nanomaterial that mimics the waterproof properties of the scales on a butterfly's wing (Figure 11.4). Dr. Yang is professor of Materials

Figure 11.4 In this intriguing image, a bead of water sits on a disk-shaped material created by Dr. Shu Yang's research group at the University of Pennsylvania. The material mimics the structure of a butterfly wing to provide color that won't fade, and a waterproof surface. Note that the wings of the butterfly sculpture in this image were also created from the same material (image/design by Felice Macera, used with permission, *Advanced Functional Materials*, University of Pennsylvania).

Science and Engineering and Chemical and Biomolecular Engineering at UPENN. Her coauthors included Jie Li, Guanquan Liang, and Xuelian Zhu. Their research was published in the April 2012 issue of *Advanced Functional Materials* [8].

In addition to waterproofing capabilities – described in science jargon as the development of *superhydrophobic surfaces* – Dr. Yang's research has achieved what she calls "structural color." She observes that color can come from periodic patterns and size of structures, while superhydrophobicity can be determined by the surface roughness and texture of a material. When water lands on a hydrophobic surface, she explains, its roughness reduces the effective contact area between water and a solid area where it can adhere, thus increasing the water repellency on the surface.

"In Nature, bioorganisms often possess a hierarchical architecture with multiple functions. For example, a butterfly wing has both structural color and superhydrophobicity, while a gecko's foot hairs are both sticky and superhydrophobic. A lot of research over the last 10 years has gone into trying to create structural colors like those found in nature, in things like butterfly wings, beetle scales and opals," Dr. Yang explains. "People have also been interested in creating superhydrophobic surfaces which are found in things like lotus leaves and in butterfly wings, too, since they couldn't stay in the air with raindrops clinging to them" [9].

Mimicking Nature requires ingenuity and technical prowess. Dr. Yang's approach involves an unconventional photolithography technique called "holographic lithography," which uses a laser to create a cross-linked 3D network from a photoresist material. The regions on the photoresist material that are not exposed to the laser are later removed by a solvent leaving submicrometer-sized "holes" periodically arranged in a 3D lattice, which provides structural color. Applying another solvent causes nanospheres to form within the lattice, which prevents the wetting of water on the surface. She said this process is both a science and an art, since each step requires fine-tuning.

Dr. Yang's colead author on the project, Dr. Guanquan Liang, is an optical physicist. While working to improve the quality of 3D photonic crystals, he experimented with many different solvents, and early in his research he often produced spongy nanospheres that gave the material a rough surface instead of a smooth surface. At first this looked like a failure and was frustrating, but when Dr. Liang showed the results to Dr. Yang, she got excited when she realized that this could be used to create a superhydrophobic surface on photonic film. Furthermore, the structural color would retain its color and would not fade from exposure to light, which happens with many pigments and dyes.

Structural color and waterproofing are features that are needed in a wide variety of applications, including semiconductors, energy-saving materials, and sensors. This technology could be used in traffic signs and consumer electronics. It could even be used as camouflage. These functions can be combined with other functions as well. For example, Dr. Yang's group is currently developing energy-efficient building skins that will integrate optical sensors. Their special focus involves interactions at surfaces and interfaces – what Dr. Yang calls the "Structure–Property Relationship." Researchers in her group are studying and fabricating a variety of properties that occur in Nature, including wetting, adhesion, biocompatibility, and optical and mechanical properties. Many of these involve multiscaled structures and unique properties. For example, the group has drawn lessons from properties exhibited by human cells and muscles to design adaptive skins for buildings, which responds to environmental changes, thus saving energy.

This fascinating convergence of biomimicry and nanotechnology is creating new types of materials as well as more efficient ways to incorporate color, waterproofing, and other capabilities. Nanoinnovations will continue to replicate biological structures observed in Nature. There is much more yet to come.

References

1 Ran, T., Kaplan, S., and Shapiro, E. (2009) Molecular implementation of simple logic programs. *Nature Nanotechnology*, **4**, 642–648.

2 Qian, L. and Winfree, E. (2011) Scaling up digital circuit computation with DNA strand displacement cascades. *Science*, **332**, 1196–1201.

3 Reif, J.H. (2011) Scaling up DNA computation. *Science*, **332**, 1156–1157.

4 California Institute of Technology (2010) Spiders at the nanoscale: molecules that behave like robots. Press Release, May 12.

5 Semenov, O., Stefanovic, D., and Stojanovic, M.N. (2012) The effects of multivalency and kinetics in nanoscale search by molecular spiders. Working paper, presented at the Italian Workshop on Artificial Life and Evolutionary Computation, February, 20–21.

6 Benyus, J. (2009) Janine Benyus on biomimicry in design on TH Radio (Part two). TreeHugger.

7 Bartlett, M.D., Croll, A.B., King, D.R., Irschick, D.J., and Crosby, A.J. (2012) Looking beyond fibrillar features to scale gecko-like adhesion. *Advanced Materials*, **24**, 1078–1083.

8 Li, J., Liang, G., Zhu, X., and Yang, S. (2012) Exploiting nanoroughness on holographically patterned three-dimensional photonic crystals. *Advanced Functional Materials*, **22**, 2980–2986.

9 Lerner, E. (2012) Penn researchers find new way to mimic the color and texture of butterfly wings. Penn News.

12
Nanotechnology: Is It Safe?

> Nanoparticles have been part of our natural environment forever, yet historically we had little grasp of them or their health effects. Today, the tools of nanotechnology allow us to characterize and understand both natural and man-made nanoparticles and thus take a more rational approach to safety.
>
> – *Youseph Yazdi, Ph.D., Department of Biomedical Engineering, Johns Hopkins University*

As we consider what happens at the convergence of bio and nano, it is important to consider safety issues, especially where and how nanotechnology – especially nanoparticles – might potentially cause harm to humans, or to the environment.

In the realm of nanotechnology, safety is an issue that impacts everyone. As a consumer, you may worry that nanoparticles in a food product or cosmetic will cause cancer or some other disease. If you're an employer at a company whose products use nanomaterials, you worry about the safety of your employees who might inhale or absorb nanoparticles in the workplace. If you work in medicine or healthcare, you know that nanoinnovations need to pass rigorous animal and human clinical trials to confirm that they are safe as well as effective. As environmentalists, we all need to ensure that nanoparticles created to solve problems don't inadvertently cause problems when they are released into the environment. If you work in city government, you may worry about health hazards if nanoparticles that find their way into the sewer system and kill bacteria needed to biologically process wastewater. If you are working in the space program at NASA, you worry that nano- and subnano particles will penetrate the hull of a spacecraft and sever DNA strands while speeding through the bodies of astronauts bound for Mars.

From its inception, the nanoinnovation community has been extremely diligent and conscientious about studying and monitoring safety issues. This attention to risk factors has helped insulate the nanotechnology sector from media attention and public criticism. However, safety is an area that requires constant diligence since unpleasant surprises can emerge unexpectedly to reveal a hidden threat.

NanoInnovation: What Every Manager Needs to Know, First Edition. Michael Tomczyk.

There is good reason to ensure that there aren't any lurking dangers embedded somewhere in nanoinnovations. In recent decades, the public learned that many products previously considered safe and marketed by reputable companies were actually unsafe, even fatal.

Asbestos was used as fire-resistant thermal insulation in construction for more than a century, and 30 years ago was confirmed to cause cancer even in minute quantities. The first generation of silicone breast implants leaked and caused debilitating diseases. Tainted livestock organs ground into a meal and fed to sheep and cattle conveyed a horrific condition called "mad cow disease" acquired by humans who ate contaminated meat.

Nanoparticles are used in materials, clothing, coatings, and even food products and packaging. Is there a hidden hazard lurking in one of these substances or applications?

In many cases, the safety issue may not be the hazard or toxicity *per se*, but the form of the material and how it is used. For example, asbestos is still used in ~3000 products today, in products that include safeguards against inhalation of particles. A thicker more stable form of silicone is used in breast implants. Long carbon nanotubes (CNTs), which have an asbestos-like form factor, are used in applications that embed or encapsulate the particles to stabilize them, so they can't be inhaled. It is difficult to validate risk factors for nanoparticles, since many nanoparticles are beneficial in one application, but may be harmful in a different application, or when those nanoparticles are disposed of.

To some extent "toxic" is a relative term. Any substance that concentrates in a human organ can have a toxic effect depending on the composition, size, and shape of the particle. Any foreign particles introduced into the body can accumulate in organs and cause inflammation. If the particles concentrate in quantity, they can cause the body to encapsulate them, forming granulomas. Some anticancer drugs are more toxic than the nanoparticles used to deliver them.

So far, a few safety issues have been identified that involve nanoparticles, but, in general, nanoinnovations appear to be safe, although we don't know everything yet since the field of nanotoxicology is very young. The term "nanotoxicology" has only been in common use since 2004. The first notable uses of the term appeared in 2004, in commentaries by R.F. Service [1] and Donaldson *et al.* [2].

So far, only a few validated safety studies have confirmed serious health or environmental problems that are unique to nanoparticles. In most cases, the same precautions used for any small particles, powders, and aerosols have helped to prevent serious problems in the workplace or marketplace for nanoparticles.

The current consensus is that most nanomaterials and nanoparticles are safe. However, most regulatory agencies and experts in the field suggest that our knowledge of nanoparticle toxicity is still limited, more research is needed, and caution should be exercised in the handling and disposal of certain nanomaterials. A full life cycle assessment is recommended for all products that incorporate nanoparticles and nanomaterials.

12.1 Early Experience with Nano Safety

Humans have been exposed to nanoparticles for centuries. Airborne nanoparticles are contained in wood smoke, eruptions from forest fires and volcanoes, exhaust from cars and trucks, and particulates from furnaces used by factories and utilities.

People around the world are being exposed to nanoparticles every day. Pollution from burning fossil fuels generates a vast amount of particles ranging in size from nano to macro – as evidenced by the visible clouds of pollution that are still suffocating major cities in the world. In some heavy traffic areas, there can be as many as 5000–3 000 000 sub-μm particles/cm^3.

Workers are exposed to nanoparticles in many industries. Miners have inhaled mineral nanoparticles as part of their jobs. Nanoparticles of carbon inhaled by coal miners contributed to black lung disease. Workers in many factories come into contact with nanopowders, aerosols, dust, and other materials that are hazardous. In most cases, workers use masks, filters, protective clothing, and ventilation systems to protect themselves.

In our daily lives, we consume millions or billions of submicrometer particles in our normal diet, including food additives such as titanium dioxide and aluminosilicates [3]. Popular condiments ranging from gum and candy to frosting can contain nanoparticles that are used to control color, consistency, or other attributes.

In the past decade, we have come in contact with hundreds of new products and materials or ingredients that are the result of nanoinnovation. We use materials that contain carbon nanotubes. We wear clothing treated with nanocoatings. We are treated with drugs that have been nanosized. We wash our clothes in appliances that have been coated with antibacterial coatings that are designed to release nanoscale silver ions. We may be exposed to nanoparticles through hair dyes, sunscreens, skin creams, and many other products.

Therefore, it makes sense to gain an understanding of the risk factors associated with nanoparticles. To do this, we need to understand the false indicators as well as the real indicators of risk hazards; we need frameworks to conceptualize and map these risks as well as ways to pool our information and collaborate across borders.

Reasons for Caution: The First Nanosafety Incidents

Public safety has always been a major consideration for nanoinnovators. For more than 25 years, *public perception* has played an important role in public debates over nanotechnology safety. The history of nanotechnology incidents contains a mix of misperceptions and false alarms, as well as verified health risks and toxicology studies.

In 1986, Eric Drexler coined the term "gray goo" in two paragraphs in his book *Engines of Creation.* The same year, the concept was featured in *Omni* magazine. This fueled a media debate about whether self-replicating "nanobots" could be unleashed on the world and spread like a virus to destroy humanity. In 2000, Robert Freitas coined the term "ecophagy" to describe how "biovorous nanoreplicators" could keep replicating themselves and ultimately consume entire ecosystems. These fears were vividly brought to life in Michael Crichton's 2002 book, *Prey*, which described nanoparticles that escaped from a laboratory to form self-aware, intelligent swarms that could take human form. The public debate and associated fears have diminished as scientists have refuted these possibilities. Today, most authorities consider the concept of nanorobots spinning out of control as somewhere between "absurd" and "cautionary."

Another safety issue that was initially feared but never materialized was a potential public backlash against nanotechnology. In the 1990s and early 2000s, there was a legitimate concern that the public would reject anything "nano" just as Europeans initially rejected genetically modified crops and foods in the 1980s and required special labeling on GMO products. However, there was no significant backlash against nanotechnology despite the hidden nature of nanoparticles and materials. Several factors have made nanotechnology "friendlier" and more acceptable to the public than genetically modified foods.

First, the reaction of government regulators was mostly favorable. In the United States and many other countries, federal agencies adopted a policy of regulating nanoparticles like other materials such as paints and coatings, devices like semiconductors, and medical drugs – all of which were early product categories that integrated nanoscale materials and structures. Most agencies treated nanoparticles as "special forms" of the bulk substances they represented and relied on existing regulations for safe handling of powders, aerosols, or known toxins, rather than the unique properties of nanoscale materials.

This light regulatory treatment was aided by the fact that nanoparticles have been used in many products for a century or more. Well-known examples include carbon black used in automobile tires and photocopier ink, nanoparticles used in cosmetics since the 1960s, and nanocoatings in stain-resistant clothing since the 1970s. Carbon black has been used in rubber products for more than a century – today, more than 10 million metric tons a year are produced with no significant safety issues.

As previously discussed, nanotechnology also benefited from the favorable legacy of microtechnology. Microcomputers and microchips launched the digital revolution with few safety concerns. Microdevices from videogames to music players, everything electronic became smaller, more compact, and "user-friendly." In 2005, Apple named its iPod music player the "Nano," which sold a million units in 17 days.

This atmosphere of user-friendliness benefited "everything nano," which was viewed by most people as an extension of "micro." The public perception was that microtechnology gave us a myriad of innovations, and nanotechnology is a logical extension. Micro is a friendly word. Nano sounds even friendlier. In popular jargon, it means tiny and ultrasmall.

Also, it was helpful that nanotechnology was not associated with food products. As the first few hundred or so commercial uses were emerging in the 1990s, it was not widely known that nanoparticles were already present in cosmetics and other consumer products, so consumer concerns were minimized.

Despite this nano-friendly climate, there were a few notable incidents that received media attention that are important to mention, that any manager should know about.

The First Nano Safety Incident Was a False Alarm

The first reported safety scare attributed to nanotechnology was actually a false alarm. In 2006, a bathroom cleaning spray called Magic Nano was blamed for sickening 70 people in Germany. Half a dozen people were hospitalized for pulmonary edema. The product had been sold for 4 years as a pump spray with no reported problems. The health problems began when the product was introduced as an aerosol spray. In the pump spray the solvent was a liquid and did not evaporate, but as an aerosol product, a corrosion inhibitor that was added to the formula made the spray more volatile and caused particles to form in the lungs when inhaled.

Because the product had "nano" in its name, the media jumped on the story as the first example of an illness caused by nanotechnology. However, the name "Magic Nano" was misleading since it did not contain any nanoparticles as ingredients.

The incident was investigated by Germany's Federal Institute for Risk Assessment. Later it was discovered that the particles in the aerosol were not nanoscale in size, so this was a case of "aerosol poisoning" not associated with nanoparticles or nanotechnology. The misleading factor was the name of the product – Magic Nano – although there was nothing "nano" about Magic Nano except the name. The aerosol version of the product was recalled as soon as the health problems occurred.

The First Nanoparticle-Linked Health Incident

In 2009, seven Chinese women working at a paint factory where nanoparticles were used fell ill with a serious lung disease, and two died. The factory manufactured polyacrylate-based paint. According to reports, cotton gauze masks were the only protection used by the women in the workplace, which had no windows and was poorly ventilated. The machinery at the factory was shut down and no other people became ill. The women experienced itchy eruptions on their face and arms and fluid build-up around their heart and lungs. The symptoms were different from symptoms normally associated with paint inhalation. Also, the symptoms persisted after the women were no longer exposed to the paint [4].

A research study detailing the incident was published in the *European Respiratory Journal* [5]. The team was led by Yuguo Song, from the Occupational Disease and Clinical Toxicology Department at Chaoyang Hospital in Beijing. The team found nanoparticles 30 nm in diameter, deep in the lungs of the women. The particle clusters had formed pleural granulomas, which are ball-like collections of immune cells in the lining of the lung. These structures formed when the immune system was unable to remove the particles. The study reported that the symptoms came from inhaling fumes produced from polystyrene boards that were heated to 75–100 °C after being sprayed with a paste-like plastic called a polyacrylate ester.

"This is the first clear case where there is an association between someone breathing in nanoparticles in the workplace and getting seriously ill," observed

Dr. Andrew Maynard, former Advisor at the Woodrow Wilson Center's Project on Emerging Nanotechnologies, now NSF International Chair of Environmental Health Sciences at the University of Michigan School of Public Health [6].

Some toxicologists have suggested that nanoparticles *per se* were not responsible. The toxicity of the plastic material and poor health and safety conditions in the facility were major, if not primary, causes. Whether this fatal incident resulted from workplace conditions, the accumulation of particles, or from properties unique to the nanoparticles, is unclear. It is known that larger particles of different sizes and shapes can also cause granulomas to form. Which factors were most responsible for these unfortunate deaths and illnesses remains unclear.

12.2
What We Know about Nanoparticle Risks

In the past half decade, hundreds of research projects and papers have focused on nanosafety issues. Researchers have studied whether short or long carbon nanotubes can cause cancer (like asbestos fibers), and whether silver ion nanoparticles kill friendly bacteria and aquatic species if released into streams and reservoirs. These studies helped confirm what we can and cannot do with nanotechnology and provided the first guidelines for safe use of nanomaterials. Most major studies have focused on the impact of nanoparticles on human health, their impact on the environment (especially aquatic ecosystems), and the potential impact of discarded nanoparticles on biological waste treatment systems.

Ongoing medical research is exploring which particles and materials are toxic to human health, since nanoparticles have the ability to penetrate the membranes of cells that can produce a toxic effect, trigger an immune response, or produce an unexpected catalytic reaction. Some substances are safe at the microscale and larger, but become reactive and potentially toxic at the nanoscale. The size and shape of the particles may also cause medical problems. For example, a long carbon nanotube can act as a nanoscale knife, slicing DNA strands, which can cause cancer. Even the surface configuration of the individual particles can have an effect.

Environmental safety research has been underway since the early 2000s. Environmental researchers have been studying what happens to nanoparticles that may be safe in industrial or consumer products, but may deteriorate and become harmful after disposal. They are attempting to identify when nanoparticles may be an unintended by-product of an industrial process. Of special concern is the impact of nanoparticles on an entire ecosystem or food chain, and its unintended consequences.

Most nanoparticle safety research has involved toxicology studies on animals. Researchers have found that most nanoparticles are harmless and/or are removed by macrophages (white blood cells). On the other hand, some particles such as long needle-shaped nanotubes are not removed. Nanoparticles can accumulate and form granulomas in the lungs and other organs.

Nanoparticles can also be absorbed by plants and accumulate in their tissues and can accumulate in the food chain, which can affect the environment. The ability of the body to absorb and concentrate nanoparticles is used to deliver therapeutic nanoparticles in nanomedicine. But which types of materials and concentrations are toxic? The challenge for science is to identify which types of nanoparticles and accumulations can be harmful, and why.

Determining the "safety" of a nanoparticle can be tricky. Some nanoparticles are both safe and toxic. For example, it is ironic that the same nanosized silver used to coat surfaces in hospitals and clinics – to prevent the growth of bacteria – can also kill friendly bacteria and harm aquatic creatures if released into streams and reservoirs.

Given these dichotomies, the relevant question for nanotechnology is not "is it safe?" but rather, "what are the safest ways to use it?" Here are some examples of findings drawn from toxicology research that provide some early "safety signals" and also illustrate how complex and sometimes confusing this type of analysis can be.

Carbon Black Nanoparticles

Carbon black is a nanopowder form of carbon that has been used in automobile tires for more than 100 years. Today, carbon black is found in computer printer ink and photocopier ink, and also in fumes from burning diesel and other fuels. In April 2011, researchers reported that carbon black nanoparticles can pose a serious threat when inhaled into the lungs [7]. In addition to inflaming the lungs – which most foreign particles do when inhaled – carbon black particles did not kill lung cells via traditional apoptosis (cell death), but rather caused the cells to explode. This kind of cell death where the cells burst apart is known as *pyroptosis.* The study showed that instead of shrinking on death, the affected lung cells burst open and released their contents, which increased the inflammation, thus producing a secondary effect that makes this contamination especially serious. Martha Monick, professor of Internal Medicine, was the lead author of the study, which was conducted by a team at the University of Iowa's Roy J. and Lucille Carver College of Medicine. The researchers also noted that the concentrations of carbon black used in the study were much higher than the levels to which most people are typically exposed. They noted that these effects would not be expected from simply walking through an area where diesel fumes are present. As in many of these early research studies, the effect on mice and humans could be different, and it is difficult to determine the amount of exposure needed to replicate these effects in humans.

Titanium Dioxide Nanoparticles

Titanium dioxide (TiO_2) nanoparticles are used in skin creams and sunscreens as well as paint, toothpaste, food colors, nutritional supplements, and other products. TiO_2 nanoparticles have been considered harmless because they do not produce a chemical reaction in the body. However, a study at UCLA's Jonsson Comprehensive Cancer Center found that TiO_2 nanoparticles had the potential to cause single- and double-stranded DNA breaks, chromosomal damage, and inflammation in mice. The mice were exposed to TiO_2 particles in their drinking water and began showing genetic damage on the fifth day. Robert Schiestl, the study's senior author, said the UCLA study is the first to show this effect from titanium dioxide. The problem, according to the study, is that the TiO_2 nanoparticles can concentrate in various organs where they can sever DNA and interfere with cellular systems [8]. In another study reported in 2011, a group of Korean researchers found that TiO_2 nanoparticles may inhibit the immune system in mice [9]. These early studies are limited in scope, use higher than normal doses, and do not confirm the extent to which particles may reach certain organs, since the body has various barriers that prevent particles from reaching many organs. This research does not confirm that titanium dioxide poses a health hazard to humans, but it is an early warning signal that suggests further investigation.

Carbon Nanotubes

One of the statistics mentioned in this book is the "world's largest carbon nanotube," which was a technical goal scientists were working on a few years ago. In the industrial world, many forms of nanotubes are used, including long, short, multiwalled, and so on.

Most types of carbon nanotubes, and other types of nanotubes, have had no safety or toxicology issues. In contrast, short and curling nanotubes appear to present minimal or no hazard, since they are typically removed by macrophages. However, several research studies have shown that the needle-like shape of long carbon nanotubes presents a potential health hazard. A seminal study published in *Nature Nanotechnology* in 2008[1] determined that long, thin multiwalled carbon nanotubes behave like asbestos fibers [10]. Asbestos is a proven cause of mesothelioma (cancer of the lining of the lungs).

Current research suggests that care needs to be taken in the handling and use of long carbon nanotubes. The potential dangers posed by long carbon nanotubes require special handling. Many manufacturers who produce and/or use long carbon nanotubes are embedding them in a matrix or substrate that holds them fast and prevents them from separating or flaking off.

For this reason, manufacturers of materials that include long nanotubes embed the CNTs in a liquid or a hard matrix or substrate that prevents the tubes from flaking off. Also, carbon nanotubes can require special handling to minimize health risks to workers. This has given rise to an "intermediaries" sector that produces and/or does the initial processing of CNTs used in industrial or consumer products. Fortunately, high-quality filters used in face masks and ventilation systems can trap most nanoparticles. The airflow and shape of the nanotubes force them against the fibers in the filter, trapping most of them. The current thinking is that carbon nanotubes require special handling, and products

1) An interesting trivia note – one of the authors of this article, Professor Ken Donaldson, from the University of Edinburgh, is credited with being among the first to use the term "nanotoxicology" in a 2004 article, along with Robert Service who first used the term in a 2004 commentary in *Science*.

that incorporate CNTs need to be evaluated with regard to the full product life cycle, including disposal, of the product.

In April 2013, the National Institute for Occupational Safety and Health (NIOSH) published Current Intelligence Bulletin that reports on occupational exposure to carbon nanotubes and nanofibers. A careful review of 54 animal studies revealed that carbon nanotubes and nanofibers can cause a variety of pulmonary problems, including inflammation, granulomas, and pulmonary fibrosis. These studies were deemed relevant to human health risks because effects observed in animals had also been observed in workers who inhaled particulates of other materials in "dusty jobs." CNTs were also found to have "similar or greater potency" to other fibers such as silica, asbestos, and ultrafine carbon black. The report estimated the risk of developing early-stage (slight or mild) lung effects over a working lifetime from 0.5 to 16%.

In addition to controlling exposures before recommended exposure levels, the report gave the following advice: ". . . it is prudent for employers to institute medical surveillance and screening programs for workers who are exposed to CNT and CNF for the purpose of possibly detecting early signs of adverse pulmonary effects including fibrosis."

It's important to note that these guidelines resulted from animal studies and authors were careful to note that "to date, NIOSH is not aware of any reports of adverse health effects in workers producing or using CNT or CNF." As recently as 2009, NIOSH had concluded that there was insufficient evidence to recommend medical tests for workers exposed to these engineered nanoparticles; however, the 2013 report reversed this view and indicated that this evidence now exists. This is indicative of how young the field of nanotoxicology is, and suggests that more evidence and safety recommendations – especially in the workplace – will be forthcoming in the coming decade.

Metallic Nanoparticles

Most nanoparticles are composed of common elements such as carbon or iron, or gold, and these elements in larger forms are routinely processed safely by the human body. In terms of delivery, metallic nanoparticles can be absorbed through the skin, breathed in, or consumed in food. They can move across membranes in the body designed to protect organs and are small enough to enter individual cells. Nanostructures can also sever DNA strands that can potentially cause cancer.

The "ecotoxicity" of nanoscale metals and metal oxide particles has been studied to determine whether nanoparticles are toxic, based on a variety of parameters including size, shape, surface charge, composition, degree of aggregation or dispersion, and associated contaminants. The most frequently studied metals are titanium dioxide, zinc oxide, and silver.

Silver, which has been extensively studied, is a well-known germicide. Silver nanoparticles are used in coatings on work surfaces in hospitals and clinics, because of their antibacterial/antiseptic properties. Ionic silver is impregnated in bandages, to treat wounds, and in other applications. Ionic silver (silver atoms with an electron removed) is also used in appliances such as washing machines, most notably in Samsung's "Silver Nano" appliances. Nanoscale silver is also used as an antimicrobial agent in toilet seats, in ice cube machines, food packaging, socks, and other clothing.

Most researchers agree that silver, gold, and other metallic nanoparticles are not especially toxic to humans or other mammals, although silver nanoparticles have been found to be toxic to fish and other aquatic creatures including beneficial microbes. Research studies have shown that silver nanoparticles can accumulate in drainage, sewage, and wastewater, presenting hazards to aquatic creatures and insects. This can seriously disrupt fragile aquatic ecosystems. The antibacterial effect of silver nanoparticles including ionized silver can kill beneficial bacteria that are needed to break down sewage sludge, and this can interfere with biological treatment systems. Silver nanoparticles that wash down drains and enter aquifers are not removed by sewage treatment systems and can persist in the environment. Most silver eventually degrades to a harmless form in the environment. A key question is how long do reactive forms persist before they become harmless, and what adverse effects might occur?

Ongoing research at Duke University has focused on bioaccumulation of nano silver particles in miniature ecosystems (mesosystems). In 2009, the researchers found that nano-silver can form new chemical compounds that are absorbed by plants, insects, and fish; these can also be passed from parent fish to embryos and can disrupt essential bacteria in aquifers and soils. In March 2013, Duke University researchers reported that even low concentrations of silver nanoparticles can cause significant disruptions to natural ecosystems [11].

Silver nanoparticles can also affect fish and other aquatic creatures. A classical 2009 study of zebrafish at the University of Wisconsin-Madison found that embryos exposed to high doses of silver nanoparticles in the laboratory developed serious malformations and congestive heart failure [12]. Studies that exposed adult trout and juvenile carp to silver nanoparticles over a period of days revealed an accumulation of silver particles in the gills and liver, but not mortality – although longer exposures and larger concentrations could have more serious effects.

Researchers have been studying metallic nanoparticles to measure and confirm whether toxicity results in laboratory tests resulted from the size and structure of the particles, the accumulation and concentration levels, and/or the gradual release of ionic silver. One concern is that silver and other metallic nanoparticles could bioaccumulate silently without being detected in aquifers and other ecosystems. These particles may not kill adult fish, but could have a devastating effect on embryos and reproduction of the species. This is what happened with the pesticide DDT – which worked its way from crops into insects and fish, and were absorbed into the eggs of birds, causing some birds to become endangered. DDT was banned in the United States in 1972 and in most (but not all) countries by 2004 under the Stockholm Convention on Persistent Organic Pollutants.

The lingering concern is that an undetected metallic nanoparticle, especially something as well known and "friendly" as silver, could accumulate in an ecosystem over time and cause environmental damage.

12.3
The Regulatory Climate and Safety Knowledge Gaps

Since the 1990s, the nanotech sector has benefited from an accommodating regulatory climate, a kind of regulatory honeymoon. The general feeling was that if something was safe and in use for many years or decades, it was probably safe if nanosized.

There was justification for early confidence in the safety of nanoparticles, since nanoparticles had been in use in some industries for >100 years. Carbon black is a nanopowder used in automobile and rubber products. Materials made of carbon fibers had been in use for years. Cosmetics used nanoparticles in their skin creams and lotions. Some paints and coatings contain nanoparticles.

A decade ago, some companies called their nanoparticle products "micro." For example, an ingredient listed as "micronized titanium dioxide" in cosmetic creams was actually a nanosized version of titanium dioxide that was ~40 nm in size. Despite early concerns that the small-sized particles could be photoreactive and generate free radicals, the FDA reviewed the available information and determined that nanoparticles of titanium dioxide were not a new ingredient, but a specific grade of the original product.

In the workplace, manufacturers had in place procedures for handling and processing fine powders and aerosols that were larger in size than nanoparticles and these procedures have been mostly deemed adequate for handling nanomaterials. Given the ability of nanoparticles to slip through even the tiniest pores, there was some concern that standard air filters, even high-tech filters, might not be sufficient to screen nanoparticles in face masks and ventilation systems. However, there is research that shows that nanoparticles even in the range of 4–20 nm get trapped in fine filters because the airflow pushes the particles against the matrix of fibers in the filter and traps them there. Still, the example of the Chinese women who died in the paint factory confirms that any workplace where aerosols, powders, and particulates are present needs to be properly ventilated. Workers need to be protected by the best available wearable filters.

Regulation and oversight of nanotechnology has been a weak link in the nanoinnovation process, not intentionally, but because so little was known about the risk factors involved in producing, handling, transporting, and commercializing nanoparticles. In the mid-2000s, the field of nanotoxicology emerged as a discipline. This gave scientists a foundation for their research and helped enable the global community to intensify its nanosafety activities. Europe and the

United States in particular developed a much more focused and integrated effort to conceptualize, map, organize, and conduct meaningful research to determine specific risk factors. Several initiatives were launched to create nanoparticle repositories that could be accessed by the international research community. Guidance was provided by government agencies on safe handling and disposal of nanomaterials. Case studies were conducted. Processes were put in place to evaluate nanotech risks and safety issues.

Conceptualizing Nanosafety

Some of the best visual diagrams portraying expert-hypothesized relationships between risk factors in nanotechnology were developed in 2005 by Kara Morgan. Dr. Morgan did this work as a Senior Advisor for Risk Analysis at the FDA [13]. The diagrams she developed provide a valuable starting point for assessing the safety risks of nanoparticles. These diagrams were developed from a series of conferences, workshops, and interviews with experts using the mental modeling approach that helped identify the individual factors and relationships. Dr. Morgan presented these diagrams as preliminary frameworks for risk analysis of nanoparticles because they did not include data. These diagrams showed how influence diagrams can be designed to show these types of relationships. They provide good templates and examples of how to conceptualize nanotechnology safety factors for any organization that is conducting a nano-safety assessment. The expert-hypothesized relationships represented by the arrows also provided topics for potential research.

Influence diagrams are used to visually portray the relationships between various factors that influence each other, in a system. For example, this framework could be used to assess the risks from carbon nanotubes embedded in a bullet-proof vest, or silver ions released into a stream. Cause and effect relationships are shown by drawing arrows between the elements. The assessment process starts with an overview diagram (see Figure 12.1) that includes a description of the product and how nanomaterials are involved. The next step is to assess the presence of nanomaterials, uptake capacity, transport and fate, and toxic effects – to determine the exposure risks and toxicity. The last step is to evaluate the human health and ecological risks. As Dr. Morgan explained, these factors and relationships are not static. These relationships can change over time, which is why the "time" dimension is shown on the diagram.

These diagrams can be layered or reconfigured to drill down to more detailed analysis, or used to describe hypotheses. For example, the accompanying toxic effects diagram (Figure 12.2) depicts a hypothetical set of relationships that could lead to health effects based on chemical composition, surface coating, surface reactivity, particle size distribution, and adsorption tendency. Linkages can include potential health effects such as inflammation, and genetic or toxic effects in various organs and tissues. Note that these are *conceptual* and preliminary diagrams to provide an idea how these frameworks can be developed as part of the safety assessment process.

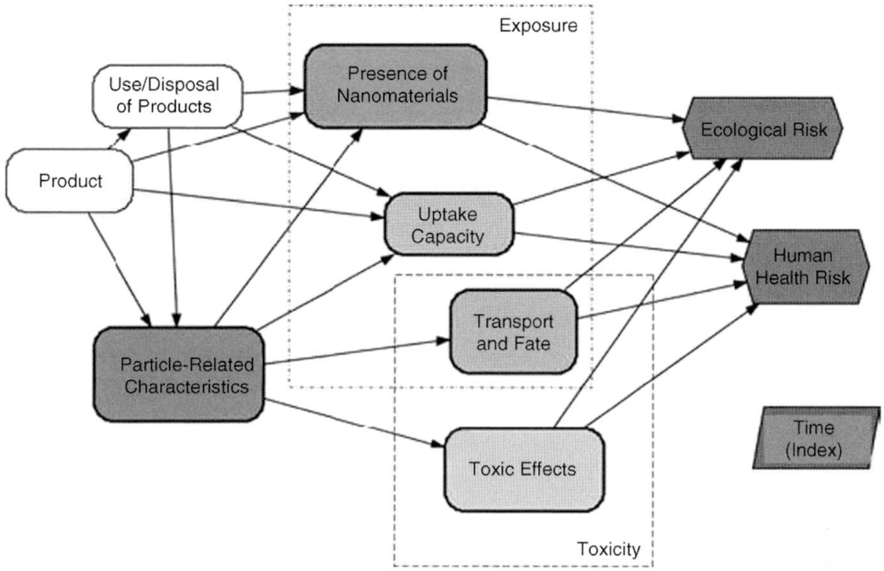

Figure 12.1 This influence diagram developed by Dr. Kara Morgan at the FDA shows the expert-hypothesized high-level relationships between a product containing nanomaterials and/ or nanoparticles and human and ecological risks (used with permission from Ref. [13]).

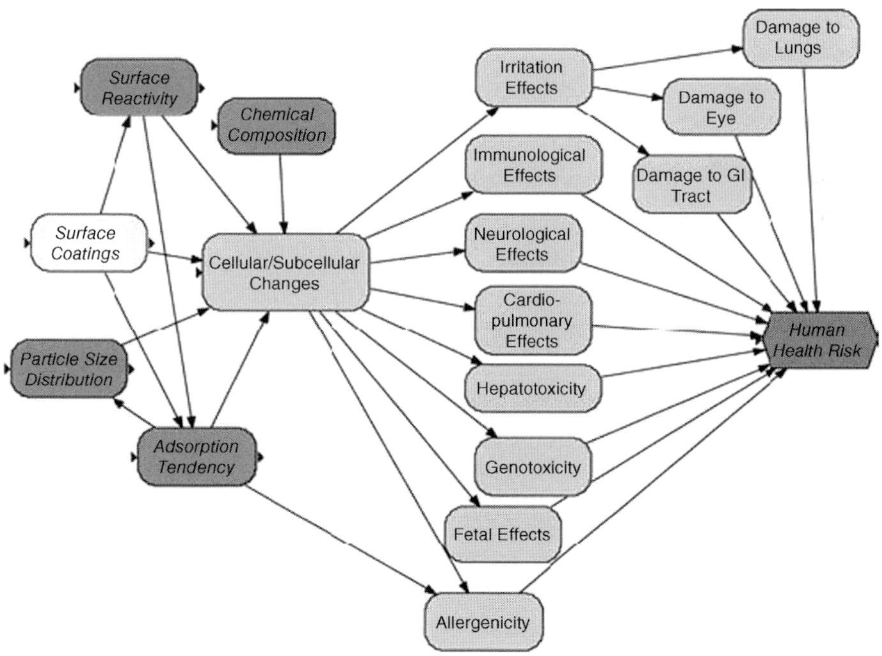

Figure 12.2 This influence diagram shows a set of potential relationships between a product containing nanoparticles and human health risks. It is a mapping tool for conceptualizing nanosafety risks (used with permission from Ref. [13]).

Another important part of safety assessment involves how to dispose of contaminants that may be released (and need to be remediated) when a product such as a cell phone reaches the end of its useful lifespan. Detection and screening of nanomaterials for harmful properties were discussed in a 2013 article [14]. The authors included researchers from Denmark, Norway, and the United States, including Steffen Foss Hansen, Kare Nolde Nielsen, Nina Knudsen, Khara D. Griegerd, and Anders Baun.

In their article, they cited a study by the UK Royal Commission on Environmental Pollution that identified four types of "novel materials" that need to be monitored for environmental impact.

1) New materials hitherto unused or rarely used on an industrial scale.
2) New forms of existing materials with characteristics that differ significantly from familiar or naturally occurring forms (e.g., silver and gold).
3) New applications for existing materials or existing technological products formulated in a new way (e.g., cerium oxide) used as a fuel additive.
4) New pathways and destinations for familiar materials that may enter the environment in forms different from their manufacture and envisaged use.

This taxonomy provides a good framework for categorizing the various types of nanoinnovations that are emerging today. The researchers also discussed five early warning signs that are useful in assessing environmental risks for new technologies and applications: novelty, persistency, whether the material is readily dispersed, whether it bioaccumulates, and whether its use leads to potentially irreversible action. These are useful frameworks, especially since there is still a lot that we really don't know about nanosafety.

A Shift in Nanotechnology Safety Regulations and Policies

In recent years, regulatory agencies have increased their scrutiny of consumer and industrial products using nanoparticles. Some countries have created repositories and inventories of the most commonly used nanoparticles to facilitate risk assessment. Workplace guidelines on safe handling of nanomaterials are being published by the NNI and other agencies.

While the United States took the early lead in development of innovative nanotechnologies, the European Community has exhibited strong leadership in monitoring and tracking nanosafety. The EU has a variety of very pragmatic programs and initiatives designed to ensure that nanosafety issues are included in the nanoinnovation process.

In Europe, nanomaterials are regulated by the European Chemicals Agency as a "chemical substance" under a regulation called REACH (Registration, Evaluation, Authorisation and Restriction of Chemicals). Nanomaterials that may be hazardous are covered by Regulation 1272, which defines the classification, labeling and packaging (CLP) of substances and mixtures. Additional documents have been published to provide more detailed guidance on handling "nanoform" substances, especially those that are considered hazardous. To provide samples

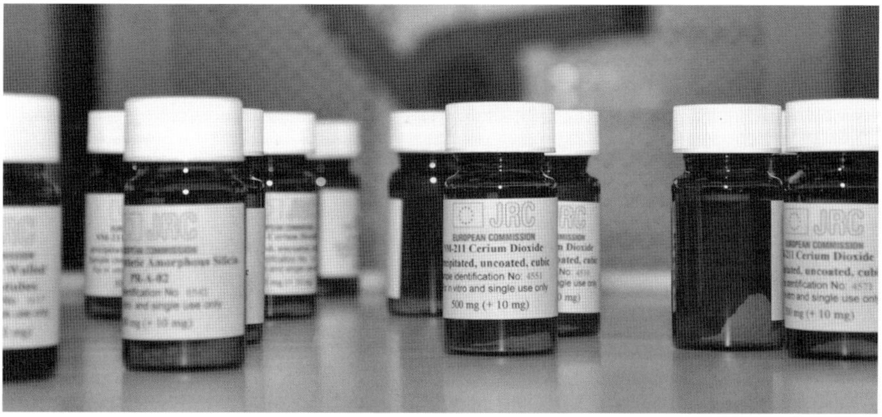

Figure 12.3 In Europe, the Joint Research Centre's nanomaterials repository has created an inventory of nanomaterials. The JRC has provided several thousand vials of nanomaterials to laboratories in France, Germany, the United Kingdom, Belgium, the Netherlands, Denmark, Spain, Poland, Italy, Austria, Slovakia, the United States, Canada, Japan, Korea, China, and Russia (image courtesy of the JRC).

of nanomaterials for testing and evaluation, the Joint Research Centre of the Institute for Health and Consumer Protection created a repository of representative nanomaterials in 2011 (Figure 12.3). A database and information platform called NANOhub hosts several data collections and is designed to consolidate information on nanomaterials that is relevant for safety and risk assessment. A 2012 compendium report [15] describes more than 30 multinational nanosafety projects.

In December 2012, the Danish Consumer Council and Danish Ecological Council, in cooperation with the Technical University of Denmark, established the first nanotechnology database of more than 1200 products that contain or are claimed to be "nano" products. A safety scale is used to rate each product. This database provides a starting point for identifying nanomaterials and ingredients that are potentially hazardous – not only for everyday use in products but also for workers who need to handle carbon nanotubes, chemical liquids or sprays, pesticides, coatings, and other nanomaterials. This rating system also provides a *de facto* incentive for nanoinnovators to make their products as safe as possible so that they can achieve a "green" rating that indicates low risk.

In the United States, a dozen government agencies fund research on nanotechnology and most of these agencies provide regulatory oversight. The agency responsible for workplace safety is the NIOSH. In early 2013, NIOSH recommended minimizing worker exposure to nanomaterials. The agency recommended that occupational exposures to carbon nanotubes and nanofibers be controlled to reduce worker's potential risk to prevent work-related lung effects. NIOSH was the first US federal agency to issue recommended exposure levels for nanomaterials. In a 2013 Current Intelligence Bulletin, NIOSH reported the results of research showing that various types of carbon nanotubes/carbon nanofibers can cause pulmonary fibrosis, inflammatory effects, and granulomas in laboratory animals exposed by inhalation. NIOSH considers these animal study findings to be relevant to human health risk based on similar lung effects observed in workers exposed to respirable particulates of other materials in dusty jobs.

While enforceable laws governing the use of nanoparticles are still very rare, there are early signs that the regulatory mood is shifting toward stronger safety policies. A notable example is the FDA's 2012 Draft Guidance, which requires companies to conduct additional safety tests and document the safety of nanoparticles used in food, food packaging, food additives, and other food-contact applications. The FDA suggests that companies conduct safety tests for nanoparticles and submit documented safety records. This is a departure from the previous approach that recognized food-related nanoparticles as "generally recognized as safe." It is becoming increasingly likely that at some point in the future, there will be tighter regulations governing the use, reporting, and labeling of nanoparticles used in foods, cosmetics, and other consumer products.

Labeling Nano

As previously indicated, nanomaterials have not faced the same labeling issues that hindered the adoption in Europe of food products from genetically modified crops. Without a major public incident or nano-related crisis, nanoinnovations have been able to move fairly smoothly into consumer products and industrial processes. The first mandatory labeling of nanotechnology products – for cosmetic products – was enacted in November 2009 by the Council of the European Union and took effect in July 2013. The European Union mandated that the designation [nano] has to be included on the label of any cosmetics that use nanomaterials. The labels are required not as a warning or to alert consumers, but rather to enhance market surveillance by member states. Despite hints or threats to expand nano labeling requirements, the widespread use of special labeling on packages for nanoparticles and nanomaterials has been very slow to develop.

12.4
Perspectives of Nano-Insiders

While interviewing nano-insiders for this book, many of the interviewees expressed insightful opinions on nanosafety issues. A few of the most interesting comments are included here to provide additional perspectives on this very important aspect of nanoinnovation.

Mark Banash (NanoComp): "One of the smartest things the nanocommunity did to allay public fears was to conduct research on safety issues such as absorption of nanoparticles, possible carcinogenic effects, and impact on the environment. A number of studies suggest nanotubes may pose an inhalation hazard and this is correlated to length. Tubes 20 to 50 microns long may be respirable and resemble asbestos type fibers. However, [at Nanocomp] we have shredded, punctured, twisted, pulled, snapped and ground them up into little balls and sat there with some of the most technological advanced air sampling instruments for air contaminants and we've found that nothing comes off of our materials. With our tubes, we don't see contaminants coming off of our products, primarily because our tubes are in the millimeter range, about 700 microns."

Youseph Yazdi (Johns Hopkins University): "Nanoparticles are ubiquitous in the natural environment. When something is burned or pulverized, you have a natural distribution of particles from large particles to micron and nano-size. In the atmosphere, there are probably nanomaterials that we're breathing all the time – fine nanoparticle soot that firefighters breathe, for instance. Now with modern nanotechnology we can start identifying, filtering, assessing and measuring them in the body. We can identify the accumulation of nanoparticles in various organs.

In the past, there was no way to find out what the size range was or what the effect might be. What's really new in terms of safety is not the existence of nanoparticles. What's new is our ability to understand, measure and manipulate things at the nanoscale. This will help make us safer when exposed to nanoparticles that naturally occur in the environment as well as engineered nanoparticles."

Fred Klaessig, Ph.D. (Pennsylvania BioNano): "We can learn from history. We are using today's methods to evaluate tomorrow's technologies. For example, Dr. Barnes introduced Argyrol, a nano-silver drug, in 1902 and epidemiology testing was done in 1934. They found that silver develops in the body like silver used in photographic paper and film. Silver turns exposed skin bluish-grey. So the method used to determine the safety of silver was the 'blueness test.' If the silver was present in sufficient quantities to turn a person blue, it was deemed unsafe. This is a safety factor using a cosmetic 'test.' Some things that were determined safe based on standards used in the past may yet be confirmed safe using modern technologies for testing nanotoxicity.

Fortunately, many commercial nanomaterials are dissolved by stomach acid or removed by white blood cells, which minimizes the health impact. Still, there is no established testing standard for nanoscale solids for a variety of reasons. Regulatory agencies have been reluctant to identify ingredients based on size. The boundary of 100 nanometers defining 'nanoscale' is arbitrary. A 90 day inhalation study can cost a million dollars. These are some of the 'in the trenches' issues that are still being worked through. We're making progress but much more needs to be done to address future new compositions."

Patrick Ennis, Ph.D. (Intellectual Ventures): "People are running around worrying about whether a nanoparticle is going to give you a disease. People are worrying about a one in a hundred trillion risk. I gave a presentation where I held up a glass of water and said, this glass is more dangerous than anything we have in nanotechnology. It's been around for thousands of years. Think how many ways it can hurt me. I can fill it up with poison. I can break it in half, cut my foot and die from the infection. I could grind it up into glass powder and put it in the air conditioning system and breathe it – that's not good for us, either.

Asbestos, if it gets in your lungs, can give you lung cancer – but if you seal it up and use it properly it has useful functions. Anything is dangerous depending how you use it and how you deploy it."

Duncan Griffiths (Nanosight): "The nanocommunity has been very conscious up front of the potential for damage on the environment or on living beings, and very proactive in studying these risks. From the beginning there have been studies on these risks, but it is difficult – we're having to come up with completely new measurement methods, and health and safety rules for nanoparticles. For example, you can't just run things through a filter because nanoparticles flow through the pores in the filter. How do we test for nanotoxins in our bodies or in the environment? What if some effects come from individual particles, while others may come from aggregations or certain concentrations?"

References

1 Service, R.F. (2004) Nanotoxicology: nanotechnology grows up. *Science*, **304**, 1732–1734.

2 Donaldson, K., Stone, V., Tran, C.L., Kreyling, W., and Born, P.J.A. (2004) Nanotoxicology. *Occupational and Environmental Medicine*, **61**, 727–728.

3 Stern, S.T. and McNeil, S.E. (2007) Nanotechnology safety concerns revisited. *Toxicological Sciences*, **101**, 4–21.

4 Gilbert, N. (2009) Nanoparticle safety in doubt. *Nature*, **460**, 937.

5 Song, Y., Li, X., and Du, X. (2009) *European Respiratory Journal*, **34**, 559–567.

6 Smith, R. (2009) Nanoparticles used in paint could kill, research suggests. *The Telegraph*, August 19.

7 Reisetter, A.C., Stebounova, L.V., Baltrusaitis, J., Powers, L., Gupta, A., Grassian, V.H., and Monick, M.M. (2011) Induction of inflammasome dependent pyroptosis by carbon black nanoparticles. *Journal of Biological Chemistry*, **286**, 21844–21852.

8 (2009) Nanoparticles used in common household items cause genetic damage in mice. *Science Daily*, November 17.

9 Moon, E.-Y., Yi, G.-H., Kang, J.-S., Lim, J.-S., Kim, H.-M., and Pyo, S. (2011) An increase in mouse tumor growth by an *in vivo* immunomodulating effect of titanium dioxide nanoparticles. *Journal of Immunotoxicology*, **8**, 56–67.

10 Poland, C., Duffin, R., Kinloch, I., Maynard, A., Wallace, W.A.H., Seaton, A., Stone, V., Brown, S., MacNee, W., and Donaldson, K. (2008) Carbon nanotubes introduced into the abdominal cavity of mice show asbestos-like pathogenicity in a pilot study. *Nature Nanotechnology*, **3**, 423–428.

11 Colman, B.P. *et al.* (2013) Low concentrations of silver nanoparticles in biosolids cause adverse ecosystem responses under realistic field scenario. *PLoS One*, **8**, e57189.

12 Bar-llan, O., Albrecht, R.M., Fako, V.E., and Furgeson, D.Y. (2009) Toxicity assessments of multisized gold and silver nanoparticles in zebrafish embryos. *Small*, **5**, 1897–1910.

13 Morgan, K. (2005) Development of a preliminary framework for informing the risk analysis and risk management of nanoparticles. *Risk Analysis*, **25**, 1621–1635.

14 Hansen, S.F., Nielsen, K.N., Knudsen, N., Grieger, K.D., and Baun, A. (2012) Operationalization and application of 'early warning signs' to screen nanomaterials for harmful properties. *Environmental Science: Processes and Prospects*, **15**, 190–203.

15 Riediker, M. and Katalagarianakis, G. (2012) Compendium of Projects in the European NanoSafety Cluster, 2012 edition, February.

13
Prologue to the Future What's Next?: Predictions and Possibilities

> It's a good thing to turn your mind upside down now and then, like an hour-glass, to let the particles run the other way.
>
> – *Christopher Morley,* The Haunted Bookshop, *1919*

Most books start with a prologue. This book ends with a prologue, because this is where the future begins – a prologue to the future.

As I was interviewing nano-insiders for this book, I asked many of them to offer predictions of radical nanoinnovations that might be possible to achieve in ~10 years. These informed opinions were combined with my own research and blended with my personal views as a technology futurist. The result is a list of radical nanoinnovations that need to be on your radar screen. If successful, these innovations have the potential to replace existing technologies and applications, transform industries, and create entirely new markets.

While these may be considered predictions, it is more appropriate to consider them as *possibilities.* With the proper resources, funding, perseverance, and luck, these nanoinnovations can be commercially available in about a decade. Which of these will succeed and which will fail depend on many factors. The important thing is that these emerging technologies and applications hold great promise. Many of these innovations are already finding their way from the research laboratory to the marketplace. This is a dynamic list and it will certainly change in the coming decade and beyond. I'm sure that you can find examples that are specific to your interests, your industry, or company. For now, these examples offer some intriguing glimpses of the future of nanoinnovation.

Aerial Drones and Robots

You may have seen YouTube videos showing quadrotors flying in different formations, in a gymnasium. This system was developed at the University of Pennsylvania. In 2012, while I was serving as managing director of the Mack Center for Technological Innovation at the Wharton School, we cosponsored the $5000 Y-Prize competition for the best business and technology plan for a robotics innovation. The winning proposal was a plan to use quadrotors with ground-penetrating radar to find improvised explosive devices (IEDs) to protect combat troops.

NanoInnovation: What Every Manager Needs to Know, First Edition. Michael Tomczyk.

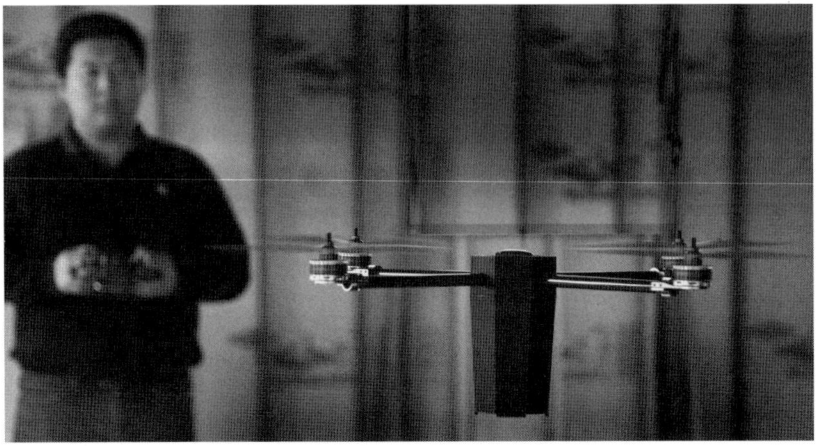

Figure 13.1 Dick Zhang, CEO and cofounder of Identified Technologies, operates a prototype quadrotor. The computer-controlled multicopter can sense and gather data where it's too dangerous or physically impossible for humans to go.

The prizewinners were engineering undergraduates Dick Zhang and Kelsey Duncombe-Smith, and doctoral candidate Andy Wu, who have subsequently launched a venture called IdentifIED Technologies Corporation (Figure 13.1). I have been privileged to advise the new venture, which is currently refining their prototype and developing a variety of industrial applications that use quadrotors to aerially inspect areas where humans cannot go. Dick indicated that their quadrotor uses lithium-polymer batteries with a LiCo nanotech substrate.

Multirotor devices (called multicopters) have been featured in games such as Call of Duty, but they also exist in real life and are being developed by real military research teams, as well as entrepreneurs and inventors. You can buy multicopter kits now on the Internet. The popularity of these devices and the implications for privacy and security have raised a variety of issues, including how and where government agencies will allow them to be flown.

Cameras of the Future

Having started my career as a photojournalist, I have keen interest in one of the most intriguing innovations in photography – the ability to change the focus of a picture *after* you take the photograph. This would eliminate all those out-of-focus pictures we took on our vacation, those blurry family photos, or a photo of a news event that was taken quickly and is out of focus. Imagine being able to adjust the focus to make everything clear *after* the photo is taken. This sounds like science fiction, but it's really true.

The Lytro Light Field Camera (Figure 13.2) allows you to adjust the depth of field (focus point) of a photograph after the photo is taken, because the camera captures the *entire light field* in the photo, instead of a two-dimensional focal plane. This allows the photographer to refocus the picture *after* it's taken using a feature Lytro calls "Perspective Shift." Not only does this capability eliminate

Figure 13.2 The Lytro Camera has revolutionized photography by allowing photographers to adjust the focus of a picture *after* it's taken (photo courtesy of Lytro – Copyright 2014, Lytro).

the problem of out-of-focus pictures, but it also allows the photographer to shift the focal area to any point in the picture. The next step for this technology, which may require nanotechnology to shrink the components, will be to integrate light field cameras in smartphones, tablets, and laptops.

Cancer-fighting Cosmetic Creams

What if a moisturizing skin cream or lotion could also be a gene therapy product, and what if this cream could turn off the genes that cause skin cancer? A research report in 2012 at the Proceedings of the National Academy of Sciences described how Chad Mirkin, Amy Paller, and their colleagues at Northwestern University used nanoparticles to turn off genes that cause skin cancer and other diseases. Their innovation involves an engineered nanoparticle consisting of small strands of RNA packed around a gold core to switch off genes that are linked to skin diseases. Their nanoparticle, which they call a "spherical nucleic acid nanoparticle conjugate," is wrapped around gold particles that are only 13 nm in diameter.

Their goal was to suppress epidermal growth factor receptor (EGFR), a gene that controls the growth of skin cells. This could provide an entirely new way to treat skin cancer, psoriasis, or other gene-related skin diseases. Some drugs exist today that target genes associated with skin disease, but these are usually administered in a pill or intravenously. The epidermis (outer layer) of human skin is structured to prevent foreign particles from penetrating the skin. Wrapping small interfering RNA (siRNA) around gold particles combined with a commercial moisturizer was able to penetrate the skin barrier. The researchers were able to get the particles to penetrate almost 100% of human skin cells *in vitro*, and to suppress the EGFR genes in various mice and human skin samples from 52 to

90% – which almost totally cleared the skin within 10 days. The EGFR gene application was developed as a proof of concept. Other applications include the potential to turn off skin genes associated with other diseases, even aging. The technology is being developed for commercial use by AuraSense Therapeutics.

Diagnostic Techwear

Nike, Samsung, Under Armour, and other innovative techwear companies are beginning to market monitoring straps, bracelets, and watches for athletes and fitness enthusiasts. These wearable computers are easily converted to medical applications that will add a new dimension and enable prevention as well as early self-diagnosis of many biometric symptoms. What are required to move to the next wave are nanocircuits, flexible electronics, nanomaterials, and nanosensors that record biometrics.

Electronic Devices Will Have New Form Factors

We've already seen the next generation of nano-enabled form factors, including phones that morph into bracelets (Nokia), bracelets that monitor fitness statistics (Adidas, Nike, and United Armour), smarter than ever smartphones (Samsung and Apple), and paper thin tablets (Apple, Asus, and Lenovo). The first flexible large screens were previewed at a private event at the January 2014 Consumer Electronics Show. We've also seen the first wave of long-life battery-powered devices that will eventually last 1 week, and then 1 month, on a single charge. Another feature that will soon be available involves transparent electronics, where the circuits are so thin (nanoscale) that they become invisible, which means they can be printed or sandwiched in paper-thin form factors, such as paper-thin computers or sensors, and functional tattoos or electronic nanoskin.

When I was helping to design the first home computers at Commodore in the early 1980s, I predicted that portable computers would eventually come in configurations that would "roll up, scroll out, bend and fold." In November 2013, Samsung previewed a prototype of an integrated phone/tablet device (phablet) that can be folded to the size of a wallet and unfolded to the size of a tablet. The device uses a new plastic substrate and an active-matrix organic light-emitting diode (AMOLED) display that uses nanoparticles. Samsung has announced plans to introduce folding phones and other futuristic devices by 2015.

Fashionable Nanowear

The integration of nanotechnology and textiles is well underway. Some of the newest innovations in this space include clothing that changes color or texture depending on the weather, temperature, or how we move. In the coming decade, fabrics that mimic the scales of a butterfly will allow the costumes of contestants on Dancing with the Stars to change color as they dance – maybe red for the jerky moves of a quick step or blue for a slow tango. A gym suit or sweatpants might change color when we run or exercise, powered by muscle tensing and relaxing as we move. On the Red Carpet at the Academy Awards, a movie actress might wear a silk dress that shimmers or sparkles, or changes dramatically as she moves, thanks to embedded nanoparticles.

Nanofabrics and nanoskins will give a future Lady Gaga some really trend-setting fashion ideas. Perhaps her tunic or dress will be made from flexible electronics that allow the costume to morph into different shapes as well as different colors and patterns. Several years ago, Phillips Lighting demonstrated fabrics with built-in LED lights that play an animated video. Future versions may be powered not by batteries but by piezoelectric energy drawn from the body's own muscle movements or pulsing arteries.

I believe that the day is not too distant when a college student walking on campus will wear a jacket that displays an animated Coca-Cola logo that lights up and keeps moving and changing as she walks to class. When her classmate asks, "Where did you get that cool jacket?" she will respond, "It's free. I got it from Coke, but I had to agree to keep the battery charged so the logo lights up when I wear it."

Google Glass: A New Dimension in Wearable Computers

Most of the functions that are currently available in smartphones are rapidly migrating to eyeglasses. This new dimension is being pioneered by Google through an awesome innovation called Google Glass – a lightweight frame with a visual display that contains all the functions currently available in a smartphone (Figure 13.3). The frame rests above the eyes and provides a visual display in the upper corner of the wearer's field of vision. It can be used with prescription or sun glasses, or by itself. The user interface responds to tapes and gestures, a touchbar, and voice commands. Google's Mirror API facilitates development of software apps that are called "glassware."

Design enhancements that may be integrated in future models include GPS, sound transmitted through bone conduction, facial recognition, and the ability to take photos simply by blinking. The game community led by Oculus Rift is exploring ways to combine reality with animation to create awesome simulations that are being called "augmented reality."

Google Glass was cocreated by Project Glass lead Babak Parviz; Steve Lee, a project manager and geolocation specialist; and Sebastian Thrun, who developed Udacity and worked on Google's self-driving car project. The first Google prototypes were introduced to early adopter "lottery winners" in 2013. Google is building a growing community of Glass users called Explorers who are helping to develop, test, and comment on this and other innovations. Google has set up Basecamps in several cities where Glass Explorers can explore the features and experiment with new apps. This is one way that anyone can become a nanoinnovator, or in this case, a Google innovator.

Of course, this technology raises privacy issues. For example, movie theatres are worried that film pirates will use the glasses to create pirated copies. Gaming authorities have issued directives cautioning casinos that this technology could be used to facilitate cheating. Banks are concerned that the glasses could film ATM users as they enter their PIN numbers to withdraw cash. Some strip clubs in the United States have already established rules requiring patrons to check

Figure 13.3 Google Glass devices can be worn with prescription glasses or sunglasses, and can be customized for special applications such as athletics and virtual games that create augmented reality (photos courtesy of Google, Inc.).

their Google Glass devices before entering the club. Regardless of these concerns, Google Glass is proving that "seeing is achieving."

Graphene and Other 2D nanomaterials

Nanomaterials that are one atom thick are called "2D materials," although most of the actual materials are given the suffix "-ene" like graph*ene*. Graphene was the first 2D nanomaterial and will be the first 2D material to be available commercially. In 2013, Samsung Electronics demonstrated a 30″ wide sheet of graphene. Other 2D materials include single atom layer tin – which is called "stanene." Stanene could be the world's first material that can conduct electricity with 100% efficiency at room temperature, according to a team from Stanford University, the US Department of Energy, Tsinghua University in Beijing, and the Max Planck Institute for Chemical Physics of Solids in Germany [1]. The researchers coined the term "stanene" from the Latin name for tin, which is "stannum" and the suffix "-ene," which is being used to describe materials that are single atom thick. Adding fluorine atoms to a layer of stanene could allow the material to conduct electricity at 100 °C (212 °F). The scientists created stanene as a simulation and still need to create it in a laboratory to show proof of concept and validate its commercial potential. Other examples of 2D nanomaterials that are being developed include phosphorene and silicene [2].

Military "Iron Man" Suits

Throughout history, wars have been won by innovations in combat technology, from the use of English longbows in 1415 at the Battle of Agincourt to stealth bombers and night combat used in the first Iraq War (1991). The next war will likely include state-of-the-art heads-up helmet displays that deliver a full spectrum of combat information to combat fighter pilots as well as on-the-ground infantry – sort of a military version of Google Glass. Infantry troops will wear powered exoskeletons that allow each soldier to carry weapons and equipment weighing hundreds of pounds – a skeletal version of this is the Lockheed-Martin HULC, which is available today. Nextgen versions will be full body combat suits that will make soldiers look like a cross between Iron Man and Robocop.

Nano-enabled composite materials are currently being incorporated in super-strength combat suits that completely enclose a soldier to support mobile combat. These suits include lightweight body armor, built-in portable weapons systems, target identification (for aircraft and artillery), communications, climate control, medical diagnostics, bioweapon filters, and more. These suits also have the ability to identify a broken bone or wound and surround the break with a cast or stabilize a bleeding wound. In 2014, RevMedX announced an innovation called XStat that treats battlefield wounds with tiny sponges that expand to stabilize combat wounds – a variation of this technology could be incorporated into a combat suit.

For troops fighting in the desert where temperatures can swing from baking to freezing, the suit stabilizes the temperature, like an astronaut's space suit. They can link to overhead surveillance drones to provide real-time over-the-horizon images of enemy positions and targets.

Heads-up displays in the soldier's helmet can show infrared images of enemy soldiers for night fighting, while using stealth technology to block visual and infrared detection by the enemy. These suits, which exist in experimental military programs, use lightweight carbon composites that are already used in body armor, flexible electronics, and solar panels enabled by nanotechnology. It's impossible to predict which country's army will use these innovations or the first war where they will be used, but it is fairly certain that at some point tactical combat suits will begin to resemble the full body suits worn by the armies of clone troopers in Star Wars.

Nanoskins for Buildings

Nanoskins for buildings are beginning to find commercial applications. Covering buildings or walls of rooms with nanoskins will increasingly allow control of climate, light, sound, and temperature and will help enable the construction of net-zero energy use buildings. The first applications of nanocoatings are designed to protect the surfaces of buildings and machines from rain, snow, sand, dust, and pollutants.

Baiku, the capital of Azerbaijan, is known as the "land of fire," so it is appropriate that the bold new Socar Tower is designed in the shape of a stylized flame. The tower is designed to withstand winds up to 190 km/h and is able to resist an earthquake up to 9 on the Richter scale. To protect the building from the elements, glass surfaces are covered with a nanoglass coating called NG-1314, which protects the outer surfaces and reduces the amount of dust that will accumulate on the glass. The weather-resistant coating was developed by Nanovations Pty Ltd., an Australian nanotechnology product manufacturer. Nanovations specializes in advanced ultrathin coating solutions for glass surfaces exposed to harsh environments. The company's glass coatings cover up to 15 times more surface area than other types of coatings, which reduces the cost while enhancing performance.

Paint-On Batteries Provide "Thin Power"

In 2012, a group of researchers from Rice University announced the development of a "paint-on" lithium-ion battery. The team included Neelam Singh, Charudatta Galanda, Andrea Miranda, Akshay Mathkar, Wei Gao, Arava Leela, Mohana Reddy, Alexandru Vlad, and Pulickel M. Ajayan. In their seminal research paper, the team noted that the use of batteries has been traditionally constrained to a cylindrical or rectangular form factor, although thin, flexible, and stretchable batteries are beginning to emerge. The Rice group developed a spray paint using a combination of single-walled nanotubes (SWNTs), lithium cobalt oxide, Super P^{TM} carbon, and ultrafine graphite into polyvinylidine fluoride and 1-methyl-2-pyrrolidone.

The group "painted" their battery material on ceramic tiles, spelling the word "RICE" in LED lights. They also painted batteries on glass, stainless steel, plastic, and on the curved side of a beer mug. One of the tiles was fitted to a solar power cell that demonstrated the feasibility of using solar power to recharge the spray painted batteries. Their early tests have successfully powered LED lights for 6 h.

It is easy to envision spray-on solar paints and coatings where solar cells are painted on instead of installed. One application involves integrating and embedding LED lights into the battery coating to produce painted-on lighting surfaces. In the future, an entire side of a building might be painted with a high-tech coating that combines painted on batteries and LED pixels integrated with motion sensors. The walls of a prison might automatically light up if an escaped prisoner is detected. An office or building might light up if an unauthorized intruder walks into a high-security facility such as a nuclear power plant or public utility.

Neelam Singh, a materials scientist, has predicted that consumers may one day use paint-on batteries just like we now paint houses and cars. The current design contains toxic materials that prevent use outdoors, however, the team is working on more environment-friendly versions that will be more suitable for use outdoors, along with improved coupling to solar cells.

Peel-and-Stick Thin-Film Solar Cells

Science is on track to developing "peel-and-stick" thin-film solar cells <1 μm thick that can be applied to almost any substrate and used to power a wide variety of products from solar cells to portable electronics and even military devices.

In January 2013, this innovation was demonstrated by Qi Wang, principal scientist, from the National Renewable Energy Laboratory of the US Department of Energy (NREL), and Xiaolin Zheng at Stanford University. The two collaborators met at a conference in 2012 where Dr. Wang was giving a presentation on solar cells and Dr. Zheng discussed her peel-and-stick technology, developed with NREL engineer William Nemeth. Less than a year later, the collaboration resulted in a demonstration of a peel-and-stick solar film powering a solar cell, which was reported in *Scientific Reports* (*Nature*) [3].

It is intriguing to think that one day we may buy solar cells in "peel-and-stick" sheets that can be applied to almost any portable electronic device, and used to power those devices by exposing them to sunlight.

Self-Healing Concrete

When we think about technological innovation, most of us don't think much about concrete. It's not as glamorous as building atomic scale computer chips or using nanoparticles to cure cancer, but it's a huge business and it's ubiquitous. Nanotechnology has already enabled improvements in the formulation of building materials, and more changes are looming on the horizon, specifically in concrete.

Imagine a highway with a lot of heavy truck traffic that begins forming cracks in the roadbed. But in this road, the cracks heal themselves. Imagine that buildings and bridges that deteriorate over time are able to resist corrosion from salt and pollution and are also able to self-heal cracks as they form.

"The Romans were using concrete 2,000 years ago – they even used iron to reinforce the concrete, which builders do today," observes Patrick Ennis, a venture capitalist at Intellectual Ventures who was an early venture capital pioneer at Arch Ventures. "If you brought a Roman engineer from 200 B.C. here and showed him concrete, he would know what it is and how to make it. Show him an i-phone and he'll think you're a magician. In another few years, we'll be seeing some nano-enabled changes in concrete that will seem like magic. These will be the first major changes in the formulation of concrete, in more than 2,000 years."

"I think we'll see a next generation of concrete that will be superior across every attribute from strength to weight," Ennis predicts. "It will have resistance to salt, which a huge issue in the winter. It will be self-healing. When a crack begins to form in the concrete, it will heal before it can proliferate. In critical applications such as concrete roads where heavy trucking damages the roadbed, an acute crack in the road will heal itself."

If we extend this metaphor to other areas where "self-healing materials" would be extremely valuable, we can point to research that is underway now to produce self-healing aircraft wings, self-healing roads, and self-healing shelters for extreme environments.

Sensors That Can Detect Anything

We are fast developing the ability to inexpensively, efficiently, and wirelessly detect almost anything – from motion, light, temperature, vibration and humidity, to chemical pollutants and poisons. In the coming decade, virtually anything that can be measured will be measured.

The best place to see this sensor revolution is in your mobile phone. If you have an Android device, you can download a plug-in from iMobLife called Sensor Box that detects sensors in the device. The number and type of sensors in your phone may surprise you. The Samsung Galaxy S4, made in South Korea and introduced in 2013, was the first major smartphone to include sensors to measure pressure, temperature, and humidity as well as a gyroscope that lets you autorotate the screen. Subsequent generations have even more sensors. New smartphones contain fingerprint sensors for biometric identification and eye scanning sensors that keep the screen turned on as long as the user is viewing it. Gestural sensors allow phone users to navigate the screen without actually touching the screen. In the coming decade, nanosensors will enhance robotic devices such as robots that mimic human movements, reactions, and intelligence as well as insect-sized flying drones and swarms of sensors.

Star Trek-Style Replicators Using 3D and 4D Printing

In just a few years, the price of a desktop 3D printer has fallen from several thousand dollars to <\$500 for a small unit used by hobbyists and students. More than a dozen companies market 3D printers. NASA has an open innovation project designed to identify new ideas for using 3D printing to engineer components, make repairs, and make food from raw materials – which will be needed by future space missions. In the future, astronauts will use a 3D printer on Mars or on the surface of the moon to create a custom bolt, a nail, or anything structural that might be needed there. The raw materials may include nanomaterials.

Medical researchers such as Anthony Atala are already using 3D printers to spray stem cells on scaffolds to regenerate human organs. In the coming decade, 3D printers will allow us to "print" products at home. A company will send us a code and maybe a raw material kit or some cartridges that we'll plug into our 3D printer and – presto – the product will be produced in our home or office, instead of being delivered.

A few years ago, while earning a master's degree in environmental studies, I took a nanotechnology course taught by Dr. Jody Roberts at the University of Pennsylvania. During one of his lectures, Jody mused, "Imagine what would happen if nanoparticles of carbon are used as the feedstock for a 3D printer – this could be used to create food, fuel, almost anything organic, like the replicators on Star Trek." I've always thought this was a terrific concept and probably one of the ways that a Star Trek replicator could be engineered. Today, hobbyists are using 3D printers to design prototypes, which are easy to design because they can vary the dimensions, colors, or materials until they reach the ideal configuration. Chefs in New York have used 3D printers to create food from raw materials. The 3D printers now use three or more materials to create such sophisticated items as working batteries.

The 3D printers are just beginning to gain a foothold among early adopters in the marketplace, and while this is happening, the technology is being pushed and expanded in new directions. One variation is called 4D printing – the "fourth dimension" means that 3D printed objects have the ability to change, mutate, expand, or somehow reconfigure themselves after they are created, when exposed to air, water, temperature, chemicals, or other change catalysts. Add nanoparticles and nanomaterials to the mix and this fourth dimension can include novel properties that are only available at the nanoscale.

Tattoos Become Functional

Hundreds of millions of people have decorative tattoos – as many as 100 million people in the United States alone. In the coming decade, tattoos will evolve from decorative adornment to functional application.

Dr. Robert Langer at MIT predicts that one day "we will have tattoos you can wear that will tell you if you're sick – basically you'd have nanoparticles that change their nature based on signals in the body and change depending if you have a specific kind of marker that signals a disease. The tattoo might change color, for example. With the right technologies, you could detect anything from heart disease to cancer."

There are many ways to incorporate nanosensors or bionanosensors in tattoos that respond to changes in skin chemistry. This capability suggests a wide variety of novel applications. For example, a tattoo may change color when a biomarker is detected that is related to a disease. Children or adults with a high risk of acquiring a disease can be given a tattoo that turns color if the disease is present. Researchers are using nanotechnology to develop biosensor tattoos that measure glucose levels, tattoos that act as lie detectors, and electronic tattoos that can communicate with wireless devices. Tattoos can deliver drugs that need to be administered gradually by absorption through the skin. One day soon, you may go to the office or to school and a friend will say, "Uh-oh, your neck tattoo is turning blue. You better go get a doctor." It could be as simple as that.

Universal Gene Testing

As discussed earlier in the book, genetic tests are well on their way to becoming a standard test for cancer patients, and are being used to diagnose patients with known hereditary risks for breast cancer and other diseases. As individual genomes become more affordable and reliable, genomic tests will become standard.

Molecular scientist David Bachinsky predicts, "I imagine that someday you'll come in sick to your doctor's office, he will take a swab, sequence the information, and you'll get a solution based on the genetic analysis." He also predicts that babies will get a genetic sequence when they're born, and this will provide a baseline that will be used to measure changes throughout their lives. Genetic analysis will be a major success story in the coming decade, although it's questionable whether the analysis will be provided by a supercomputer in a laboratory, a desktop system, or a portable "lab on a chip."

13.1
Keeping Nanoinnovation on Your Radar Screen

These are only a few examples of the many hundreds of nanoinnovations and nano-enabled innovations that are looming on the near horizon. It is truly amazing how many of these innovations have the ability to dramatically transform industries and markets, and improve our lives. Nanoinnovation is changing the world.

Whoever you are, whatever your role in life, you need to know about this technological revolution. You need to keep nanoinnovation on your radar screen. Find ways to get involved and support this technological renaissance. Together, we can help nanoinnovation to fulfill its remarkable potential.

References

1 Xu, Y., Yan, B., Zhang, H.-J., Wang, J., Xu, G., tang, P., Duan, W., and Zhang, S.-C. (2013) Large-gap quantum spin hall insulators in tin films. *Physical Review Letters*, **111**, 136804.

2 Barras, C. (2014) Graphene rival 'phosphorene' is born to be a transistor, *New Scientist*, January.

3 Lee, C.H., Kim, D.R., Cho, I.S., Nemeth, W., Wang, Q., and Zeng, X. (2012) Peel-and-stick: fabricating thin film solar cell on universal substrates. *Scientific Reports*. doi: 10.1038/srep01000.

Appendix A
Answers to the Nano I.Q. Quiz

1. Which of these consumer products use nanotechnology?

   ☐ Dockers jeans
   ☐ Sunscreen
   ☐ Toyota Prius
   ☐ Computers and Cell phones
   ✓ All of the above

All of the Above use nanotechnology. Dockers "Go" khakis were reported to use protective nanocoatings in 2004; today, several clothing manufacturers use nanocoatings. Sunscreens are probably the largest user of nanoparticles; more than 300 sunscreen products use nanoparticle formulations of titanium oxide or zinc oxide to make their creams, which help make their lotions clear instead of white. Most cars and trucks use nanocomposites in automotive lubricants, exterior paints, engine coatings, seat covers, batteries, capacitors, computer systems, and much more. Computers and other electronic devices have used nanoscale architectures in semiconductors for more than a decade; computers and flat-screen monitors use nanocoatings. These are only a few examples.

2. In 2014, approximately how many consumer products used nanotechnologies (according to the Woodrow Wilson project)?

   ☐ <10
   ☐ 100
   ☐ 560
   ☐ 1300
   ✓ >1800

The Emerging Nanotechnologies Project of the Woodrow Wilson Project/Pew Charitable Trusts inventory of consumer products reported to contain or utilize nanotechnology reached 1317 products in March 2011, the last year the consumer products inventory was updated. Based on the trend peline, the site projected that the total would exceed 1800 products by 2013. The real number is undoubtedly much higher, given the very large number of materials, cosmetics, packaging, composites, catalysts, and other categories that have emerged just in the past few years.

NanoInnovation: What Every Manager Needs to Know, First Edition. Michael Tomczyk.

3. The term "nanotechnology" was first coined by

   - ☐ Richard Smalley
   - ✓ Norio Taniguchi
   - ☐ K. Eric Drexler
   - ☐ Richard Feynman

Although Eric Drexler has been credited for independently coining and promoting the term nanotechnology *in the United States* (in his 1986 book *Engines of Creation: The Coming Era of Nanotechnology*), the term was actually first coined in 1974 by Norio Taniguchi of Tokyo Science University. Technically, Professor Taniguchi can be credited for being first in the world to originate the term, while Dr. Drexler is credited for introducing the term in the United States.

4. This Nobel Prize winning scientist gave a Caltech lecture in 1959 entitled "There's Plenty of Room at the Bottom" – setting the scientific community on a course of discovery that gave rise to the field of nanotechnology.

   - ☐ Richard Smalley
   - ☐ Carl Sagan
   - ☐ K. Eric Drexler
   - ✓ Richard Feynman

Richard Feynman (1918–1988) is credited with sparking the movement to manipulate and control matter at the atomic and molecular scale – the field we now call nanotechnology. He issued a challenge in his speech to a meeting of the American Physical Society at Caltech in December 1959. He shared the Nobel Prize in 1965 for his work on quantum electrodynamics and was one of the first scientists to conceptualize quantum computers. As part of his presentation, he offered two $1000 prizes: one for building a tiny motor (which was achieved using conventional technology not nanotech) and the other, to reproduce information from a book page on a surface at 1/25 000 the original scale. The second prize was won in 1985 by a Stanford graduate student who reduced the first paragraph from *A Tale of Two Cities.*

5. The scanning tunneling microscope (SEM) capable of imaging the nanoscale was invented in

   - ☐ Germany
   - ☐ Japan
   - ✓ Switzerland
   - ☐ The United States

It is wonderfully ironic that the scanning tunneling microscope was invented in Switzerland, home of precision instruments and clocks. The scanning tunneling microscope was invented in 1981 at IBM's Zurich Research Laboratory by Gerd Binnig and Heinrich Rohrer, who shared the Nobel Prize in 1986 for their invention. The STM can image individual atoms on a metal or semiconductor surface by scanning the tip of a needle over the surface. This invention opened the entire field of nanotechnology by using electron beams to view structures and processes at the nanoscale, which could not be viewed using optical microscopes. Dr. Binnig is still on the research staff at the Zurich laboratory; Dr. Rohrer retired from IBM in 1997.

6. What was the first recognizable image created by manipulating individual atoms?

   ✓ The IBM logo
   ☐ A Mickey Mouse cartoon from 1929
   ☐ The yin and yang symbol
   ☐ A yellow "happy face"

On November 11,1989, Donald Eigler became the first to manipulate individual atoms with nanoscale precision, a truly historic achievement in the history of physics and nanoinnovation. Using a low-temperature, high-vacuum scanning tunneling microscope that he built, he positioned 35 xenon atoms to spell the letters "IBM" – the world's smallest logo. Dr. Eigler is a physicist and IBM Fellow, and corecipient of the Kavli Prize, the highest scientific award in nanoscience.

7. The first US president to have his image rendered at the nanoscale is

   ☐ George W. Bush
   ☐ William Clinton
   ✓ Barack Obama
   ☐ Ronald Reagan
   ☐ None of the above

The image of President Barack Obama on the cover of this book is the first image of a US president to be rendered at the nanoscale – although technically, President Obama was still a candidate in 2008 when the image was created. Anastasios John Hart at the University of Michigan used nanolithography to create the image by growing carbon nanotubes on a semiconductor substrate, using a masking process similar to how semiconductor chips are made. Each nanoscale image is comprised of 150 million carbon nanotubes. The nanoimage was adapted from a graphic by artist Shepherd Ferry based on a photo taken at the National Press Club in 2006 by AP photographer Mannie Garcia. These images and a detailed explanation of the process can be found at www.nanobama.com.

8. Nanoparticles of gold (below 100 nm) are what color?

   ☐ Gold
   ☐ Silver
   ✓ Red or purple
   ☐ Black

Particles of gold appear red at 100 nm or less, and can also exhibit other colors such as green or purple, depending on the size that determines the wavelength of light that is reflected to the viewer. This is an excellent example of how materials exhibit novel properties at the nanoscale that are different from their macro properties. Historically, nanobits of gold were used in stained glass windows to create vibrant red images. Nanobits of colloidal gold suspended in a fluid such as water may appear red or purple. Colloidal gold and silver in the Lycurgus Cup (fourth century) causes a red/green color-shifting effect (the glass is green when viewed normally, but turns red when a light such as a candle is shone through the glass). The Lycurgus Cup can be seen in the British Museum in London.

9. Nanoparticles of iron oxide – called "nanorust" – can be used to

☐ remove bacteria from washing machines
☐ desalinize seawater
✓ remove toxic arsenic from drinking water
☐ kill cockroaches

Iron oxide – more commonly known as "rust" – has long been known to bind particles of arsenic, which is a heavy metal that contaminates drinking water in many regions of the world – most notably in Bangladesh where as much as 80% of all groundwater may be tainted. Researchers at Rice University have discovered that nanoparticles of iron oxide, known as "nanorust," can bind with arsenic in drinking water, which can then be removed using an ordinary magnet. The team, led by Prof. Vickie Colvin, is part of the Center for Biological and Environmental Nanotechnology (CBEN) at Rice.

10. Buckyballs and nanotubes are forms of

☐ gold
☐ silicon
✓ carbon
☐ semiconductor circuits

Buckyballs are a nickname for "buckminster fullerenes," which are a form of carbon molecule comprised of 60 carbon atoms arranged in what looks like a soccer ball or geodesic sphere. Buckyballs were discovered by the late Richard Smalley, a professor at Rice University who was awarded the Nobel Prize for his discovery in 1985. The molecule was named after famed architect Buckminster Fuller, who used geodesic shapes in his designs.

11. Which country holds the most nanotechnology patents?

☐ Japan
☐ Germany
☐ India
☐ South Korea
✓ The United States

The United States is in the leading patent position with over 40% of all applications, followed by Japan (17%) and Germany (10%) (*Source:* OECD, 2009). The total number of US patents that include "nano" in the title and/or text exceeds 24 000 – this excludes nanotechnology patents that do not include the term "nano," which includes patents that may involve quantum dots, fullerenes, biomarkers, and other nanoscale inventions. A record 4400 of these "nano" patents were granted in 2009, a year to year increase of more than 40% from 2008. At the end of 2009, there were more than 49 000 US patents pending that referenced nano or nanotube. Alton Parrish, who tracks these statistics on his Nano Patents and Innovations blog, estimates that there may be more than 73 000 nanotechnology patents at year-end 2009 (*Source:* Nano Patents and Innovations, December 29, 2009).

12. Which country is the nanotechnology research leader in Latin America?

- ☐ Argentina
- ✓ Brazil
- ☐ Chile
- ☐ Mexico
- ☐ Uruguay

Brazil, which has Latin America's largest economy, was the first nation in Latin America to provide public funding for nanotechnology (in 2000) and by most metrics has the largest nanotechnology research infrastructure. Mexico and Argentina, which are generally ranked second and third in nanotechnology in Latin America, also benefited from providing early public funding to develop their nanotechnology infrastructures.

13. Which of these atoms are smaller than 1 nm in diameter?

- ☐ Oxygen
- ☐ Cesium
- ☐ Uranium
- ☐ All atoms are >1 nm
- ✓ All atoms are <1 nm

Atoms range in diameter from about 0.1 to 0.3 nm, which means that all atoms are <1 nm in diameter. As you add protons, neutrons, and electrons to an atom, the atom increases in size, but as more particles are added, the force holding the electrons also increases, which condenses the atom and keeps even the largest atoms under 1 nm. The simplest, smallest atom is hydrogen, comprised of one proton and one electron. Measuring the "radius" of an atom is complicated by the fact that electrons "orbit" the nucleus, so the atom is not really "solid." Electrons often behave more like waves than particles, and form a "cloud" around the nucleus.

14. Which of the following has been created by scientists in nanotech laboratories?

- ☐ A nanoscale box with a lid, made of synthetic DNA
- ☐ A paper battery that you can fold or cut
- ☐ A hollow gold nanosphere capable of delivering cancer fighting drugs
- ☐ Carbon sheets one atom thick
- ✓ All of the above

All of these very interesting technical achievements have been demonstrated in research laboratories, by extraordinary teams of nanoinnovators. DNA origami – the building of structures using synthetic DNA molecules – has progressed to the point that a box with "locks" has been created. Paper batteries have been created by growing carbon nanotubes on a substrate of silicon, peeling off the carbon sheet, and using it to create a battery that you can cut, fold, and bend, while retaining the battery's properties. In several medical laboratories, biomedical researchers have demonstrated that hollow gold nanospheres can either be delivered to tumor sites and heated to kill just the tumors or used as nanocarriers to deliver drugs to disease sites. Carbon sheets one atom thick – called graphene or graphane – also exist and are moving from research to commercialization.

15. Does Europe spend more or less than the United States on nanotechnology research?

    ☐ More
    ☐ Less
    ✓ Approximately the same

Europe spends approximately the same money on nanotechnology research, as the United States.

16. Which of the following is an example of nanoinnovation?

    ☐ High-performance baseball bats and golf clubs
    ☐ Flexible thin-film solar cells
    ☐ Digestible computers for tracking pills
    ☐ Cell phones that bend and twist like plastic
    ✓ All of the above

These nanoinnovations reveal the broad spectrum of technologies and applications that are already being developed and commercialized. Baseball bats and golf clubs already benefit from integrated nanoparticles. Flexible thin-film solar cells still have relatively low efficiency levels, but experimental nano-solar prototypes are increasing these levels to more commercial performance levels. Proteus Biomedical's digestible computer chip is in clinical trials. Nokia's concept phones use nanotechnology to bend and twist. These are just a few examples.

17. Which of the following are nanoscale (<100 nm in diameter)?

    ✓ West Nile Virus
    ☐ Red blood cells
    ☐ White blood cells
    ☐ Human hair
    ☐ *Escherichia coli* bacteria

The West Nile Virus is a type of *Flavivirus*, ranging from 40 to 60 nm. Viruses in general range from 20 to 200 nm. White blood cells are ~10 000 nm, and red blood cells are about 7000 nm. Most bacteria range from 1000 to 10 000 nm. A human hair can vary in thickness, from 50 000 to 100 000 nm in diameter.

18. The world's largest carbon nanotube manufacturing facility is located in which country?

☐ Germany
☐ The United States
☐ Japan
✓ China
☐ India

CNano Technology, headquartered in Santa Clara, California, operates a 500 metric ton carbon nanotube production facility, which is located in China. Showa Denko K.K. (SDK) in Japan operates a 400 tonne/year nanotube plant, primarily focused on resin composite applications. The Belgian company Nanocyl operates a 460 metric ton plant which is the second largest.

19. Which of the following companies holds the most nanotech patents?

☐ Fujitsu
☐ GE
✓ IBM
☐ Nantero, Inc.
☐ Siemens

IBM, which holds more patents than any United States-based company, holds more than 200 patents, including an early fundamental patent on single-walled carbon nanotubes. Of the 1180 companies that own nanotechnology patents in the United States, half of them are held by just 95 entities.[1]

20. Which products in Europe are required to identify nanosized ingredients with the word "nano" in brackets on their labels?

☐ Food products sold in supermarkets
✓ Cosmetics
☐ Candy
☐ Clothing
☐ All consumer products

In November 2009, the European Union adopted new product labeling rules that require cosmetics sold in Europe to include the word "nano" in brackets after any ingredient <100 nm in size. The rules went into effect in 2013.

1) Chen, H. *et al.* (2008) Trends in nanotechnology patents. *Nature Nanotechnology*, **3**, 123–125.

Appendix B
Carbon Nanotubes: Company List

	Company name	Products	Location	Web sites
1	Advanced Nanopower	MW	Taiwan	http://www.anp.com.tw/English/main01.htm
2	Ahwahnee, Inc.	MW	CA	www.ahwahneetech.com/
3	Alfa Aesar	MW	MA	www.alfa.com
4	AlphaNano Technology	SW, MW	CHI	http://www.nanotubes.cn/index-en.htm
5	American Elements	SW, MW	CA	www.americanelements.com/
6	ApNano Materials	CNT	NY	www.apnano.com
7	Apex Nanomaterials	SW, MW	CA	www.apexnanomaterials.com/
8	Applied Nanotech	CNT	TX	www.appliednanotech.net
9	Applied Science	Fiber	OH	www.apsci.com/home.html
10	Arkema	MW	FRA	www.arkema-inc.com/index.cfm
11	ARRY International	SW, MW	Hong Kong	www.arry-nano.com
12	BASF	MW	GER	http://www.plasticspotal.com/products/ultraform.html
13	Bayer (exited 2013)	MW	GER	www.baytubes.com/
14	Bio-Nano International	CNT	Singapore	www.bionano.com.sg
15	Biophan	CNT	NY	www.biophan.com
16	BuckyUSA	SW, MW	TX	buckyusa.com
17	Buhler Partech		GER	www.buhlergroup.com
18	Canatu	CNT	Finland	www.canatu.com
19	CarboLex, Inc.	SW	PA	carbolex.com/
20	Carbon Design Innovation	CNT tip	CA	www.carbondesigninnovations.com
21	Carbon Nanomaterials Technology	SW, MW	KOR	http://www.carbonnano.co.kr/english.htm
22	Carbon Nanotube and Fiber 21	SW, MW	Austria	www.carbon-nanofiber.com
23	Carbon NT&F21	CNT, CNF	Australia	www.carbon–nanofiber.com
24	Carbon Solutions	SW	CA	www.carbonsolution.com/

(*continued*)

NanoInnovation: What Every Manager Needs to Know, First Edition. Michael Tomczyk.

(Continued)

	Company name	Products	Location	Web sites
25	Catalystic Materials	MW	NC	www.catalysticmaterials.com
26	Cheap Tubes, Inc.	SW, MW	VT	www.cheaptubesinc.com/
27	Chengdo Organic Chemicals	CNT	CHI	www.timesnano.com
28	Cnano	SW, MW	CA	www.cnanotechnology.com/
29	Cnanoz	CNT	NC	cnanoz.com/index.html
30	C-Polymers	CNT, CNF	Austria	www.c-polymers.com
31	Dupont	CNT	DE	www.dupont.com
32	E-City Nano Technologies	SW	MD	www.ecitynanotech.com/
33	Eikos -		MA	www.eikos.com
34	Electrovac	CNF	Austria	www.electrovac.com
35	Fujitsu Laboratories	CNT	JPN	http://jp.fujitsu.com/group/labs/en
36	Future Carbon GmbH	MW	GER	www.future-carbon.de
37	Graphene Solutions		WI	www.graphenesolutions.com/home
38	Hanwha Chemical	SW, MW	KOR	http://hcc.hanwha.co.kr/English/index.html
39	HeJi	SW, MW	CHI	www.nanotubeseu.com
40	Helix Material Solution	SW, MW	TX	www.helixmaterial.com
41	Hodogaya Chemicals		JPN	http://www.hodogaya.co.jp/English2/web/
42	Honjo Chemical	CNT	JPN	www.honjo-chem.co.jp
43	Hyperion Catalyst	MW	MA	www.fibrils.com
44	IBM	CNT	NY	http://www.research.ibm.com/Nanoscience/nanotubes.html
45	Innovations Unified Technologies	CNT	India	www.iutechnologies.com
46	Ionic Liquid Technologies	SW, MW	GER	www.nanomaterials.iolitec.de
47	Liftport		NJ	http://www.liftport.com/nanotech.htm
48	Lockheed Martin	CNT	CA	www.lockheedmartin.com/
49	MER Corporation		AZ	http://www.opus1.com/~mercorp/index.htm
50	MicroTechNano	SW, MW	IN	www.microtechnano.com/
51	Mitsui		JPN	mitsuichemicals.com
52	MK Nano	SW, MW	Canada	www.mknano.com
53	Molecular Nanosystems Inc.		CA	www.monano.com/
54	mPhase Technologies	CNT	NJ	www.mphase.com
55	Mystic MD	CNT	CT	www.mysticmd.com
56	Nanergy	CNT	NJ	www.nanergyinc.copm

57	Nano Factor Materials Tech.	SW, MW	IND	www.nanofactortubes.com
58	Nano NB	SW, MW	Canada	www.nanonb.com
59	NanoRidge	CNTcomp	TX	www.nanoridge.com/index.html
60	Nanostructure & Amorphous	SW, MW	TX	www.nanoamor.com
61	Nano-C	SW	MA	www.nano-C.com
62	Nanocarb Lab	SW	Russia	www.nanocarblab.com
63	Nanocomp Technologies		NH	www.nanocomptech.com/
64	Nanocomposites	CNT	TX	www.nanocompositesinc.com
65	NanoCompound	MW	GER	www.nanocompound.de
66	NanoCS	SW, MW	NY	www.nanocs.com
67	Nanocyl	SW, MW	GA	www.nanocyl.com
68	NanoIntegris	SW	IL	www.nanointegris.com/
69	Nano-Lab	SW, MW	MA	www.nano-lab.com
70	Nanoledge	CNT	can	www.nanoledge.com
71	NanoMas Technologies	CNT	NY	www.nanomastech.com
72	Nanomaterials Research Unit	SW, MW	Thailand	physcis.science.cmu.ac.th/nano
73	Nanomix Inc.		CA	www.nano.com/
74	Nanoselect	CNT	DE	www.nanoselect-sensors.com
75	Nanoshel	CNT	India	www.nanoshel.com
76	Nanotailor	SW	TX	www.nanotailor.com/
77	NanotechLabs	CNT	NC	www.nanotechlabs.com
78	Nanothinx	SW, MW	Greece	www.nanotubesx.com
79	Nantero	CNT	MA	www.nantero.com
80	Natural Nano		NY	www.naturalnano.com
81	N-Tech	CNT	Norway	www.n-tech.no
82	NTP	SW,	MW CHI	www.nanotubes.com.cn
83	Optomec	CNT	NM	www.optomec.com
84	PlasmaChem	MW	GER	www.plasmachem.de
85	Polytech & Net	MW	GER	www.polytech-net.de
86	Porifera	CNT mem.	CA	poriferanano.com/
87	Pyrograph Products	CNF	OH	www.apsci.com/ppi-markets.html
88	Raymor Nanotech	SW	Canada	www.raymor-nanotech.com
89	Rosseter Holdings	SW, MW	Cyprus	www.e-nanoscience.com
90	Samsung Electronics	CNT	KOR	www.samsungelectronics.com
91	SELAH Technologies	SW	SC	www.sealhtechnologies.com
92	Seldon Laboratories	CNT	Vermont	www.seldontechnologies.com
93	SES Research	SW, MW	TX	www.sesres.com
94	Shenzhen Nanotechnologies	CNT	CHI	www.nanotubes.com.cn
95	Showa Denko	CNT	JPN	http://www.sdk.co.jp/html/english/index.html
96	Smart Nanomaterials	SWCNT	TX	www.smartnanomaterials.com/
97	Solarno	CNT	TX	www.solarno.com

(*continued*)

(Continued)

	Company name	Products	Location	Web sites
98	SouthWest NanoTechnologies,	SW	OK	www.swnano.com/index.php
99	Stanford Materials	SW, MW	CA	www.stanfordmaterials.com/
100	Sun Nano	SW, MW	CA	www.nanomaterialstore.com
101	Sun Nanotech Co.	MW	CHI	www.sunnano.com
102	ST Microelectronics	CNT	Swiss	www.st.com
103	Surrey Nanosystem	CNT	UK	www.surreynanosystems.com
104	Teco Nanotech	MW	Taiwan	www.teconano.com.tw
105	Thomas-Swan	SW, MW	UK	www.thomas-swan.com
106	Tokyo Chemical Industry	CNT	JPN	www.tci-asiapacific.com
107	Unidym, Inc.	SW,	MW CA	www.unidym.com/
108	Vorbeck Materials	CNT	MD	www.vorbeck.com/
109	Xintek	MW	NC	www.xintek.com
110	YTC America	CNT	CA	http://www.ytca.com/carbon_nanotubes
111	Zyvex	MW	TX	www.zyvexpro.com

Adapted with permission from Appendix II, *Review of the Emerging Nanotechnology Industry: Materials, Fabrications, and Applications* by H.Y. Kang, California Department of Toxic Substances Control, Pollution Prevention and Green Technology, September 2010.

Appendix C
University Nanotechnology Research and Educational Centers

US Universities: A Representative List

Arizona State University
The Nanostructures Research Group
Applied NanoBioscience Center

Brown University
Computational Nanotechnology
Nano and Micromechanics Laboratory

California Institute of Technology
Roukes Research Group

Central Michigan University
The National Dendrimer & Nanotechnology Center

Columbia University
Center for Electronic Transport in Molecular Nanostructures

Cornell University
Cornell Nanoscale Science and Fabrication Facility
Center for Nanoscale Systems, Nanoscale Science and Engineering Center (NSEC)
Nanobiotechnology, Science and Technology Center

Department of Defense
Department of Defense Research Laboratories, Nanoscience and Technology
Multidisciplinary University Research Initiatives
Defense University Research Initiatives on NanoTechnology

Department of Energy, Office of Science
Nanoscale Science Research Centers

NanoInnovation: What Every Manager Needs to Know, First Edition. Michael Tomczyk.

Georgia Institute of Technology
Nanoscience + Nanotechnology @ Georgia Tech

Harvard University
Nanoscale Systems and Their Device Applications (NSEC)

Howard University
Keck Center for the Design of Nanoscale Materials for Molecular Recognition

Lawrence Berkeley National Lab
National Center for Electron Microscopy

Massachusetts Institute of Technology
Institute for Soldier Nanotechnologies
Nanostructures Laboratory
Space Nanotechnology
Nanoscience and Nanotechnology in Depth

Materials and Process Simulation Center (MSC)
Applied Physics at Caltech

National Aeronautics and Space Administration
Ames Center for Nanotechnology
Institute for Nanoelectronics and Computing
Nano-and Opto-Electronics Team
Nanotechnology Home Page

NASA-Nanotechnology Team
Nanotech Home Page

National Institute of Standards and Technology
Center for Nanoscale Science and Technology
Center for Neutron Research
National Nanofabrication Infrastructure Network (NNIN)

Naval Research Laboratory
Nanoscience Institute
Nanoscience Research Laboratory

New York University
Ned Seeman's Laboratory Homepage

North Carolina State University/University of North Carolina
North Carolina Center for Nanoscale Materials

Northeastern University
Center for High-Rate Nanomanufacturing

Northwestern University
Electronic Transport in Molecular Nanostructures (NSEC)
Institute for Nanotechnology (NSEC)
Center for Transportation Nanotechnology
Integrated Nanopatterning and Detection (NSEC)
Center for Nanofabrication and Molecular Self-Assembly
National Center for Learning & Teaching in Nanoscale Science & Engineering

Oregon Nanoscience and Microtechnologies Institute (ONAMI)
Center for Advanced Materials Characterization
Center for Electron Microscopy & Nanofabrication
Microproducts Breakthrough Institute
OSU Electron Microscopy & Imaging Facility

Pacific Northwest National Laboratory
*Institut*e for Integrated Catalytics

Penn State University
Industrial Opportunities from Nanotechnology Manufacturing Technologies

Princeton University
Biologically Inspired Materials Institute (BIMat)
Institute for the Science and Technology of Materials
Center for Complex Materials

Purdue University
Nanotechnology Simulation Hub

Rensselaer Polytechnic Institute (NY)
Directed Assembly of Nanostructures (NSEC)
Materials Science and Engineering Department

Rice University
Center for Biological and Environmental Nanotechnology
Center for Nanoscale Science and Technology
Nanoscience in Biological and Environmental Engineering (NSEC)
The Smalley Group
Professional Master's Program in Nanoscale Physics

Stanford University
Stanford NanoFabrication Facility

State University of New York at Stony Brook
USB Buckyball Homepage, Physics Department

Texas A&M University
Institute for Intelligent Bio-Nanomaterials & Structures for Aerospace Vehicles

University of Arkansas, University of Oklahoma
Center for Semiconductor Physics in Nanostructures

University of California-Davis
Nanophases in the Environment, Agriculture, and Technology (NEAT)

University of California-Irvine
Integrated Nanosystems Research Facility

University of California-Los Angeles
Institute for Cell Mimetic Space Exploration
California NanoSystems Institute

University of California-Santa Barbara
California NanoSystems Institute
Nanotech Fabrication Facility

University of Delaware
Multiuniversity NSF NIRT Project

University of Florida
The Center for Structural Biology

University of Illinois at Urbana-Champaign
Beckman Institute for Advanced Science and Technology
Center for Nanoscale Science and Technology

University of Michigan
Life Sciences at Michigan

University of Michigan, Ann Arbor
Center of Experience in Nano Mechanical Science & Engineering (NAMSE)
Michigan Nanotechnology Institute for Medicine and Biological Sciences http://nano.med.umich.edu/index.html

University of Minnesota
Center for Nano-Energetics Research

University of North Carolina
Center for Computer Integrated Systems for Microscopy and Manipulation
(CISMM) at UNC Chapel Hill
Nanoscale Science Education Center
Nanoscale Science Research Group
North Carolina Center for Nanoscale Materials

University of Notre Dame
Center for Nanoscience and Technology

University of Pennsylvania
Krishna P. Singh Nanotechnology Center
Nano/Bio Interface Center

University of South Carolina
NanoCenter at the University of South Carolina

University of Southern California
The Laboratory for Molecular Robotics

University of Washington
The Center for Nanotechnology

Index

o

p